T0305681

TECHNOLOGIES IN DECLINE

The central questions of this book are how technologies decline, how societies deal with technologies in decline, and how governance may be explicitly oriented towards parting with 'undesirable' technology.

Surprisingly, these questions are fairly novel. Thus far, the dominant interest in historical, economic, sociological and political studies of technology has been to understand how novelty emerges, how innovation can open up new opportunities and how such processes may be supported. This innovation bias reflects how in the last centuries modern societies have embraced technology as a vehicle of progress. It is timely, however, to broaden the social study of technology and society: next to considering the rise of technologies, their fall should be addressed, too. Dealing with technologies in decline is an important challenge or our times, as socio-technical systems are increasingly part of the problems of climate change, biodiversity loss, social inequalities and geo-political tensions. This volume presents empirical studies of technologies in decline, as well as conceptual clarifications and theoretical deepening. *Technologies in Decline* presents an emerging research agenda for the study of technological decline, emphasising the need for a plurality of perspectives.

Given that destabilisation and discontinuation are seen as a way to accelerate sustainability transitions, this book will be of interest to academics, students and policy makers researching and working in the areas of sustainability science and policy, economic geography, innovation studies, and science and technology studies.

Zahar Koretsky is post-doctoral researcher in STS and transition studies at the French Research Institute for Agriculture, Food and the Environment (INRAE) and Gustave Eiffel University in Paris. He works on sustainability transitions, focusing on a critical take on it. His current research lies within sustainability transitions in the extractive sector. Zahar has published on the history of various declined technologies across the world and on the EU's climate change mitigation.

Peter Stegmaier is assistant professor at the Department of Technology, Policy and Society, section Science, Technology and Policy Studies (STePS) of the University of Twente, the Netherlands. Peter has worked as researcher and lecturer at the Universities of Dortmund, Hagen and Düsseldorf, carrying out at the latter two the project 'Law as a social practice—from everyday legal methods to a legal methodology for everyday work' (2000–2004); and as lecturer and researcher in criminal and police sociology at the Ruhr University of Bochum, focusing especially on public security institutions (2004–2006).

Bruno Turnheim is INRAE Research Scientist at the Interdisciplinary Laboratory on Sciences, Innovations and Societies (LISIS, Université Gustave Eiffel, France) and Honorary Research Fellow at the Manchester Institute of Innovation Research (MIoIR, University of Manchester, UK). His research focuses on socio-technical innovation and transition processes, particularly in relation to grand societal challenges.

Harro van Lente is full professor of Science and Technology Studies at the Faculty of Arts and Social Science, Maastricht University. He is one of the founding fathers of the Sociology of Expectations, which studies how representations of the future shape current socio-technical developments. He has published more than 100 journal articles, book chapters and edited volumes on technology dynamics, innovation policy and knowledge production. He is recipient of the 2018 EASST Freeman Award with the co-edited book (with Marianne Boenink and Ellen Moors), *Emerging Technologies for Diagnosing Alzheimer's Disease: Innovating with Care.*

TECHNOLOGIES IN DECLINE

Socio-Technical Approaches to Discontinuation and Destabilisation

Edited by Zahar Koretsky, Peter Stegmaier, Bruno Turnheim and Harro van Lente

Cover image: L.E. Baskow www.lefteyeimages.com

First published 2023
by Routledge
4 Park Square, Milton Park, Abingdon, Oxon OX14 4RN

and by Routledge
605 Third Avenue, New York, NY 10158

Routledge is an imprint of the Taylor & Francis Group, an informa business

British Library Cataloguing-in-Publication Data
A catalogue record for this book is available from the British Library

Library of Congress Cataloging-in-Publication Data
A catalog record has been requested for this book

ISBN: 978-1-032-10102-6 (hbk)
ISBN: 978-1-032-10098-2 (pbk)
ISBN: 978-1-003-21364-2 (ebk)

DOI: 10.4324/9781003213642

Typeset in Bembo
by Taylor & Francis Books

CONTENTS

ILLUSTRATIONS

Figures

Tables

Box

CONTRIBUTORS

Wouter Achten, bio-engineer in land and forest management by training, is Professor of Environmental Management at the Université Libre de Bruxelles. His research activities are mainly focused on life cycle thinking, both applied (environmental impact assessment) and methodological (LCA, LCSA, consequential LCA, S-LCA, IO-LCA, Territorial LCA, etc.).

Tom Bauler is ecological economist and Assistant Professor at the Université Libre de Bruxelles, where he holds the chair of "Environment & Economy." In his research Tom explores the role of policy instruments and tools in the environmental policy domain.

Ela Callorda Fossati is Researcher at SONYA/ULB (SOcio-eNvironmental dYnAmics research group—Université Libre de Bruxelles) and an associate member of CIRTES/UCLouvain (Interdisciplinary Research Centre on Work, State and Society—Université Catholique de Louvain). With a background in Development Economics (PhD from the University of Bordeaux), her research interests revolve around plural and substantive understandings of the economy. Her work focuses on institutional and organisational transformations of the social and solidarity economy, mainly in Belgium, and encompasses sustainability transitions approaches. Currently, Ela investigates 'exnovation challenges' in transitions towards a sustainable economy in Brussels (see the GOSETE research programme that she coordinates).

Frédéric Goulet is Researcher at the Centre de Coopération Internationale en Recherche Agronomique pour le Développement (CIRAD, France), Visiting Professor at the Universidade Federal Rural do Rio de Janeiro (UFRRJ), and currently Visiting Researcher at CIMMYT, Mexico. His research interests include the withdrawal and emergence of technologies in agriculture, transformations in agricultural research in response to societal challenges, and the production and sharing of knowledge among farmers.

Karoliina Isoaho works for the Environmental Services at the City of Helsinki, focusing on promoting sustainable mobility and reducing the environmental impacts of the transport sector. Isoaho holds a PhD in environmental policy from the University of Helsinki and an MSc in Environment and Development from the London School of Economics and Political Science.

Zahar Koretsky is Postdoctoral Researcher in STS and transition studies at the French Research Institute for Agriculture, Food and the Environment (INRAE) and Gustave Eiffel University in Paris. He works on sustainability transitions, focusing on a critical take on it. His current research lies within sustainability transitions in the extractive sector. Zahar has published on the history of various declined technologies across the world and on the EU's climate change mitigation.

Jochen Markard works as Senior Researcher at the University of Applied Sciences Winterthur and is a Privatdozent (habilitation and venia legendi) at ETH Zurich. He holds a PhD in Innovation Studies and degrees in Electrical Engineering and Energy Economy. In his research, Jochen focuses on sustainability transitions: large-scale sectoral changes of for example energy and transport systems to more sustainable modes of production and consumption.

Peter Newman is professor of Sustainability at Curtin University in Perth, Australia. He is the Co-ordinating Lead Author for the UN's IPCC on Transport, and the author of 23 books and over 400 papers on sustainability. Peter has worked in local and state government as an elected councillor and advisor. In 2014 he was awarded an Order of Australia for his contributions to urban design and sustainable transport.

Bonno Pel is Researcher/Lecturer at SONYA/ULB. Having a background in environmental planning and political philosophy, his work focuses on the dynamics, governance and politics of sustainability transitions and social innovations. In recent years he has published extensively on the social innovation dimensions of transitions, both conceptually as well as empirically.

Adrian Rinscheid is Assistant Professor of Environmental Governance and Politics at Radboud University, the Netherlands. His research focuses on 1) comparative environmental politics and public policy, and 2) public perceptions of environmental challenges and policies. His work has been published in leading interdisciplinary journals such as *Environmental Innovation and Societal Transitions* and *Nature Energy* as well as more traditional Political Science outlets like *Journal of European Public Policy* and *Politics & Governance.*

Daniel Rosenbloom is an SSHRC Postdoctoral Fellow at the University of Toronto, Canada. Drawing on transition and political perspectives, his research explores the intersection of climate change, energy and societal transformation.

Philipp Scherer works in the area of innovation management in Berlin, Germany. He holds a Master's degree in Public Economics from the Freie Universität Berlin and a Bachelor's degree in Economics from the University in Trier, Germany.

Peter Stegmaier is Assistant Professor at the Department of Technology, Policy and Society, section Science, Technology and Policy Studies (STePS) of the University of Twente, the Netherlands. He studied Sociology, Social Psychology, Law and Economics at the University of Munich and at Goldsmiths' College in London. Peter has worked as a researcher and lecturer at the Universities of Dortmund, Hagen, and Düsseldorf, carrying out at the latter two the project 'Law as a social practice—from everyday legal methods to a legal methodology for everyday work' (2000–2004); and as a lecturer and researcher in Criminal and Police Sociology at the Ruhr University of Bochum, focusing especially on public security institutions (2004–2006). From 2007 to 2009 he held a postdoc position at the Centre for Society and Genomics, Radboud University Nijmegen, investigating 'society and genomics' research, education and dialogue activities ('ELSA', 'ELSI') in the Netherlands and in the United Kingdom as science governance in action. His research interests include social theory and methodology, knowledge and normativity, theories of action and institutionalisation, phenomenology and neurology, and the sociologies of science, governance, innovation and citizenship, as well as that of music.

Solène Sureau is Researcher at SONYA/ULB. With a background in Macroeconomics and Environmental Management, she worked during her PhD on the methodological development of Social Life Cycle Analysis. She works currently on the analysis of exnovation processes, from the perspective of their sustainability impacts.

Gregory Trencher is Associate Professor at the Graduate School of Global Environmental Studies, Kyoto University. He obtained a PhD in Sustainability Science from the University of Tokyo and previously held posts at Clark University (USA) and Tohoku University (Japan). His research specialises in 1) policies and processes that drive or hamper the transition to sustainable energy and mobility systems and cities, 2) phase-out of environmentally unsustainable substances, technologies and processes, and 3) divestment and decarbonisation strategies in fossil fuel-centric industries. He serves as an editor for *Energy Research & Social Science* and *Frontiers in Sustainable Cities* and has published in leading journals like *PLOS One, Energy Policy, Energy Research & Social Science, Global Environmental Change* and *Renewable and Sustainable Energy Reviews*.

Bruno Turnheim is INRAE Research Scientist at the Interdisciplinary Laboratory on Sciences, Innovations and Societies (LISIS, Université Gustave Eiffel, France) and Honorary Research Fellow at the Manchester Institute of Innovation Research (MIoIR, University of Manchester, UK). His research focuses on socio-technical innovation and transition processes, particularly in relation to grand societal challenges. Since 2016, he is a member of the Sustainability Transition Research Network (STRN) Steering Group. Since 2020, he leads the WAYS-OUT project, financed by the Programme d'Investissements d'Avenir (PIA) under the Make Our Planet Great Again (MOPGA) programme. The WAYS-OUT project focuses on destabilisation as a process, on the formulation of destabilisation pathways, on the comparison of destabilisation case studies in various sectors, and on the modalities for governing the discontinuation of established socio-technical regimes.

Dirk van de Leemput is collaborative PhD student at Maastricht University, Faculty of Arts and Social Sciences (Netherlands) and Tate Gallery (London, UK). He has a background in Science and Technology Studies and Archival sciences. In his PhD research, he investigates how conservators of contemporary art and technical experts care for declining technologies—such as film, CRT TVs or software—that are used in time-based media works of art.

Harro van Lente is Professor of Science and Technology Studies at the Faculty of Arts and Social Science, Maastricht University. He is one of the founding fathers of the Sociology of Expectations, which studies how representations of the future shape current socio-technical developments. He has published more than 100 journal articles, book chapters and edited volumes on technology dynamics, innovation policy and knowledge production. He received the 2018 EASST Freeman Award with the co-edited book (with Marianne Boenink and Ellen Moors) *Emerging Technologies for Diagnosing Alzheimer's Disease: Innovating with Care.*

Daniel Weiss is Researcher and doctoral candidate at the Chair of Innovation Management at Freie Universität in Berlin, Germany. He completed his Master's degree in Economics at the Freie Universität Berlin and his Bachelor's degree in Economics at Ruprecht-Karls-University in Heidelberg, Germany.

Linda Widdel works as Researcher at the Competence Center Policy and Society at the Fraunhofer Institute for Systems and Innovation Research ISI, Karlsruhe. She writes her dissertation on social justice in sustainability transitions at the Laboratoire Interdisciplinaire Sciences Innovations Sociétés (LISIS). Linda Widdel holds a Master's degree in Democracy and Global Transformations from the University of Helsinki and a Bachelor's degree in Political Science from Ludwig-Maximilians-University Munich.

PREFACE

We started writing this book in a different world—one before the Covid-19 pandemic and before Europe saw war on its soil for the first time in the 21st century. Since that time the topic of technologies in decline has expanded its possible empirical subjects: from studying how to urgently address the climate crisis (movements to stop subsidising and to ban fossil fuels, for instance) to discontinuations of some industries during the pandemic lockdowns (for instance, aviation due to shortage of demand, food production due to a shortage of workers and the breakdown of supply chains) and war (namely, destruction in Ukraine, and, for that matter, what could be the beginning of a decline in the Russian aviation, military and IT technological fields as a result of international sanctions and an exodus of specialists). By the time this book is published, the readers will be in a better position to assess how these topics have evolved in real life. We, as editors, hope the lessons offered by this edited volume will be of use.

This book started with the conference panel on 'How are technologies abandoned?' initiated by Zahar Koretsky, Harro van Lente and Peter Stegmaier, at the European Forum for Studies of Policies for Research and Innovation (Eu-SPRI) 2020 conference. Having received ample first signs of interest from other science and technology studies and transitions researchers, we were approached by Routledge to consider turning the panel submissions into an edited volume. We decided to accept the invitation. As the book proposal was taking shape, Bruno Turnheim got increasingly involved and became a key part of the editors' team.

We may equally count the beginnings of this book from Zahar's PhD project on decline and phase-out (2017–2021) at Maastricht University; or from Peter's work on the governance of discontinuation of socio-technical systems in the Open Research Area project 'DiscGo' (2013–2016) with research partners from France (LISIS), the United Kingdom (SPRU), the Netherlands (STePS, University of Twente), and Germany (TU Dortmund), where the discontinuation governance framework was developed. Likewise, the book may be said to have started several years prior when Bruno published his work on destabilisation. Or when Harro started engaging with the topic of Responsible Research and Innovation (RRI) and questions of sustainability and needs. The point is, for the past decade or so the topic of declining technologies, systems and practices has been with us in some way or another, in our experiences of the material world, in the expression of social claims related to impending crises, in our conceptual thinking, and in our puzzling about how the significant variety of approaches to decline may be brought together in a meaningful way. The four of us were impressed and happy to find like-minded scholars, the contributors to this volume, who were engaging with this important topic and grappling with similar puzzles, by their eagerness and by the quality of their contributions. With this book, we hope to have contributed to the identification of a community of scholars interested in the study of technologies in decline, and we look forward to seeing how this community will develop over time, including among the readers of this book.

The Editors
Zahar Koretsky, Peter Stegmaier, Bruno Turnheim, Harro van Lente

ACKNOWLEDGEMENTS

We would like to thank the contributors for their dedication and patience during the various commenting and editing stages. Big thanks go to the participants of the 2020 Eu-SPRI panel 'How are technologies abandoned?' for the general positive reception and constructive discussion of the ideas and, for many of you, for following up on these ideas in this volume. Likewise, we thank the participants of the WAYS-OUT workshop in Paris, 2021, for their engagement and constructive feedback.

Various institutional funding sources have supported or indirectly enabled the production of the volume. This work, partly inscribed in the WAYS-OUT research project led by Bruno Turnheim, has benefited from State assistance managed by the French National Research Agency under the 'Programme d'Investissements d'Avenir' under the reference ANR-19-MPGA-0010. The research for this volume has also partly been enabled through results from the Open Research Area Scheme (ORA) Grant no. 464-11-057 from NWO, DFG, ANR and ESRC within the project called 'Governance of the discontinuation of socio-technical systems' (DiscGo). Equal acknowledgements go to the funding from the Horizon 2020 grant for the project 'Fostering improved training tools for responsible research & innovation' (FIT4RRI), Grant no. 741477, as well as funding by Maastricht University.

We thank our publisher, Routledge, and especially the editors Jyotsna Gurung and Annabelle Harris for picking up on this important topic, their patience with us, and all of the production work that went into this book, as well as Tom Bedford for his pertinent copyediting work.

1

INTRODUCTION

The relevance of technologies in decline

Zahar Koretsky, Harro van Lente, Bruno Turnheim and Peter Stegmaier

The central question of this book is *how technologies decline*. Surprisingly, this question is fairly novel. The dominant interest in historical, economic and sociological studies of technology has been to understand how novelty emerges and how innovation can open up new opportunities. This 'innovation bias' in the disciplines studying technology reflects how in recent centuries modern societies have embraced technology as a vehicle of progress. Indeed, the development and use of technologies have brought remarkable improvements in health, mobility and standards of living. In the last two centuries, technologies were figured as solutions to address societal problems. Yet, in a time of growing concerns related to the challenges of climate crisis, biodiversity loss, social inequalities or geo-political tensions, technologies increasingly figure as part of the problem, too (cf. Beck 1992; Douglas 1970). Technologies that once embodied progress, such as pesticides or coal-fired power production, now embody problems and stand in the way of better directions. It is timely, therefore, to broaden the horizon of technology dynamics and the technology-society relationship: next to considering the rise of technologies, we should also consider their fall, too. These are two sides of the same coin when it comes to how the relationship to technology is constantly re-negotiated in a social context. In this volume we present some outlines for the study of technological decline.

1.1 Limits to innovation

After WWII most industrialised countries adopted a techno-optimistic approach. The idea was that by stimulating scientific research and technological development, society would benefit from the boons of technology. This idea became known as the 'linear model' as it assumed a direct line from scientific discovery to the implementation and diffusion of technologies (Godin 2017). The techno-optimistic view of the linear model was emblematically captured in the 1933 Chicago World Fair motto: 'Science Finds, Industry Applies, Man Conforms'.

DOI: 10.4324/9781003213642-1

But, of course, men and women do not readily conform. From the 1970s onwards, critical discourses about the problems of technology became more prominent. A landmark event for the public recognition that technologies can also put society at risk was the publication of the 1972 Club of Rome report on the *Limits of Growth*. The report indicated the depletion of resources and accumulation of pollution. It also announced that the general idea that economic growth would bring progress was misguided and not necessarily consensual. While such criticisms have changed the discourses and policies on technology, it is also clear that now, 50 years later, the problems flagged have only deepened. The climate crisis and the ongoing loss of biodiversity—as corroborated by a series of IPCC and IPBES[1] reports—indicate that widely used technologies, like the internal combustion engine, coal-fired power generation or the routine preventive use of pesticides, can pose serious threats for current and future generations.

The question of technology was broadened with the issues of reducing risks and increasing democratic control—questions linking to literature on critical theory of technology, where technology is seen as neither value neutral nor universal, and opposing the privileging of technical manipulation over other relations to reality (Feenberg 2017). How to stimulate technologies while avoiding unwanted side-effects? Can unintended effects be anticipated and avoided?

1.2 Critical discourses on technology

The set of questions broadened further with the economic crises of the 1980s that saw hampered industries and painful economic reshufflings in many parts of the world. Technology appeared as the stake in intensive global competitions; national industrial policies were set up to gain a favourable position in the innovation races. As with prior techno-optimism, as if part of a cyclical pattern, this was the era of strategic research and innovation, and of significant investment programmes in ICT, biotechnology and new materials. New regional specialisations emerged under technological and competitive pressures of global capitalism. Western countries faced the social consequences of deindustrialisation in heavy industry and sought to retain competitive edges by investing in R&D for advanced technological sectors. As emerging economies in the East and South became innovation powerhouses and expanding mass consumption markets, promises of growth and prosperity provided a counterpoint to problems associated with the decline of crafts and traditional industries (e.g., hand weaving in India, cf. Mamipudi 2016). Technology and innovation remained intimately and, again, techno-optimistically, tied to economic and social prosperity, but this time there was a more widespread awareness of the downsides, and particularly the social costs of regional decline in technological races.

During the last few decades, various political initiatives have been proposed to address the removal of certain technologies. Compared to efforts to stimulate technology these initiatives are modest, but there have been some successes: for instance, the ban of chlorofluorocarbons (CFCs) under the Montreal Protocol in

1987, the chemical substances responsible for ozone depletion which were used, for instance, in refrigerators. Various forms of technology assessment are now being used in policy settings and in firms to evaluate the desirability of technologies. In EU research, the notion of 'responsible research and innovation' has become commonplace, indicating an ambition to have more public control on technologies. Currently, there is a renewed emphasis on technological sovereignty (Edler et al. 2020), to gain independence from energy sources from warring states on the occasion of the war in Ukraine, or to be less dependent on supply chains that suffered from the Covid-19 pandemic.

In recent decades, environmental, health and social concerns have begun to assume more central roles in studies on technology, innovation and economic development. The UN Millennium Goals and their extension into the UN Sustainable Development Goals are important milestones signalling new long-term global orientations for economic and technological development. The rising interest in managing the concerns relating to technologies might be stemming from the economic stagnation of the global North in the last two decades (Streeck 2014; Albertson 2020) and dissatisfaction with the way institutionalised decision-makers have been handling economic and environmental issues (Oreskes & Conway 2010; Wille 2010). We can also observe a fatigue from current hyper consumerist societies and the unresolved environmental concerns (Gibson-Graham 2008; Escobar 2015; Hossain 2018; de Saille et al. 2020; Hickel & Kallis 2020) coming from both rich and poor regions of the world. Both types of concerns manifest in a new style of protest, one that not only frames a problem explicitly, but also articulates the need to change systems by taking them down. The actions of Fridays for Future, Black Lives Matter or Extinction Rebellion are cases in point. More societal and environmental problems such as climate change, biodiversity loss, unhealthy lifestyles, redistributive justice, privacy breaches or the spread of fake news are galvanising a return to critical discourses about technology. Against this background, calls for more desirable alternatives (e.g., eco-innovation, responsible innovation) are being complemented by calls for deliberately discontinuing existing systems deemed undesirable. Phasing out coal and fossil fuels has, for instance, become an important priority for climate action. Similarly, we are witnessing the emergence of policy objectives and programmes seeking to shift food production systems towards pesticide-free agriculture. Problems around nuclear decommissioning have been around for decades, but haven't yet found widely accepted solutions.

One of the difficulties is that attempts to discontinue technologies tend to remain largely translated into new agendas and horizons for innovative activity without fundamentally challenging its underlying logics (e.g., the 'green growth' oxymoron) or established systems. Significant R&D funding is being spent to invent ways to maintain and improve lifestyles without causing deterioration of the environment. Such eco-innovation optimism is, however, struggling to deliver fully on its promises: despite significant deployment of renewable energy, electric mobility or organic agriculture in some countries, these remain a far cry from the 'fundamental system transformations' called for (IPCC 2018; EEA 2019; UNEP 2022). Meanwhile, existing technologies and underlying systems remain relatively stable (e.g., empty passenger planes flying during pandemic) or even expand (e.g., SUVs, re-opening of coal mines in Europe due to the war in Ukraine, LNG extraction plus terminals to replace natural gas shortages), while new industries that are neither ecologically sustainable nor economically necessarily viable continue to emerge (e.g., space tourism) (Markard et al. 2021). Such examples can also show how much discontinuation has to contend with contradictory or competing rationales, interests, opportunities and framings (Turnheim 2023; Stegmaier 2023; Koretsky 2023).

In short, the prominent techno-optimistic discourses are under pressure, and existing socio-technical systems, ranging from energy production, to mobility, to agri-food, are increasingly under critique. As a result, many questions come to the fore: Is it possible to do away with undesirable or unsustainable technologies? If so, how? Does this necessarily involve substitution or does it involve other shifts, too? What societal, political and industrial strategies may help to reduce our dependence on harmful technologies and socio-technical systems? Should specific products or larger systems be targeted? How can investment patterns related to harmful and polluting production be discontinued? These questions require another approach to technology: exits and divestments, destabilisation and discontinuation are high on the agenda.

1.3 Studying technology beyond innovation

In the scholarly fields of innovation studies and science and technology studies, the emergence of technologies has traditionally been the focus of study. Even a decade ago Elizabeth Shove noted that '[w]ithin the fields of innovation studies and transitions theory, processes of emergence and stabilisation are better documented and more widely discussed than those of disappearance, partial continuity and resurrection' (Shove 2012: 363). Yet, the attempts to abandon undesirable technologies have been hampered by insufficient insights into how such processes unfold—whether they are deliberately pursued or not. The question of how technologies decline, which we are concerned with in this book, is timely and differs markedly from earlier questions about technology in society.

Of course, the recognition that technologies may be disruptive is not new. A century ago, the founder of innovation studies Joseph Schumpeter coined the phrase *creative destruction* to characterise the role of technical change in economies. Technologies do not just bring an accumulation of improvements, he argued, but will necessarily destabilise economic sectors, too. This still holds today: think about the woes of the postal services, which suffer from the popularity of e-mail. Before Schumpeter, Karl Marx analysed the exploitative and alienating nature of capitalism and its mobilisation of technology for this, and pointed to the fundamental disruption of social structures. The destructive character of technology, the destabilisation and eventual decline of industries or organisations, the social and environmental costs of capital accumulation around technology and its use have a certain regularity and genericity that makes them observable in different geographical, sectorial or temporal settings.

The main starting point for this book is the observation that (desirable) exit or reduction objectives informed by critical discourses on technology are qualitatively different from fostering desirable innovation. They involve a different kind of phenomenon, requiring different skills, different interventions and different kinds of thinking: *decline is not just the reverse of innovation*. Moreover, deliberate decline is likely to face resistance from significant vested interests, which may be powerful incumbents as well as more vulnerable populations and communities facing to lose significantly from the end of systems they depend on. Deliberate decline entails significant challenges, such as those associated with regulating or restricting activities and livelihoods associated with 'undesirable' technologies, related political contestation and struggles, but also dealing with the fact that there will be winners and losers as a result of decline. Decline is likely to be as much about setting directions and objectives as it is about managing a process and handling its aftermath—including loss (Elliott 2018). The difficulty and, at the same time, opportunity is that technologies neither persist, nor disappear into oblivion automatically: they require work to do so (Callon 1987; MacKenzie and Spinardi 1995; Russell and Vinsel 2018). Moreover, formerly established systems may still be needed for very specific purposes (DDT for vector control, special purpose incandescent light bulbs, special purpose vehicles with internal combustion engines that are fossil-fuelled). At least for a transitional period, they may leave traces that outlive the discontinuation of their active use (e.g., dealing with nuclear waste long after the disconnection and dismantling of nuclear power plants) and require dedicated infrastructures. Sometimes they also threaten to come back as zombie technology because strong interests want to push them back into the market and effectively revert phase-out programmes (e.g., the revival of coal power generation) or critical discourses are shifting (e.g., nuclear energy framed as green and CO2-neutral or dirty and life-threatening). Thus, decline is neither a linear nor an irreversible process. This is new territory for policy and research alike, and calls for revisiting concepts, methods, capabilities and means for intervention.

1.4 Perspectives and concepts

This book proposes to ask what kind of processes are involved, what forms of decline can be observed, what lenses and concepts can be usefully applied and what questions remain unanswered. In this volume on technologies in decline, we seek to draw on a rich empirical base, which is diverse in terms of technologies, geographic locations and political settings. We intend to explore and use various intellectual starting points and concomitant concepts. The study of technologies in decline is necessarily *interdisciplinary*, drawing from multiple disciplines such as sociology, history, management and economics. Yet, this diversity can only be productive when there is also some common ground, which allows us to compare and contrast empirical findings and to connect conceptual claims.

Overall, the meta-theoretical lens of this book is *socio-technical* and we should clarify what this means, what it entails and what it requires. Firstly, 'socio-technical' refers to the insight that the social and the technical are deeply interwoven. They do not exist in separate domains, but are *mutually embedded* in tight relationships (Hughes 1983; Callon 1984; Latour 1999). As a consequence, socio-technical decline includes at least a partial dis-embedding of society-technology relationships. Secondly, a 'socio-technical' perspective implies that technological artefacts do not exist in and of themselves but only as a part of networks, configurations or systems. They can be seen as *configurations that work*, as Rip and Kemp (1998) phrased it. Consequently, technology removal or technology substitution is not the appropriate unit of analysis: technologies in decline is a matter of transformations in networks, configurations or systems. Finally, the socio-technical perspective points to the *interplay of social and technological dynamics*. The processes of technologies in decline then involve co-evolving social processes (political, cultural, psychological) and technical processes (in design, standardisation, manufacture, etc.).

The notions and terms mobilised in this volume showcase the varying foci and interests of its various contributors. Some of the notions point to emergent and long-term processes and mechanisms, such as 'destabilisation'. Others, such as 'discontinuation' or 'phase-out', help describe policy or policy goals and the related policy processes. Using past and contemporary examples, the contributions put forward different kinds of explanations, illustrate which strategies might work and which might not, and how decisions to turn away from a questionable technology could be initiated and navigated.

Judging just by the topics of the contributions and the theoretical constructs employed, we may sketch a preliminary frame of reference for various forms of technology decline, see Figure 1.1. The contributors of this volume are preoccupied with

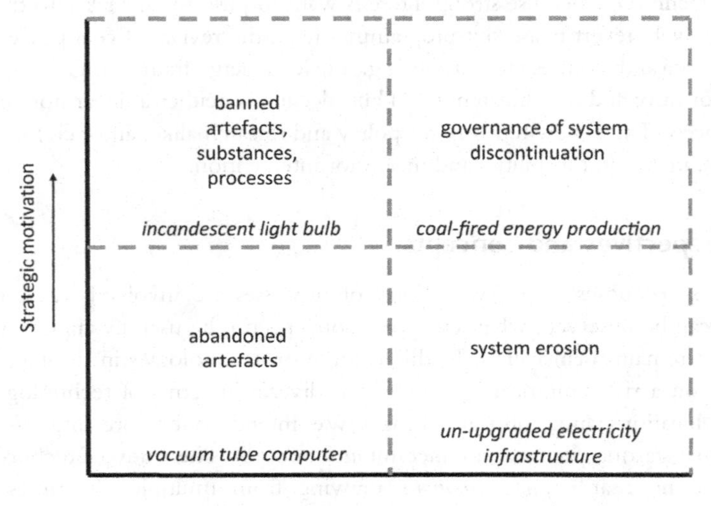

FIGURE 1.1 Varieties of technological decline

the particular processes of decline and adjacent concepts and how they unfold over time, their relevant characteristics, phases, trade-offs and dimensions. Two dimensions appear particularly relevant to make sense of the variety of perspectives: the scope of analysis (i.e., the perimeter of the technology, system or configuration in question) and the strategic motivations vested in a decline process (i.e., the extent to which decline is actively pursued or on the contrary a more emergent process). Of course, these dimensions are to be seen as gradients rather than binary categories, but they already allow us to qualify some archetypical foci along which *technologies in decline* are being thought up. Decline can range from the abandonment of a particular artefact or substance (say, a specific product model) or the erosion of a socio-technical system's relevance and centrality, to the more purposive ban of specific substances or products, or attempts to actively discontinue entire systems.

1.5 Outline of the book

We structure the volume in three Parts: from conceptual explorations, to empirical ones and, finally, to governance explorations. Table 1.1 provides an overview of the chapters, and bears witness to the significant variety of key notions deployed, analytical scale and contexts, focal context and primary research focus.

The conceptual Part of the book deals with three notions. The first one is decline, or *technological decline*, which in Zahar Koretsky's chapter is reconceptualised away from its colloquial, umbrella-term status in the book's title (and this introductory chapter) to a more specific and empirically supported characterisation: a measurable trend and a socio-material process of scaling down of production and/or use of a given product or process. In the chapter, Koretsky presents an overview of literature trying to understand the mechanisms of technological decline and offers, based on it, a socio-material characterisation. From a distinct conceptual starting point, Bruno Turnheim in his chapter focuses on the notion of *destabilisation* next to decline and phase-out. Destabilisation is understood as an emergent process of exposure of socio-technical orders to pressures significant enough to threaten their continued existence and 'normal' functioning, but also strategic responses of affected actors to this exposure and changing commitment to core productive engagements. Turnheim maps the theme of destabilisation in the transitions literature and proposes a research agenda. The third notion, central for Peter Stegmaier's chapter, is *discontinuation*, seen as a property of a technological trajectory in which its constituting relations become misaligned to such an extent that its distinctive character is lost, and also seen as a possible result of various permutations of distributed agency, contingency, emergence or deliberate governance. In this chapter we find a discussion of how the actions of groups of actors affect both the discontinuation of a trajectory itself and of governance practices that help stabilise it. These three notions—decline, destabilisation, discontinuation—provide lenses for the exploration of the problems around passive and active withdrawal of technologies from societies. As is discussed throughout the book and returned to in the Conclusions (Koretsky et al. 2023), these notions are interrelated and provide entry points for dealing with the challenges that come with re-negotiating our relationships with technology.

A Part focusing on empirical studies continues the book's inquiry. To characterise their focal phenomena, some contributors have mobilised a range of additional notions and frameworks. Jochen Markard, Karoliina Isoaho and Linda Widdel (2023) study discursive destabilisation in a comparative setting, by examining framings of coal phase-out in mainstream press outlets across Europe. They adopt the conceptual lens of Technological Innovation Systems (TIS) to study the case of coal power generation in three countries. Daniel Weiss and Philipp Scherer (2023) mobilise the notion of 'phase-out' and approach it as an outcome of processes of decrease in production or consumption. They study the phase-out of the internal combustion engine also through the TIS framework, discussing the role of geographical context in phase-out and decline, focusing on divergent responses to phase-out pressures in the US and the EU. Frédéric Goulet (2023) studies decline as a process of innovation through withdrawal, and examines its relationship with processes of novelty creation. Using an illustrative case of bio-pesticides as alternatives to synthetic pesticides, his chapter explores how the development of substitutes can contribute both to the decline and continuity of problematic technologies. Dirk van de Leemput and Harro van Lente (2023) study the duality of decline of and care for a technology, framing care as 'aftercare' (cf. Stegmaier et al. 2014). Using the example of the 16 mm film as an object of art, they draw from museum studies and care studies. They show how actors can preserve and care for declining technology in pockets of resistance to decline, and how these processes are often invisible.

In the third Part of the book, we turn to governance-related perspectives. Adrian Rinscheid, Gregory Trencher and Daniel Rosenbloom (2023) focus on phase-out, which they see as a policy intervention for a stepwise decrease and termination of production or consumption of a product or process. They offer a systematic review of academic literature on the notion of 'phase-out' since the 1970s. They observe the changing attention to this notion in the literature and comment on the travels of this concept across environmental and societal challenges, policy efforts and instruments. The chapter by Ela Callorda Fossati, Bonno Pel, Solène Sureau, Tom Bauler, and Wouter Achten (2023) mobilises the concept of 'exnovation', where the authors seek to advance empirical knowledge on this notion adjacent to decline and discontinuation. The study focuses on the efforts of the Brussels-Capital region to establish and maintain a low-emission zone. The authors discuss political, jurisdictional and epistemic issues with the implementation of an exnovation policy, and highlight contestations and concomitant discourses of discontinuation in a complex governance setting. A captivating activist/academic testimony by Peter Newman (2023) on the end of leaded petrol complements the prior chapters with a powerful and, ultimately, hopeful account of, and reflection on, the possibility and reality of decline of an undesirable technology.

The concluding chapter by Zahar Koretsky, Bruno Turnheim, Peter Stegmaier, and Harro van Lente (2023) brings the three Parts together by returning to the

TABLE 1.1 Overview of the contributions to the present volume

		Core notions O Side notions •						Analytic scale			Focal context					Research focus					
Book section	**Book chapter**	Decline	Destabilisation	Discontinuation	Phase-out	Exnovation	Withdrawal	Domain	Country	Number of	Organisational	Municipal	Regional	National	Supranational	Transition	Governance	Innovation	Socio-technical	STS	Sectorial
Conceptual explorations	Koretsky	O			•			Energy, IT, Meteorology	USA, EU, RU	3	•			•	•	•			•	•	
	Turnheim	•	O		•			-	-	-	•		•	•		•			•		
	Stegmaier	•	•	O	•	•	•	-	-	-	•	•	•	•	•	•	•		•	•	
Empirical explorations	Markard et al.	O	•		•			Coal	UK, D, FIN	3	•			•		•			•		•
	Weiss & Scherer	O			•			Car	EU, JP, USA	1				•	•	•		•			
	Goulet	O		•				Agriculture, Pesticides	ARG, BRA	2				•		•		•			•
	Van de Leemput & van Lente	O						Media	UK	1	•									•	•
Governance explorations	Callorda Fossati et al.		•		•	O		Mobility	BE	1		•					•	•			•
	Rinscheid et al.				O			Sustainability	-	-				•	•	•	•				•
	Newman				O			Fuel	Several	1	•			•	•	•	•				•

question what the study of technologies in decline entails and requires. The chapter reflects on the progress made in this volume and delineates a research agenda for further study and reflection. In this way, with the book we hope to inspire more efforts to move to a next step in the relationship of current societies to technology as questions of *decline* become more prominent.

Note

1 The Intergovernmental Panel on Climate Change and the Intergovernmental Science-Policy Platform on Biodiversity and Ecosystem Services, both organisations of the United Nations.

References

Albertson, K. (2020) The problem with markets. In *Responsibility Beyond Growth*. Bristol University Press. doi:10.2307/j.ctv13qfvjk.10.

Beck, U. (1992) *Risk Society: Towards a New Modernity*. Sage.

Callon, M. (1984) Some elements of a sociology of translation: Domestication of the scallops and the fishermen of St Brieuc Bay. *Sociological Review*, 31(S1), 196–233.

Callon, M. (1987) Society in the making: The study of technology as a tool for sociological analysis. In Bijker, W.E., Hughes, T.P., and Pinch, T. (eds) *The Social Construction of Technological Systems: New Directions in the Sociology and History of Technology: Anniversary Edition*. MIT Press.

Callorda Fossati, E., Pel, B., Sureau, S., Bauler, T. and Achten, W. (2023) Implementing exnovation? Ambitions and governance complexity in the case of the Brussels Low Emission Zone. In Koretsky, Z. *et al.* (eds) *Technologies in Decline: Socio-Technical Approaches to Discontinuation and Destabilisation*. Routledge.

de Saille, S., Medvecky, F., van Oudheusden, M., Albertson, K., Amanatidou, E., Birabi, T. and Pansera, M. (2020) *Responsibility Beyond Growth*. Bristol University Press. https://doi.org/10.1332/policypress/9781529208177.001.0001.

Douglas, J. D. (1970) *Freedom and tyranny: Social problems in a technological society*. Knopf.

Edler, J., Blind, K., Frietsch, R., Kimpeler, S., Kroll, H., Lerch, C., Reiss, T., Roth, F., Schubert, T., Schuler, J. and Walz, R. (2020) *Technology Sovereignty: From Demand to Concept* (No. 02/2020). Fraunhofer Institute for Systems and Innovation Research (ISI). www.isi.fraunhofer.de/content/dam/isi/dokumente/policy-briefs/policy_brief_technology_sovereignty.pdf.

EEA (2019) *Sustainability Transitions: Policy and Practice*. Publications Office of the European Union. doi:10.2800/641030.

Elliott, R. (2018) The sociology of climate change as a sociology of loss. *Archives Europeennes de Sociologie*, 59, 301–337. https://doi.org/10.1017/S0003975618000152.

Escobar, A. (2015) Degrowth, postdevelopment, and transitions: A preliminary conversation, *Sustainability Science*, 10, 451–462. doi:10.1007/s11625-015-0297-5.

Feenberg, A. (2017) A critical theory of technology. In Felt, U. *et al.* (eds) *The Handbook of Science and Technology Studies*, 4th edition. MIT Press.

Gibson-Graham, J.K. (2008) Diverse economies: Performative practices for 'other worlds', *Progress in Human Geography*. doi:10.1177/0309132508090821.

Godin, B. (2017) *Models of Innovation: The History of an Idea*. MIT Press.

Goulet, F. (2023) The role of alternative technologies in the enactment of (dis)continuities. In Koretsky, Z. *et al.* (eds) *Technologies in Decline: Socio-Technical Approaches to Discontinuation and Destabilisation*. Routledge.

Hickel, J. and Kallis, G. (2020) Is green growth possible? *New Political Economy*, 25(4), 469–486. doi:10.1080/13563467.2019.1598964.

Hossain, M. (2018) Frugal innovation: A review and research agenda, *Journal of Cleaner Production*, 182, 926–936. doi:10.1016/j.jclepro.2018.02.091.

Hughes, T.P. (1983) *Networks of Power: Electrification in Western Society, 1880–1930.* Johns Hopkins University Press. doi:10.2307/3104214.

IPCC (2018) Summary for policymakers. In Masson-Delmotte, V. *et al.* (eds) *Global Warming of 1.5°C: An IPCC Special Report on the Impacts of Global Warming of 1.5°C above Pre-Industrial Levels and Related Global Greenhouse Gas Emission Pathways, in the Context of Strengthening the Global Response to the Threat of Climate Change.* IPCC. www.ipcc.ch/sr15/chapter/spm/.

Koretsky, Z. (2023) Dynamics of technological decline as socio-material unravelling. In Koretsky, Z. *et al.* (eds) *Technologies in Decline: Socio-Technical Approaches to Discontinuation and Destabilisation.* Routledge.

Koretsky, Z., Stegmaier, P., Turnheim, P. and van Lente, H. (2023) Conclusions and continuations: Horizons for studying technologies in decline. In Koretsky, Z. *et al.* (eds) *Technologies in Decline: Socio-Technical Approaches to Discontinuation and Destabilisation.* Routledge.

Latour, B. (1999) *Pandora's Hope: Essays on the Reality of Science Studies.* Harvard University Press.

MacKenzie, D. and Spinardi, G. (1995) Tacit knowledge, weapons design, and the uninvention of nuclear weapons. *American Journal of Sociology*, 101(1), 44–99. doi:10.1086/230699.

Mamipudi, A. (2016) *Towards a Theory of Innovation in Handloom Weaving in India.* Maastricht University. https://cris.maastrichtuniversity.nl/portal/files/4083287/c5295.pdf.

Markard, J., van Lente, H., Wells, P. and Yap, X.S. (2021) Neglected developments undermining sustainability transitions. *Environmental Innovation and Societal Transitions*, 41, 39–41.

Markard, J., Isoaho, K. and Widdel, L. (2023) Discourses around decline: Comparing the debates on coal phase-out in the UK, Germany and Finland. In Koretsky, Z. *et al.* (eds) *Technologies in Decline: Socio-Technical Approaches to Discontinuation and Destabilisation.* Routledge.

Meadows, D. H., Meadows, D. L., Randers, J., and Behrens, W. W. (1972) *The Limits to Growth. A Report for the Club of Rome's Project on the Predicament of Mankind.* Universe.

NESTA (2012) *The Art of Exit: In Search of Creative Decommissioning.* NESTA. www.nesta.org.uk/report/the-art-of-exit/.

Newman, P. (2023) The end of the world's leaded petrol era: reflections on the final four decades of a century-long campaign. In Koretsky, Z. *et al.* (eds) *Technologies in Decline: Socio-Technical Approaches to Discontinuation and Destabilisation.* Routledge.

Oreskes, N. and Conway, E.M. (2010) *Merchants of Doubt: How a Handful of Scientists Obscured the Truth on Issues From Tobacco Smoke to Global Warming.* Bloomsbury Press.

Rinscheid, A., Trencher, G. and Rosenbloom, D. (2023) Phase-out as a policy approach to address sustainability challenges: A systematic review. In Koretsky, Z. *et al.* (eds) *Technologies in Decline: Socio-Technical Approaches to Discontinuation and Destabilisation.* Routledge.

Rip, A. and Kemp, R. (1998) Technological change. In Rayner, S. and Malone, E.L. (eds) *Human Choice and Climate Change.* Battelle Press.

Russell, A.L. and Vinsel, L. (2018) After innovation, turn to maintenance. *Technology and Culture*, 59(1), 1–25. doi:10.1353/tech.2018.0004.

Shove, E. (2012) The shadowy side of innovation: Unmaking and sustainability. *Technology Analysis and Strategic Management*, 24(4), 363–375. doi:10.1080/09537325.2012.663961.

Stegmaier, P. (2023) Conceptual aspects of discontinuation governance: An exploration. In Koretsky, Z. *et al.* (eds) *Technologies in Decline: Socio-Technical Approaches to Discontinuation and Destabilisation*. Routledge.

Stegmaier, P., Kuhlmann, S. and Visser, V.R. (2014) The discontinuation of socio-technical systems as a governance problem. In Borrás, S. and Edler, J. (eds) *The Governance of Socio-Technical Systems*. Elgar. https://doi.org/10.4337/9781784710194.00015.

Streeck, W. (2014) How will capitalism end? *New Left Review*, 87, 35–64.

Turnheim, B. (2023) Destabilisation, decline and phase-out in transitions research. In Koretsky, Z. *et al.* (eds) *Technologies in Decline: Socio-Technical Approaches to Discontinuation and Destabilisation*. Routledge.

UNEP (2022) Presidents' final remarks to plenary: Key recommendations for accelerating action towards a healthy planet for the prosperity of all. Stockholm+50. https://wedocs.unep.org/bitstream/handle/20.500.11822/40110/Key%20Messages%20and%20Recommendations%20-%20Formatted.pdf.

van de Leemput, D. and van Lente, H. (2023) Caring for decline: The case of 16mm film artworks of Tacita Dean. In Koretsky, Z. *et al.* (eds) *Technologies in Decline: Socio-Technical Approaches to Discontinuation and Destabilisation*. Routledge.

Weiss, D. and Scherer, P. (2023) Mapping the territorial adaptation of technological trajectories—The phase-out of the internal combustion engine. In Koretsky, Z. *et al.* (eds) *Technologies in Decline: Socio-Technical Approaches to Discontinuation and Destabilisation*. Routledge.

Wille, A. (2010) Political-bureaucratic accountability in the EU commission: Modernising the executive. *West European Politics*, 33(5), 1093–1116. doi:10.1080/01402382.2010.486137.

PART I

Conceptual explorations

2

DYNAMICS OF TECHNOLOGICAL DECLINE AS SOCIO-MATERIAL UNRAVELLING

Zahar Koretsky

2.1 Introduction

In both mainstream policy-making and academic literature on technologies and innovations, new technologies seem to be both more desirable and more interesting phenomena than their decline. The topic of the end of life of technologies has been periodically appearing in economics and management literature (e.g., Ayres 1987; Tushman and Rosenkopf 1992; Meckling and Nahm 2019). This literature, however, I argue, suffers from a lack of granularity in terms of timescales and unit of analysis, thus remaining somewhat abstract regarding technological decline. Furthermore, typically, processes of decline tend to be conflated with substitution, e.g. in literature on disruptive innovation (e.g., Ayres 1987; Yu and Hang 2010; Taylor and Taylor 2012) and in climate models (as critiqued by Anderson and Peters (2016)). Decline is also seen as an unappealing topic altogether (Smith, Voß and Grin 2010; Stegmaier, Kuhlmann and Visser 2014) because of perceived connotations with failure and a greater interest in markets and competition (Polanyi 1944; McCarraher 2019; Vinsel and Russell 2020) as the nexus of both social life and technological development.

In this chapter,[1] in contrast with prior economic and political economic literature on decline (see Stegmaier 2023), I qualify technological decline as reducing production and/or use of a given technology to a niche degree or to complete abandonment. Technological decline can be a result of emergent or coordinated processes or (more realistically) a mix of both. When a result of phase-out or discontinuation (see Stegmaier 2023), decline is an effective policy intervention which has proven instrumental for stopping wide exposure to carcinogenic DDT gas (Levain et al. 2015) and ozone-depleting aerosols (Simmonds et al. 1993), improving the energy efficiency of indoor lighting with the ban on the incandescent light bulb (Kierkegaard 2014; Stegmaier, Visser and Kuhlmann 2021) and

DOI: 10.4324/9781003213642-2

more. Many countries have instigated the decline of coal and nuclear power, and policies promoting decline increasingly appear on political agendas as countries are faced with the dangerous mid- and short-term prospects of climate crisis (Rogge and Johnstone 2017).[2]

Decline, however, can and will generate uncertainties in economies, environments, labour relations, national security, democracy and more (Lieu et al. 2019; Turnheim 2023). The literature, to date, has not focused on explaining why and how this occurs, nor how to navigate it. An elaborated conceptualisation of the processes involved with decline is needed, in which there would be a higher level of granularity of analysis (Murmann and Frenken 2006) than in existing literature. Thus, in this chapter, I seek an answer to the question: What are the key mechanisms of the socio-material dynamics of technological decline? When answering the research question I will conduct a theoretical comparison of three cases, and formulate conceptualisations by way of middle-range theorising.

The idea of a middle-range type of theorising originates with Merton (1949). It implies an analytical focus on a specific phenomenon, rather than on the social system broadly, as in a "grand theory". Such a narrower focus allows middle range theories to close in on the observed phenomenon, while being abstract enough to explain the observed variations between its instances across cases. According to Merton (1949), the benefit of middle range theories is that they are not derived from a larger theory of social systems (though they may be consistent with them); rather, their starting point is empirical observation and basic generalisation, which are then complexified after subsequent rounds of observation and abstraction of patterns.

The importance of middle range theory has been recognised in science and technology studies (STS) and transition studies (e.g., Frickel and Moore 2006; Geels 2007; Hamlin 2007; Wyatt and Balmer 2007) as part of ongoing self-reflection on own societal value. Frickel and Moore (2006) call STS scholarship to refocus from critique to theorising, and Wyatt and Balmer (2007) ask whether STS has prematurely given up on theorising altogether in its scepticism of generalising. More recently, middle range theorising has been picked up in studies of healthcare technology implementation (Lyle 2021), urban transitions (Mora et al. 2021), social innovation (Pel et al. 2020) and sustainability transitions (Geels 2018), among others. I aim to contribute to this trend and formulate an approach helpful for future analyses of technological decline.

Methodologically, I follow Christensen's (2006) approach of descriptive and predictive theory building. Christensen suggests to start inductively, with a careful description and documentation of the observed phenomena, and to continue with formulating abstractions to rise above the messy empirical detail and to understand key operations of the phenomena. This, he asserts, makes it possible to inductively classify the abstractions into a typology to simplify and organise the studied phenomena. The final task is to identify associations between the category-defining attributes of the phenomena and the observed outcomes. Similarly, Eisenhardt and Graebner (2007) argue that the theory-building process occurs via recursive cycling among rich empirical data, emerging theory and extant literature. The hope is that the result will be able to produce theory that is accurate, parsimonious, interesting and testable (Eisenhardt and Graebner 2007).

Following this logic, I start with a literature review in section 2.2 and formulate a conceptual framework in section 2.3 to analyse and compare empirical cases of decline in section 2.4. Based on this analysis, in section 2.5 I discuss the outlines of a middle-range theory of the dynamics of technological decline, and in the concluding section 2.6 I make final remarks and propose possible future research directions.

2.2 Decline in theories of technological change

In this section I will outline some of the most prominent theories[3] in the adjacent fields of transition studies and STS on technological change that offer insightful groundwork for the study of decline, even if they tend to not discuss decline explicitly. A more detailed review can be found in Koretsky (2023). To conduct this theoretical review I searched the Web of Science for the terms "technology" in combination with (variants of) "decline", "destabilisation", "control", "discontinuity", "dismantling", "convergence", "extinction", "loss", "disappearance", "death", "abandonment", "termination" and "phase out". I compiled a list of literature of the first 600 most cited items, and manually screened their titles and abstracts for relation to the topic of the chapter, after which 43 relevant entries remained. The number of dropped entries was so high because the search terms are rather popular words found in articles that had nothing to do with the topic of technological decline. In fact, additional searches using the snowballing method, where I identified key sources and bodies of literature with the help of other literature reviews and research agendas (e.g., Köhler et al. 2017, 2019; Kanger, Sovacool and Noorkõiv 2020; Rosenbloom and Rinscheid 2020; Turnheim and Sovacool 2020), revealed more entries: namely, 107. The final list included exactly 150 academic sources. As I closely read them, I used an Excel table to note if and how the authors define and address technology, technological change, and technological decline and adjacent concepts. The literature searches were conducted in 2018 and 2022.

The literature shares conceptualisations of the object of analysis and key mechanisms of change. First, configurations (Rip and Kemp 1998; Shove, Pantzar and Watson 2012), networks (Callon 1984; Latour 1999) and systems (Hughes 1983; Geels 2002; Geels and Schot 2007; Hekkert et al. 2007; Bergek et al. 2008) appear as focal entities in the analysed literature. While understood and used quite differently, these terms stem from one ontological source—a relational understanding of the world. In these three concepts, the emphasis is on heterogeneous ensembles of artefacts, infrastructures, individuals, scientific knowledge, cultural categories, cultural/symbolic meaning, consumption patterns, industry structures, natural resources, markets, and norms and laws (Geels 2004; Shove, Pantzar and Watson 2012; Hess and Sovacool 2020). Here, society and technology, humans and things are "mutually constitutive and hopelessly mangled" (Garud and Gehman 2012: 983), i.e. co-constructed or co-produced, "with no single dimension dictating change by itself" (Sovacool and Hess 2017: 31). In this literature decline tends to be understood as break of reproduction and of ties within and shrinkage of the network/system/configuration. The concepts of configuration, network and system inform us that it could be productive to look for their disruptions when tracing and conceptualising decline.

A lot of attention in literature is given to meanings and sense-making, drawing from symbolic interactionism and structuration theory (Giddens 1984). The key observation is that there is a dichotomous relationship between structures of rules and meanings on the one hand and, on the other hand, the agency of actors, who are free to interpret and enact these structures and rules. Bijker et al. (1987) discuss these interpretations with their concept of "interpretive flexibility". Interpretive flexibility foregrounds that scientific knowledge and technological design require a negotiation between interpretations and the concomitant social processes. Van Lente and Rip (1998) further argue that expectations and promises structure actions: researchers, engineers and firms propose technology options and claim promises. When accepted, such promises are typically taken as requirements to be fulfilled. Thus, in transitions theory, expectations, promises and symbolic/cultural meanings of technology are performative forces of a socio-technical system (Geels 2002; Geels and Kemp 2007). They must, then, play a key role in decline.

A key mechanism of change in the literature is the co-evolution and co-design of the elements within the configuration, network and system. Material objects are often key here (Latour 1992; Bijker 1995; David and Schulte-Römer 2021). According to Geels (2020), STS scholarship has brought the discussion of the pervasive influence of material and technological dimensions to the mainstream sociological discussion about agency and structure. Callon (1987) famously suggested that technology is a way of studying "society in the making", and Latour (1991) saw technology as "society made durable", arguing that artefacts provide ways of anchoring new routines and practices (Geels 2020). In synthesising approaches such as the MLP or TIS materiality is included as infrastructures and resources. If materiality plays as big of a role in change as society, it must also be traced to study decline.

Next, in the literature, networks, systems and configurations are typically seen as dynamically, actively and continually forming and reforming according to the intentionality of human actors. They are constrained by an accumulated "momentum", inertia or "frame". In ANT this (re)formation work is referred to as performance, in the sense of reproduction, "making and remaking groups" (Latour 2005: 35). Similarly, Law writes about the processes of "heterogeneous engineering" in which actors try to "maintain some degree of stability in the face of the attempts of other entities or systems to dissociate them into their component parts" (Law 1987: 248). Some authors in theories of practice also refer to performance, seeing it as the reproduction of the associations between material, cognitive and symbolic dimensions of a practice (Shove, Pantzar and Watson 2012; Welch and

Warde 2015). For studies of decline it might be productive to mobilise the concept of performance to partly describe the dynamics of the configuration.

An important process in performance is learning, or creation, retention and transfer of skills, knowledge and competences. Learning is emphasised in transition theory as an indicator of success of an innovation, and is something that is protected in niches (Loorbach, Frantzeskaki and Avelino 2017). In STS, learning is involved in processes of appropriation and domestication of technology (Sørensen 2006). Learning involves processes of reframing that may result in a change of perspective among stakeholders on the problem, upon which they will try to collectively find a solution (Grin, Rotmans and Schot 2010). In practice theory, activities related to learning and other forms of competences co-constitute configurations, together with materials and meanings (Shove, Pantzar and Watson 2012). They must be key in decline as well.

Prior literature already started pushing these shared concepts of configuration/system/network, co-evolution, sense-making, materiality, performance, competences into analyses of decline. Much of the literature discusses decline of a configuration as processes of delegitimation or competition with other configurations, to which the configuration may readjust (reconfigure) or "drift" into a declined, "dormant" state. Such mechanisms are discussed across studies of industries and firms in LTS, technology life cycle and regime destabilisation literature, TIS, social practices and discontinuation (Table 2.1). Case studies of dissociations, disconnections and de-alignments have been emerging lately with work on withdrawal. There, more attention is being paid to framing and creation of competing configurations. Other approaches, such as MLP and SCOT, have not yet developed a body of empirical work on decline and remain somewhat theoretical in their descriptions of decline mechanisms.

Overall, I draw five insights from the reviewed literature for studies of decline:

- Decline is a systems process constituted by technical and sense-making dynamics.
- These technical and sense-making dynamics tend to involve a weakening of an element of the system (i.e., it becomes less stable, reliable, resilient, flexible compared to its previous condition).
- In addition, system connections may weaken, too.
- When declining, systems shrink and their role in the environment decreases.
- Due to an emergence bias that has dominated studies of technological change for so long, but also due to differing units of analysis and of observation (with, I would claim, a superfluous focus on industry and firms), there is not enough empirical granularity in existing studies of technological decline (yet).

Based on these insights, I claim that a good characterisation and middle-range theorisation of the dynamics of technological decline will need to meet the following criteria. They will need to: 1) be grounded in empirical studies of detailed granularity of analysis (Murmann and Frenken 2006), 2) appreciate the complexity and messiness of the socio-material dynamics of decline (i.e., sense-making, materiality, competences, expectations and cultural meanings), and 3) appreciate the developments outside of market and industry, as not the only arenas of socio-technical change.

Based on the insights from the existing literatures on technology and technological change, in the next section I formulate a conceptual framework to structure the analysis of the phenomenon of technological decline.

2.3 A conceptual framework and methodology to study the dynamics of decline

From the literature on technological change it appears that the basic starting point of an analysis of technologies is the acknowledgment of a strong interrelatedness of artefacts and society, economy and culture, institutions and practices. In other words, the association of material and socio-cognitive entities and their relations in a configuration are often the unit of analysis, and so they are in the present chapter.

Within this unit of analysis units of observation need to be established,[4] without which the configuration will be very difficult to study with a high degree of granularity. In an attempt to avoid a critique that analysing the intertwined socio-technical dynamics amounts to concluding that "everything is in everything" (Sovacool and Hess 2017: 15 quoting Gingras 1995), the units of observation need to be analytically distinguished from each other and analytically separated to the extent that this is possible. I draw from STS literature that analytically separates the "strands" of the entangled socio-technology: Shove and colleagues (2012) who discuss materials, meanings and forms of competence; literature on domestication (e.g., Lie and Sørensen 1996; Sørensen 2006) where symbolic, practical and cognitive work is distinguished; MacKenzie and Spinardi (1995) where "tacit knowledge, control over materials, and the translation of interests form… a necessary three-sided approach to… uninvention" (MacKenzie and Spinardi 1995: 88); and Latour and Callon, who distinguish between human and non-human actors (Callon 1984; Latour 1999). Borrowing the terminology used by Shove and colleagues (2012), I thus operationalise the "strands" that co-constitute the configuration as *materials*, *meanings* and *forms of competence*. When talking about these I will mean, respectively:

- objects, tools, hardware, infrastructures, production facilities, resources
- laws, rules, public discourses, competing narratives of supporters and opponents before and after the decline and changes in cultural meanings and regulatory regimes they may lead to (or have led to)
- tacit, codified and other knowledge on design, manufacture and use, as well as labour relations[5] (see Table 2.1).

Such separation into three strands is an analytical move and not an ontological claim, and a simplification that I commit to as an operationalisation strategy (cf. Klein and Kleinman 2002; Geels 2007).

The three strands are not monolithic and are themselves networks of different entities that co-exist and compete (Callon's "punctualisation" (1984)). The dynamics

TABLE 2.1 Units of analysis and units of observation in the conceptual framework

Unit of analysis (the 'strands')	*Units of observation*
Materials	Objects, tools, hardware, infrastructures, production facilities, material resources
Meanings	Public discourses, narratives, laws
Forms of competence	Tacit, codified and other knowledge on design, manufacture and use, labour relations

Table reprinted from Koretsky (2022b)

within the strands can be expected to be driven by the same rules as other networks, e.g. they are dynamically stable as long as there is alignment (Goulet 2021) and they can become unstable if there is too much internal contestation. I will consider that meanings, competences and materials are associated, entangled in a configuration as long as the configuration is performed, i.e. the technology is participating in manufacture or routine use. Every time the given technology is manufactured or used, the associations are reproduced and entangled tighter. This is a state of continuity, dynamic stability and the absence of decline. The reverse processes I propose to term "unravelling".

In empirical cases I follow a reactive sequence methodology (Mahoney 2000) commonly used, implicitly or explicitly, in historical sociology (Clemens 2007; Haydu 2010). Reactive sequences are a useful way to study trajectories and paths narratively, i.e. as a sequence of causally linked and temporally ordered events. Key in reactive sequence analysis is to demonstrate the causality of event chains from a key antecedent event to the studied final event (Mahoney 2000). Analysis of reactive sequences is possible with narrative explanation (Abell 2004; Clemens 2007; Haydu 2010), which goes beyond mere chronological documentation of events but orders them into a causally linked storyline. Mahoney (2000) quotes Goldstone's (1998) case of invention of a first coal-powered steam engine, developed to pump water out of flooded coal mines, as an example of a reactive sequence: "it was just chance that England had been using coal for so many centuries, and now needed a way to pump clear deep mines that held exactly the fuel needed for the clumsy Newcomen pumping machine" (Mahoney 2000: 535). In my cases the final "event" of the event chain is a (quantifiable) sign of decline of the technology in question. A narrative approach is also productive for comparison of historical cases, which is a non-trivial issue for historical studies using variables, but is more amenable in narrative explanations.[6]

In the next section I will use three historical cases on which I have elaborated elsewhere (Koretsky et al. 2022; Koretsky and van Lente 2020; Koretsky 2021) to analyse the dynamics of technological decline.

2.4 Comparing cases

The first case is a geoengineering technology of cloud seeding. Cloud seeding projects started in the USA (as well as the USSR and the UK) right after World War II. As huge amounts of government and state funding were spent in various projects over 25 years, the technology came under fire after it was used in Vietnam, alongside other disastrous uses of chemicals there. When this was revealed in the press, the negative narratives on cloud seeding ("dangerous technology", "unethical", etc.) clashed with the reinforcing ones ("new weapon", "forest fire relief", "water supply control for agriculture", etc.). This clash intensified the discussion on what cloud seeding actually is for, peace or war, with the former winning. A key event for closing the controversy was the signing of an international treaty to stop using cloud seeding for military purposes (the 1977 ENMOD treaty). Cloud seeding funding quickly dropped thereafter. Many federal-funded projects were shut down, but a lot of private and state cloud seeding endeavours continued. There cloud seeding was seen as a fire relief and water supply tool. Much cloud seeding skill and knowledge was preserved in those small-scale and often intermittent projects. In effect, the configuration of "cloud seeding" did not disappear, it just contracted. To grow it, actors would have had to revive and strengthen it. This is exactly what happened in the 1990s as new actors associated a new hyped-up meaning of "geoengineering as solution to global warming" to the pre-existing configuration. As a result, by 2010, the seeded area in the US was at least twice as large as during previous peak years.

The key events in the history of cloud seeding's incomplete unravelling were 1) the formulation of a critical narrative by the press and its quick accentuation in Congress and by civil society, who were already agitated by the pacifist and environmentalist movements, 2) the governance decisions to withdraw funding and sign the ENMOD treaty, and 3) the reinvigoration of academic and political debates (e.g. it was routinely brought up in Congress) on global warming, and cloud seeding as a possible response. Cloud seeding is a case of strongly linked configuration, whose unravelling started with internal contestation and weakening in the meanings strand, followed by a similar weakening in the materials strand. The unravelling interrupted when internal contestation within the meanings strand stopped. If no publication about the use of cloud seeding appeared in the press, it would not have or would have much later been picked up by Congress. The timing for the actual discussions was opportune as they coincided with existing environmental and pacifist societal critique. Without (or with a later) awareness in the press and Congress, cloud seeding would have gone unnoticed and would have likely continued to be funded by the government. No international treaty would have been signed or even proposed, or it would have happened much later.

The case of cloud seeding's incomplete and reversed unravelling differs from a more successful unravelling of the incandescent light bulb (ILB) in Europe. The ILB was specifically targeted in the mid-2000s by a coalition of stakeholders which included the industry, the EU and environmentalist groups. Other relevant social groups, the end users at homes and factory floor workers, were not, or only belatedly, consulted. In fact, the latter two bore a high cost for the phase-out, ranging from health effects caused by the replacement lamps, to layoffs, retraining and jobs switches of the workers involved. Although in some countries labour unions seemed to have cushioned the hit for workers, the numbers of job cuts were still two to three times larger than anticipated by the European Commission.

Similar to the cloud seeding case, the unravelling of the ILB started with an attack of a group of actors on the stabilised meanings of the ILB. The view of the ILB as a provider of pleasingly warm, "natural" light was challenged by a critical narrative of energy inefficiency, resonant with a powerful larger discourse of climate change. Within five years, similar to the cloud seeding case, instability in competences and materials followed: massive layoffs, drop of patenting activity, and drop in sales and installation capacity. What is dissimilar to the re-emerged cloud seeding is that the ILB remains in use, care (see van de Leemput and van Lente, 2023) and even production in small pockets of enthusiasts in a downscaled and ever-shrinking state.

Thus, the key events in the history of the ILB were 1) the formation of a negative narrative by a powerful social group, 2) their lobbying for anti-ILB (i.e. phase-out) policy, 3) the imitation of the ILB by the LED, and 4) the withdrawal of the ILB to small pockets of production and use. Had the key social group not formulated a negative narrative of the ILB, the European Commission would not have to draft the ban, or at least would do it later. As a consequence, without the sales ban, the ILB would remain on the shelves for much longer.

A final case I have studied is the Russian/Soviet original computer series Ural. Ural was one of the most popular computers used in industrial design, medicine, meteorology and banking in the Soviet Union in the 1960s, until it was all but gone by the 1980s (so, well before the dissolution of the country). One of the key events in the unravelling of Ural was Bashir Rameev leaving his post as the lead designer whom other, junior engineers followed. This happened during a notable ministerial rerouting of resources, thanks to a competing priority computer series ES ÈVM. As a result, materials also weakened. Namely, due to a shortage of good quality hardware, Ural designers had to downgrade their initial designs from magnetic disks to inferior punch cards and "punch tapes". This weakening caused the unravelling. The remaining team resisted the unravelling by attempting to integrate the Ural in ES ÈVM, but were not successful. Because of the difficulties already suffered, there were significant delays with production. Eventually, clients and sponsors withdrew. Subsequently, Ural-related competences slowly dispersed as team members moved on or retired. Without the continuity of the configuration, users rather quickly replaced Urals with ES ÈVMs. The only legacy it had left was the very idea of a Soviet mass computer, adopted by the ES ÈVM.

Perhaps, had Rameev remained the leader of the design team, judging from his documented leadership qualities and commitment to the Ural, he could have managed to secure resources for continual development and production of Ural. If so, it would have remained popular and could eventually be integrated into the ES ÈVM. As a consequence, the Ural may have survived until the collapse of the USSR.

When comparing these three cases, it is curious to observe that a competing/alternative configuration was present in the ILB and Ural cases, but absent in the cloud seeding case, and this was the case of re-emergence. This may be strongly indicating the importance of alternatives in decline (see Goulet, 2023). To complexify, in the ILB and Ural cases the competing configurations played slightly different roles. For both the ILB and the alternative LED, continuity was key: the competing LED was deliberately made to look like ILB in shape and colour output, and the controversy over the ILB did not end until the substitute LED technology had met all the informal standards set by the ILB (safety, full light spectrum, ease of use, familiar appearance, affordability). In Ural's case, the Ural was trying to align with the alternative, not the other way around. This may be due to a relatively low embeddedness and visibility of the Ural compared to the ES ÈVM. Thus, transformations and exchange of elements of the strands (such as visual imitation) can sometimes contribute to the speed of the unravelling, but can also be used to resist it.

Cloud seeding and the ILB followed similar pathways up to a point when cloud seeding started growing in geography and regularity of use. The decline of the ILB is, in principle, reversible as long as all strands (lamp parts, the knowledge of their construction and use, the need for lighting, etc.) are available. Such availability was there in the case of cloud seeding, and was not in the case of the Ural, and that is, I argue, why the former is re-emerging and the latter has declined.

2.4.1 Key observations

Space limitations do not allow me to compare additional cases from the unravelling perspective, although I do it elsewhere (Koretsky 2022). The cases above differ in the outcome of unravelling, with one declining, one returning, and yet another stuck in-between, contracted into a pocket. But how can we distinguish how and why unravelling differs? As a next step I will discuss three apparent aspects in which the cases of unravelling differ.

Besides the obvious observation that these are different kinds of technologies situated in differing contexts and on different scales, a first observation is that technological decline is a phenomenon in time *and* space. In all of the discussed cases, the geographical and temporal intensity of associations changed during unravelling (Figure 2.1). Configurations which are performed hardly ever and hardly anywhere (relative to before) are the declined ones. This usefully specifies the phenomenon of decline, first sketched at the start of this chapter as decrease in use and production, further as radical decrease of intensity of performance of the technology over time and space.

A second observation relates to the starting point of technological decline. In my three case studies contestation in meanings tended to catalyse unravelling of the configuration. This could be because in my cases intentional processes of decline (e.g., state driven), rather than emergent (e.g., market driven), are more prominent. Additional empirical research will be needed to draw stronger conclusions, but the leading role of meanings in unravelling could indicate that the three strands may not be equally important in *catalysing* unravelling. Of course, catalysing is only part of the unravelling processes, as I have shown in the cases.

A third observation, however, confirms an earlier STS and transition studies finding that there is no strict dichotomy between emergent and coordinated decline, however obvious it might seem. Indeed, in some cases it seems that technologies are "just" gone, i.e. turn obsolete and succumb to the forces of innovation, such as film cameras (generally) losing the competition to digital cameras. But technologies may also decline forcibly, purposefully, as a result of public or industrial policy, even if their use value is still high. There are clear examples where different fields of application in different contexts in a literal sense *declined* the usefulness of the DDT, nuclear plants and Teflon pans. The dichotomy between emergent (unintentional) and coordinated (purposeful) decline seems obvious.

However, there are also signs that such a strict dichotomy is misleading. Instances of mixtures of the two are well-known. On the one hand, some of the outcompeted technologies were planned to be outcompeted. Consider anti-competitive practices, planned obsolescence (where a company's older product is made to be outcompeted by a newer one), and, more generally, the historic role of the state in creating and supporting free markets (Polanyi 1944). From this perspective, a fully emergent quality of decline can be seriously questioned. On the other hand, top-down decline is usually first triggered by some sort of emergent destabilisation in the system. Geels and colleagues (2019) observe that decline can unfold through an external shock, technical advances or shifted demand that lead to gradual regulatory tightening or a rapid phase-out decision. Thus, it appears that instead of an "emergent-coordinated" dichotomy it may be more accurate to speak of a much messier interaction between emergent and coordinated processes. This idea is already present in the term "governance" which, in contrast to the word "govern*ment*" (Rhodes 1997), is aimed to reconcile coordination at the systems level with an emergent character of the system, arising from the interaction between multiple societal groups (EEA 2017).

In seemingly clear-cut cases of either emergent or purposeful decline (e.g., computers replacing typewriters or the DDT ban, new drugs replacing older versions or the phase-out of ozone-depleting aerosols) the key may lie in methodology: perceptions of emergence or purposefulness depend on the part of the historical event chain one looks at (Pettigrew 1990). For example, in the case of British coal power (Turnheim and Geels 2012, 2013), one could study coordinated decline within a period starting in the year 1975 when the Margaret Thatcher government closed down the coal mines, or one could study emergent decline within a period starting in the year 1925 and see falling competitiveness of British coal and public protests against it. Similarly, in all three cases, emergent dynamics led up to a phase-out decision. From the STS perspective, the scale of analysis itself tends to skew the conclusions: emergence, autonomy and technological determinism are more (albeit deceivingly) apparent if larger timeframes are considered (Misa 1988; also Wyatt 2008). Thus, attempting to identify some essential emergent or purposeful character of decline becomes meaningless, and it is more meaningful to speak of a messy interaction of emergent or coordinated processes.

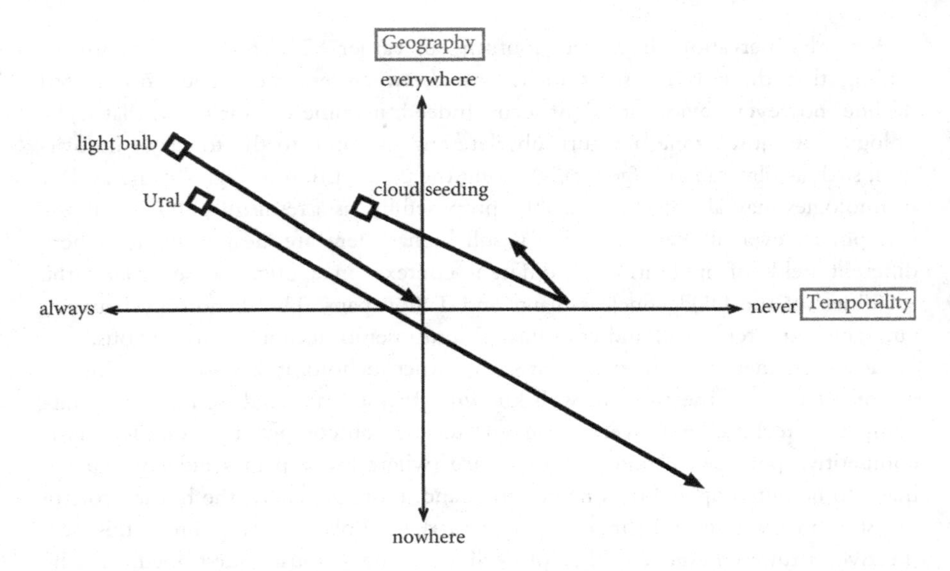

FIGURE 2.1 Sketching a relationship between relative geographic and temporal intensity performance of configuration, with decline in the bottom right quadrant
Figure modified from Koretsky (2022b)

Similar to causes, processes of decline may also be said to have more of the emergent or more of the coordinated processes, while, in fact, both types are always present. Depending on cases where either emergent or coordinated processes matter more, one may observe differences between actors' actions. With emergent decline a firm may try increasingly desperate strategies to preserve the status quo (Turnheim 2012), and with coordinated decline a firm may have time to prepare, or it even supports the decline, so the phase-out of the technology in question can go on faster and relatively orderly (Stegmaier et al. 2023; Stegmaier, Visser and Kuhlmann 2021). It was not among the aims of the present chapter to compare emergent and purposeful decline processes. It is, however, a fruitful topic for future research, for instance, to discuss solutions that could avoid the (political) gap between proponents of strictly market and strictly regulatory solutions.

Based on these observations, I am now in a position to formulate a generic representation of technological decline and to present a working approach for studying future cases of decline.

2.5 A working unravelling approach of technological decline

As discussed in section 2.2, a theorisation of the dynamics of technological decline will need to: have empirical grounding of detailed granularity, appreciate the complexity and messiness of the socio-material dynamics of decline, and be middle range. I take the socio-technical configuration as the main ontological entity to start outlining a working perspective on the dynamics of technological decline. I proposed in section 2.3, the three strands of the configuration are not monolithic and are themselves networks of different entities that co-exist and compete. I propose that the strands remain dynamically stable as long as there is alignment of these constituting entities, and they turn unstable when there is misalignment, i.e., too much internal contestation. When this dynamic stability is significantly decreased even in one strand, the configuration cannot be performed properly (i.e., manufactured or routinely used), and the process of the unravelling of the configuration starts. Shove and colleagues (2012) describe analogous dynamics in practices with the metaphor of an electric circuit to emphasise the sequential character of changes. In the following I will formulate propositions on the processes of misalignment within the strands, and then turn to the unravelling dynamics.

2.5.1 Misalignment in materials

The discussed cases have shown that the material strand may become misaligned due to changes in materialities (broadly understood), so changes in physical availability, technical and design characteristics, material degradation, loss of funding, etc. The alignment in the material strand may be linked with the concept of "obduracy" of the focal technology: i.e. the inflexibility of a technology as a result of materialised past decisions (Hommels 2005). Obduracy inherits from prior debates on agency of smaller mundane artefacts, such as seat belts or doors (Latour 1992). In both small- and large-scale objects, obduracy represents continuity; it both stands in the way of the new and preserves the potentiality of the old to be revived. Although these studies do not make such direct connections themselves, STS literature discusses repairability (Sims and Henke 2012; Cohn 2016; Russell and Vinsel 2018) as a way to maintain obduracy.

The misalignment in materials may start an unravelling of the configuration by triggering misalignment in forms of competences, as in the case of the ILB: shutting down the production facilities forced the workforce to transfer, seek new jobs or retire. Another example is the Walkman, related competences of which (such as knowing how to manually rewind a cassette) were replaced by the CD player-related competences.

Meanings may be affected by the misalignment in materials as well. Sims (2009) uses the term "slippage": a mismatch between the expectations of how technology should be performing with perceptions of how it is actually performing. As a result of slippage, either degradation or obsolescence occurs. In the Ural's case, for instance, I observe obsolescence: the last Urals in the late 1970s were perceived as years behind the new technical advances. Maintenance work is required to counteract slippage (Hommels 2005; Howe et al. 2016; van de Leemput and van Lente, 2023).

What, then, becomes of the actual artefacts as a result of misalignment and unravelling? My cases show that they get dismantled (Ural) or used in small pockets (the ILB and cloud seeding). Shove et al. (2012) note that in such cases they disappear with little or no trace, become dormant (i.e., not gone, but also not performed), or connect to other configurations. Insights from sociological studies of debris (e.g., Edensor 2005; Qviström 2012; Stoler 2013a; Schopf and Foster 2014) are useful to expand on the fates of materials. In particular, Stoler notes that abandoned materiality can be more than inert, can "hold and spread… toxicities and become poisonous debris" (Stoler 2013b: 13), their locations can become cherished or dreaded history.

2.5.2 Misalignment in meanings

The discussed cases indicate that misalignment in meanings may be caused by the presence of strong competing negative meanings. Such misalignment may be seen on a societal level as delegitimation, and on a more individual level as the inverse of "domestication" (Lie and Sørensen 1996; Sørensen 2006) of a given technology. This can be seen, for example, in the clash of positive and negative interpretations of cloud seeding right after the revelations of the seedings in Vietnam. According to the European Environment Agency, "[m]edia campaigns, public debates, (scientific) publications and reports can advance particular frames, discourses and metaphors that erode the cultural legitimacy of technologies or practices" (Geels et al. 2019). Those are what Roberts calls "negative storylines" (Roberts 2017), the inverse of what Levidow and Upham (2017) describe as "anchoring", i.e., as the loss of recognition of the given technology and the resulting decline of related communication and coordination activities across social groups. The end result of misaligned, contested meanings is a discredited, marginalised and/or forgotten technology.

Misaligned meanings may affect materials, forcing resources to be rerouted away, as seen in the cases of Ural, ILB and cloud seeding, as well as the decommissioning of nuclear reactors in Germany, Italy and other places. Forms of competences may misalign due to a misalignment in meanings as well: consider a decline in smoking after strong contestation of prior meanings of smoking as fashionable and even healthy (Oreskes and Conway 2010).

2.5.3 Misalignment in forms of competences

Misalignment in forms of competences is harder to conceptualise. In the discussed cases, misalignment in forms of competences was largely caused by shortage of competences embodied in the workers, but also codified in manuals. Thus, loss of competences is much more observable in the cases than their misalignment and contestation. This loss can be seen in the closure of ILB manufacture facilities and the slow decline of the Ural team of designers and their expertise. Loss of competences can affect meanings via reframing of demands, standards and perceptions of performance (Borup et al. 2006; van Lente et al. 2013), as well as materials via degradation due to lack of repair and maintenance.

Michener and colleagues (1997) propose a model that could be useful for the study of loss of competences. They study degradation of methodological instructions in scientific organisations: from the fullest amount of content shortly prior to and at the time of publication of the instructions, to losses of specificity with repeated revisions, to accidental loss of data, to losses due to retirement and career changes, to, finally, losses of remaining records due to death of persons who created the information. Such unravelling may be called "collective forgetting" (Shove, Pantzar and Watson 2012) or "unlearning", the process of discarding knowledge (Becker 2005). In the Ural and the ILB cases most of the competences were forgotten as workers retired or moved on and codified knowledge became more cryptic, whereas in cloud seeding much of the competences travelled to the adjacent field of meteorology.

The case of unlearning in nuclear power in the post-Soviet context is informative in illustrating the effects of loss of competences in materials. After the collapse of the Soviet system, many kinds of specialists in Eastern Europe could not find work anymore and moved to more economically stable places. This had negative repercussions for the often high-tech industries that they had left, such as nuclear power. Sturm (1993) documents the unlearning of nuclear energy operations on a societal scale that led to neglect and degradation in materials (increased frequency of service interruptions, mainly due to equipment failure). Sturm concludes that the reason was indeed the loss of competences: "The departure of… operators, for example, left the… station without enough skilled personnel to operate all its reactors" and "[t]he lack of domestic experience and comprehensive manufacturing capabilities in Eastern Europe—each country specialized in specific aspects—have exacerbated difficulties in obtaining spare parts and maintaining plants" (Sturm 1993: 188).

2.5.4 Dynamics and pathways of unravelling of the configuration

As seen above, if we conceptualise technological decline as socio-material dynamics of a configuration of three strands, materials, meanings and forms of competences, we may shed more light on the mechanisms of said decline. The mechanisms are: misalignment in the strands, their dissociation, and the "unravelling" of the configuration.

One of the key insights of STS is that holding the configuration together requires work (Law 1987; Latour 2005). In fact, the stability of a socio-technical configuration is an impressive and rare achievement of actors in the configuration (Latour 2005). Bourne suggests that configurations "consist of trajectories that seem to want to pull apart or decay" (Bourne 2016: 18). Invention, innovation, investment, routine maintenance, continual manufacture and/or use are needed to delay that pulling apart because they help maintain the resilience of the strands to internal contestation.

As seen in some cases, unravelling is stimulated by coming in contact with competing configurations that put pressure on the strands of the focal configuration. Such was the case of the light bulb and the LED, the Ural and the ES ÈVM, the Polaroid and the digital camera (Minniti 2016), writing ink and the ballpoint pen (Shove, Pantzar and Watson 2012), coal-based power generation and oil-based power generation (Turnheim and Geels 2012), and probably many more—perhaps most—technologies.

As the cases demonstrate, decline is not a binary switch, but a spectrum between continual performance (i.e., use and/or production) and a lack thereof. In section 2.4 I visualised decline as a radical decrease of intensity of performance across time and space, where unravelling is a vector, or directionality, of technology towards decline. I can now be more specific and demarcate two types of decline indicated by the degree of performance intensity: weak and strong. Weak decline indicates a state where technology is simply not used or manufactured anymore, while all of the strands remain intact. For instance, it may be said that the technology of manually prepared writing ink is in a weak decline because all of the components continue to exist: codified knowledge to manually prepare the ink is noted down in books, the necessary ingredients can be found, and the meaning of "ink" as a coloured liquid material for writing and printing persists. Or in the UK in the 1970s and 1980s (Shove 2012), cycling may be said to have been in weak decline because at least one meaning of the bicycle persisted (transport or exercise), the competences to make and ride bicycles had not disappeared, and bicycles and the lanes were still around. Strong decline indicates the completeness of misalignment in strands and unravelling between them (Figure 2.2), and is thus discontinuous decline. The difference between weak and strong decline amounts to how likely the technology is to return.

Figure 2.2 is a rough sketch for illustrative purposes, and "everywhere", "nowhere", "always" and "never" should be interpreted not as quantifiable states,

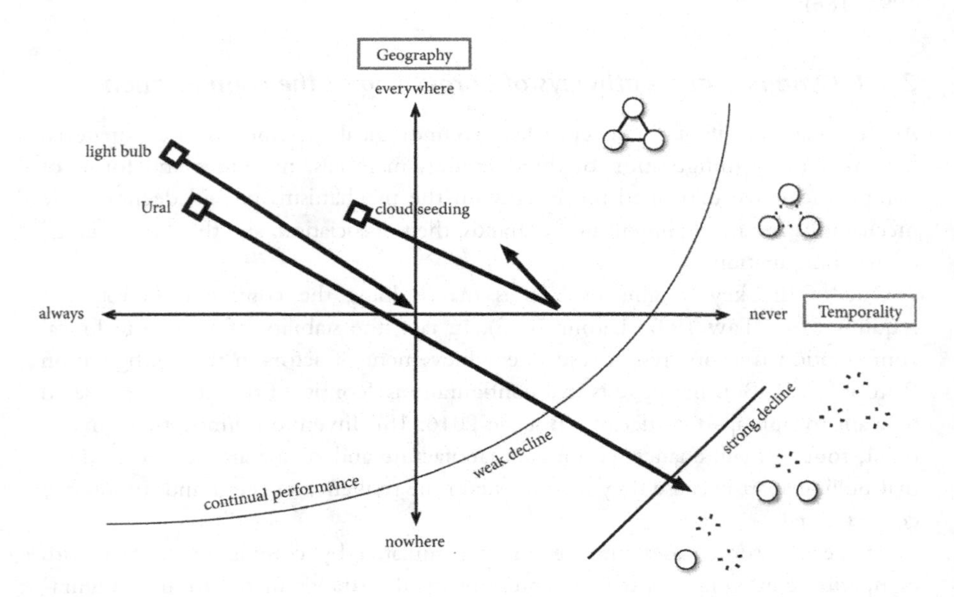

FIGURE 2.2 Sketching decline as a relationship between relative geographic and temporal intensity of performance of configuration and types of unravelling

Source: Figure modified from Koretsky (2022b)

but as general indicators of high or low geographic or temporal performance *in a given geographic and temporal context.* In this sense, weak and strong decline, in a given geographic and temporal context, are ideal-types rather than empirical evaluations. Other dimensions, such as the scale of industry and the policy environment, matter, of course; they are not foregrounded in the figure, but are indicated by the different states of the configuration (the three circles).

As a next step, the distinction between weak and strong decline allows me to portray unravelling as part of certain (ideal-type) pathways of socio-technical development (Figure 2.3). Note that in the figure the axes are geography and temporality, and the exact shape of the arrow is only an indication.

1. Fall of intensity of performance.

The technology is used and/or manufactured less than before, its trajectory is aiming downward. Such is the case of the incandescent light bulb (e.g., in the EU), but also of for example coal-fired power plants (e.g., in the UK) and nuclear power in Germany.

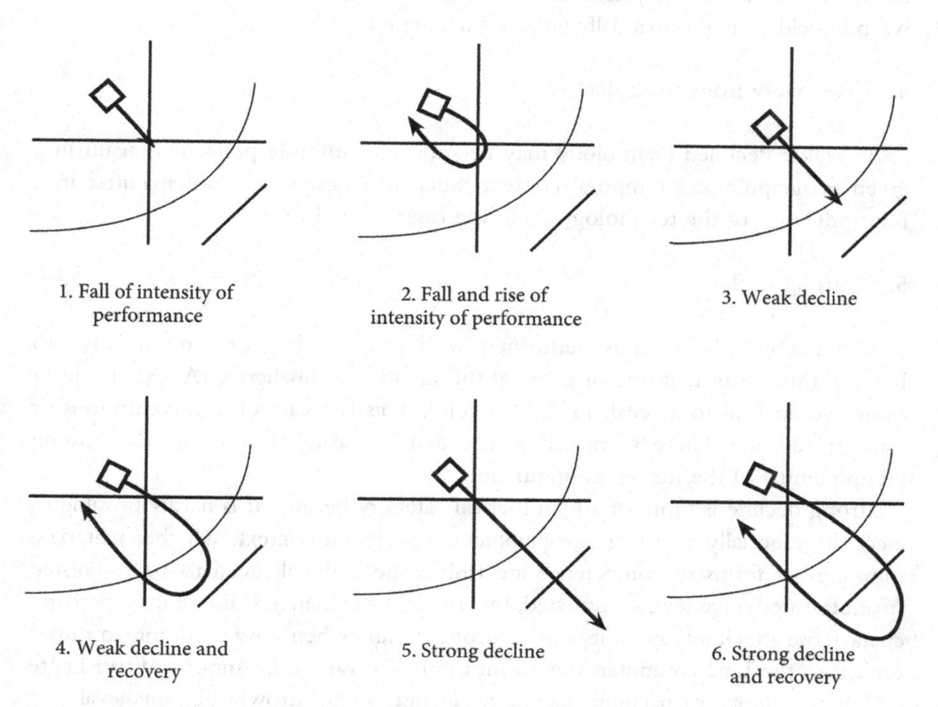

FIGURE 2.3 Six non-sequential ideal-type pathways of unravelling overlaid on the axes of geography-temporality (shape of the arrows are for illustration only and will differ per case)

Source: Figure modified from Koretsky (2022b)

2. Fall and rise of intensity of performance.

Here the configuration remains in the "continual performance" area of Figure 2.2 and no decline takes place. Configurations are in a state of reduced performance, fluctuating between higher and lower performance intensity (in a given geographic and temporal context). One example is cloud seeding which has been seeing a comeback (e.g., in the 2008 and 2022 opening ceremonies of the two Beijing Olympics).

3. Weak decline.

A configuration which continues in the direction of decreased performance intensity ends up in a state of "weak" decline in a given geographic and temporal context. Here, the technology is performed so rarely and in such a geographically dispersed way that the associations between the strands are barely there at all. The networks of producers, users and/or maintainers have significantly shrunk. One could say such technologies are on "standby", although not gone yet. Most technologies, "declined" in the colloquial use of the word, can be placed in the category of "weak" decline, such as nuclear weapon field testing or manually prepared writing ink.

4. Recovery from weak decline.

A weakly declined technology may become continuously performed again in a given geographic and temporal context. Such re-emergence could manifest in a reintroduction of the technology, as in the case of vinyl records.

5. Strong decline.

When a configuration is performed with less and less intensity (relative to before), this means that one or more of the strands are misaligned. A loss of one or more would lead to a break in the "circuit". This is a case of a discontinuous or "strong" decline where work will be needed to create or re-create the missing component(s) of the former configuration.

Strong decline is more of a hypothetical category because it is hardly possible to ascertain, especially in a large geographic and temporal context, whether materials, meanings or forms of competence are truly gone and will need to be re-created (manufactured, invented, constructed, theorised, learned, etc.). But one may, perhaps, characterise a technology on its way to strong decline when it is found only in museums, e.g., the Ural computer, the 16 mm film (see van de Leemput and van Lente 2023), the linotype machine, kamikaze aircraft, stone arrowheads, medieval and antique construction technologies and other traditional crafts. Strong decline occurs when associations are impossible because slippage, de-anchoring and/or unlearning are too profound (e.g., all materials are destroyed, carriers of competences or knowledge are gone, specific parts of the technology are banned from use or manufacture). After a strong decline, all that remains are dissociated materials, meaning and competences "debris" in the form of "fossilised" (Shove and Pantzar 2005) rust, hints and secrets.

6. Recovery from strong decline.

Even after a strong decline, a technology can hypothetically make a U-turn and re-emerge. From the perspective of the actors involved in the re-creation of materials, meanings and/or forms of competences, this pathway will probably look like the process of invention. Re-emergence in this case occurs when work has been done to (re)construct the strands and re-associate, "re-ravel" them. A weak decline is easier to recover from as to reintroduce a strand (back) to the configuration is less costly.

All three cases I have explored above have passed through the decrease of intensity of performance. The decrease has stopped in the case of the light bulb because the ILB has found its niche in industrial applications and horticulture, as well as among people with preferences for the ILB. The light bulb can be categorised as a pathway 1 exemplar. The intensity of performance has actually reversed for cloud seeding, so it is a pathway 2 exemplar. The Ural's status could be categorised as weak decline because it is not used nor produced as prominently as before. However, the Ural has been delegitimised and for a long time forgotten, and the associated competences are gone with the people who retired and can hardly be found in a codified form. The only intact Ural can be found in a museum and its (cultural) legacy has been largely forgotten. Thus, the Ural may be better categorised as a pathway 5 exemplar, or, bearing in mind the difficulties of ascertaining strong decline, at least on its way to it.

2.6 Conclusions

Typically, policy-makers' focus of nurturing novelties overshadows a focus on initiatives and the ongoing decline of technologies. Such overshadowing might be fuelled by a utilitarian logic of rational progression towards more efficiency. The assumption is that superior technology substitutes the inferior one and the latter then proceeds to fade away, or, as Lindqvist hyperbolised: "[as if] all bicycles [were] thrown in the ditch and all horses shot when the first Model T rolled off the assembly line" (Lindqvist 1994). However, there has been a rise in deliberate decline policies since the late 2010s (Green 2018). Perhaps, actors start recognising that, according to Heyen and colleagues, a "regulated phase-out can offer legal as well as planning certainty for business, workers, consumers and infrastructure planning" (Heyen, Hermwille and Wehnert 2017: 327).

My conceptual starting point, drawn from the literature, was to view technology as a socio-technical configuration of entangled socio-material "strands" (materials, meanings and forms of competences). Their dissociation, or unravelling, is proposed as a key mechanism behind decline. Decline, I found, is not, in principle, irreversible, the strands can be re-associated by actors, and this can lead to re-emergence. From this starting point I compared my cases and gained additional insights on how to characterise decline and how to adjust and further develop the conceptual framework. Based on the framework and backed up by an analysis of three cases, I proposed a working approach to study future cases of decline and represent the constituting processes.

The proposed unravelling working approach identifies processes within each strand and offers six ideal typical pathways. The proposed working approach will need further fleshing out and adjustment, but already holds, I suggest, four benefits for the field of STS and sustainability transitions. First, in alignment with the symmetry principle, I propose with this working approach to focus more explicitly on a given technology's "end of life" to counteract the prevalence of studies on emergent technology. Second, I propose that a pragmatic analytical distinction between social and material processes within decline is useful to show their entangled dynamics (cf. Klein and Kleinman 2002; Geels 2007). Third, I propose to focus more on high empirical granularity of analysis and on generation of rich case descriptions that are used towards middle-range theorisation (cf. Frickel and Moore 2006; Wyatt and Balmer 2007). And fourth, I propose to focus more on the analysis of processes outside of market and industry dynamics, since labs, public places, parliaments, newsrooms, kitchens and other sites are equally important (cf. Polanyi 1944; Hughes 1994; Latour 2007; Soete 2011; Upham et al. 2015).

Since the presented unravelling approach stems from the STS and transitions literature, it draws, just like this literature, from ecology, complex systems theory, resilience research, ecosystem services, adaptive governance, institutional theory and more. There is room for more connections and adjustments from those literatures for empirical work, theorisation and design of policy recommendations.

This chapter is not without limitations. The separation of technology into three "strands" can be questioned, as well as the ontological status of these strands. Specifically, my finding regarding the prominent role of meanings in nearly all presented cases could imply that the separation was invalid in the first place. I would challenge such possible critique with a reminder that a distinction between the three strands is analytical; they were never supposed to be exclusive and non-overlapping. Of course technology and society are deeply implicated, but the separation is no more than an analytical attempt to navigate the complexities in *what causes* unravelling and in *how* it *proceeds*. Similarly, the somewhat uncompromising collating of different things into the three strands can be questioned. Indeed, this is a radical simplification inherited from the literature on the dynamics of social practice theory. I tried to alleviate this by clearly operationalising what these strands are constituted of and maintaining internal consistency between cases by strictly adhering to this operationalisation. Additional research could no doubt improve the proposed approach on both accounts.

Second, the role of actors, structures, infrastructures, levels, knowledge paradigms and ideologies could have been studied more systematically in this work. Future research could go beyond the implicitly flat ontology of the present work and analyse how the different levels of aggregation matter in processes of decline. It could also be productive to study local situations from the unravelling perspective (e.g., city or regional bans). Transition studies stem from a systems perspective. By definition, commodities, or end products, are not resilient to decline, and are not obdurate; to phase out a commodity is fruitless because it can be, by definition, replaced. This is why destabilisation and phase-out of *systems* is needed. While I have considered configurations in this chapter, I have not focused on what is typically understood as systems in transitions literature, i.e. configurations that provide societal functions such as mobility, food or energy. The proposed approach would benefit from exploration of such cases, but decline of configurations smaller than systems is (also) an occurring phenomenon worth studying.

Although I have not studied these larger socio-technical systems, the presented unravelling perspective might be applicable in studying them too, bringing it closer to the MLP. It is then the responsibility of the analyst to delimit the unit of analysis and maintain it throughout the study. Unruh (2002) writes that discontinuity on a sub-system level (e.g., internal combustion engine and fuel cells) can look like continuity on a system level (transportation). This line of thought may be continued by observing that higher levels of aggregation (e.g., transportation) have more inertia and thus are significantly less susceptible to decline. A specific model of a smartphone is easier to phase out (even though it, being a commodity, will probably be quickly replaced) than the smartphone as a communication technology.

Third, the proposed working approach is limited in its ability to prescribe how to trace the "same" technology. A technology changes with time, thus how to know when change stops and unravelling begins? In addressing this limitation, I have, again, followed the rationale in transition studies where the responsibility to define the object of analysis and keep it in focus lies with the analyst (Geels 2011; Stirling 2019).

Lastly, although I was motivated to understand decline as a problem of sustainability and decarbonisation within the larger challenge of the climate crisis, the discussed cases only indirectly relate to sustainability. This, however, should not disqualify the observed patterns, and, if anything, expands the programme of understanding and engaging with decline into issues beyond sustainability and decarbonisation. There is no reason why phasing out established systems cannot be equally useful for dealing with ethical questions, for instance responsibility in deliberate decline, nuclear weapons field testing, mass production, surveillance and more (Escobar 2015; Bostrom 2019). Further research could investigate the connection of decline and phase-out with (post-)colonialism, imperialism and other forms of exploitation, both to study the caused decline and the ways to cause decline of these forms. It is a task of future research (see other contributions to this volume) to study more cases of decline of various technologies for the sake of sustainability and topics beyond it.

Notes

1 This chapter is based on parts of my PhD dissertation (Koretsky 2022) titled "Unravelling: The Dynamics of Declining Technologies", where it appears in a more expanded version. This research was supported by Horizon 2020 project FIT4RRI (grant agreement

741477). I thank my supervisors Harro van Lente and Ragna Zeiss, without whose generous support and reviews this work would not have been possible. I also thank Peter Stegmaier and Bruno Turnheim for the review and useful suggestions on the present chapter.

2 Among such prospects, which are reminders of the urgency of the problem, is Paterson's (2020) bleak forecast of civilizational collapse (absolute declines in human populations, collapse of food production systems, collapse of social institutions) in this century as a result of the climate crisis.
3 I take theory broadly here, including approaches and conceptual frameworks, or as "a statement of relations among concepts within a boundary set of assumptions and constraints" (Bacharach 1989: 496).
4 The distinction between unit of analysis (key focal entity of the study (Sheppard 2020)) and unit of observation (entities to be observed, measured or otherwise studied to learn more about the unit of analysis (Sheppard 2020)) is more typical in mainstream sociology, and here I find it useful for the clarity of methodology.
5 Forms of competence can also be understood from the economics perspective as "intellectual capital".
6 A theoretical explanation of this point may be found in Haydu (1998) and Mahoney (2000).

References

Abell, P. (2004) Narrative explanation: An alternative to variable-centered explanation?, *Annual Review of Sociology*, 30, 287–310.

Anderson, K. and Peters, G. (2016) The trouble with negative emissions, *Science*, 354(6309), 182–183. doi:10.1126/science.aah4567.

Ayres, R.U. (1987) *The Industry-Technology Life Cycle: An Integrating Meta-Model?*Laxenburg.

Bacharach, S.B. (1989) Organizational theories: Some criteria for evaluation, *The Academy of Management Review*. doi:10.2307/258555.

Becker, K.L. (2005) Individual and organisational unlearning: Directions for future research, *International Journal of Organisational Behaviour*, 9(7), 659–670.

Bergek, A. *et al.* (2008) Analyzing the functional dynamics of technological innovation systems: A scheme of analysis, *Research Policy*, 37(3), 407–429. doi:10.1016/j.respol.2007.12.003.

Bijker, W.E. (1995) *Of Bicycles, Bakelites, and Bulbs: Toward a Theory of Sociotechnical Change*. MIT Press. doi:10.2307/2077312.

Bijker, W.E., Hughes, T.P. and Pinch, T. (1987) *The Social Construction of Technological Systems: New Directions in the Sociology and History of Technology*. MIT Press.

Borup, M. *et al.* (2006) The sociology of expectations in science and technology, *Technology Analysis & Strategic Management*, 18(3–4), 285–298. doi:10.1080/09537320600777002.

Bostrom, N. (2019) The vulnerable world hypothesis. *Global Policy*. doi:10.1111/1758-5899.12718.

Bourne, M. (2016) Invention and uninvention in nuclear weapons politics. *Critical Studies on Security*, 4(1), 6–23. doi:10.1080/21624887.2015.1106427.

Callon, M. (1984) Some elements of a sociology of translation: Domestication of the scallops and the fishermen of St Brieuc Bay. *Sociological Review*, 32(S1), 196–233. doi:10.1111/j.1467-954X.1984.tb00113.x.

Callon, M. (1987) Society in the making: The study of technology as a tool for sociological analysis. In Bijker, W.E., Hughes, T.P. and Pinch, T. (eds) *The Social Construction of Technological Systems: New Directions in the Sociology and History of Technology: Anniversary Edition*. MIT Press.

Christensen, C.M. (2006) The ongoing process of building a theory of disruption. *Journal of Product Innovation Management*, 23, 39–55. doi:10.1111/j.1540-5885.2005.00180.x.

Clemens, E.S. (2007) Toward a historicized sociology: Theorizing events, processes, and emergence. *Annual Review of Sociology*, 33, 527–549. doi:10.1146/annurev.soc.33.040406.131700.

Cohn, M.L. (2016) Convivial decay: Entangled lifetimes in a geriatric infrastructure. In *Proceedings of the ACM Conference on Computer Supported Cooperative Work, CSCW*. doi:10.1145/2818048.2820077.

David, M. and Schulte-Römer, N. (2021) Phasing out and in: System transition through disassociation in the German energy transition – The case of light and coal. *Energy Research and Social Science*, 80, 102204. doi:10.1016/j.erss.2021.102204.

Edensor, T. (2005) The ghosts of industrial ruins: Ordering and disordering memory in excessive space. *Environment and Planning D: Society and Space*, 23(6), 829–849. doi:10.1068/d58j.

EEA (2017) *Perspectives on transitions to sustainability*. EEA.

Eisenhardt, K.M. and Graebner, M.E. (2007) Theory building from cases: Opportunities and challenges. *Academy of Management Journal*, 50(1), 25–32. doi:10.5465/AMJ.2007.24160888.

Escobar, A. (2015) Degrowth, postdevelopment, and transitions: A preliminary conversation. *Sustainability Science*, 10, 451–462. doi:10.1007/s11625-015-0297-5.

Frickel, S. and Moore, K. (2006) Prospects and challenges for a new political sociology of science. In *The New Political Sociology of Science: Institutions, Networks, and Power*. University of Wisconsin Press.

Garud, R. and Gehman, J. (2012) Metatheoretical perspectives on sustainability journeys: Evolutionary, relational and durational. *Research Policy*, 41(6), 980–995. doi:10.1016/j.respol.2011.07.009.

Geels, F.W. (2002) Technological transitions as evolutionary reconfiguration processes: A multi-level perspective and a case-study. *Research Policy*, 31(8–9), 1257–1274. doi:10.1016/S0048-7333(02)00062-8.

Geels, F.W. (2004) From sectoral systems of innovation to socio-technical systems: Insights about dynamics and change from sociology and institutional theory, *Research Policy*, 33(6–7), 897–920. doi:10.1016/j.respol.2004.01.015.

Geels, F.W. (2007) Feelings of discontent and the promise of middle range theory for STS: Examples from technology dynamics, *Science Technology and Human Values*, 32(6), 627–651. doi:10.1177/0162243907303597.

Geels, F.W. (2011) The multi-level perspective on sustainability transitions: Responses to seven criticisms. *Environmental Innovation and Societal Transitions*, 1(1), 24–40. doi:10.1016/j.eist.2011.02.002.

Geels, F.W. (2018) Disruption and low-carbon system transformation: Progress and new challenges in socio-technical transitions research and the multi-level perspective. *Energy Research and Social Science*, 37, 224–231. doi:10.1016/j.erss.2017.10.010.

Geels, F.W. (2020) Micro-foundations of the multi-level perspective on socio-technical transitions: Developing a multi-dimensional model of agency through crossovers between social constructivism, evolutionary economics and neo-institutional theory. *Technological Forecasting and Social Change*, 152, 119894. doi:10.1016/j.techfore.2019.119894.

Geels, F.W. and Kemp, R. (2007) Dynamics in socio-technical systems: Typology of change processes and contrasting case studies. *Technology in Society*, 29(4), 441–455. doi:10.1016/j.techsoc.2007.08.009.

Geels, F.W. and Schot, J. (2007) Typology of sociotechnical transition pathways. *Research Policy*, 36(3), 399–417. doi:10.1016/j.respol.2007.01.003.

Geels, F.W. *et al.* (2019) *Sustainability Transitions: Policy and Practice*. Publications Office of the European Union. doi:10.2800/641030.

Giddens, A. (1984) *The Constitution of Society: Outline of the Theory of Structuration*. University of California Press. doi:10.1007/BF01173303.

Goldstone, J.A. (1998) The problem of the "early modern" world. *Journal of the Economic and Social History of the Orient*, 41(3), 249–284. doi:10.1163/1568520981436246.

Goulet, F. (2021) Characterizing alignments in socio-technical transitions: Lessons from agricultural bio-inputs in Brazil. *Technology in Society*, 65, 101580. https://doi.org/10.1016/j.techsoc.2021.101580.

Goulet, F. (2023) The role of alternative technologies in the enactment of (dis)continuities. In Koretsky, Z. *et al.* (eds) *Technologies in Decline: Socio-Technical Approaches to Discontinuation and Destabilisation*. Routledge.

Green, F. (2018) The logic of fossil fuel bans. *Nature Climate Change*, 8, 449–451. doi:10.1038/s41558-018-0172-3.

Grin, J., Rotmans, J. and Schot, J. (2010) Introduction: From persistent problems to system innovations and transitions. In *Transitions to Sustainable Development: New Directions in the Study of Long Term Transformative Change*. Routledge. doi:10.4324/9780203856598.

Hamlin, C. (2007) STS: Where the Marxist critique of capitalist science goes to die? *Science as Culture*, 16(4), 467–474. doi:10.1080/09505430701706780.

Haydu, J. (1998) Making use of the past: Time periods as cases to compare and as sequences of problem solving. *American Journal of Sociology*, 104(2), 339–371. doi:10.1086/210041.

Haydu, J. (2010) Reversals of fortune: Path dependency, problem solving, and temporal cases. *Theory and Society*, 39(1), 25. doi:10.1007/s11186-009-9098-0.

Hekkert, M.P. *et al.* (2007) Functions of innovation systems: A new approach for analysing technological change. *Technological Forecasting and Social Change*, 74(4), 413–432. doi:10.1016/j.techfore.2006.03.002.

Hess, D.J. and Sovacool, B.K. (2020) Sociotechnical matters: Reviewing and integrating science and technology studies with energy social science. *Energy Research and Social Science*, 65, 101462. doi:10.1016/j.erss.2020.101462.

Heyen, D.A., Hermwille, L. and Wehnert, T. (2017) Out of the comfort zone! Governing the exnovation of unsustainable technologies and practices. *GAIA*, 26(4), 326–331. doi:10.14512/gaia.26.4.9.

Hommels, A. (2005) *Unbuilding Cities: Obduracy in Urban Sociotechnical Change*. MIT Press.

Howe, C. *et al.* (2016) Paradoxical infrastructures: Ruins, retrofit, and risk. *Science Technology and Human Values*, 41(3), 547–565. doi:10.1177/0162243915620017.

Hughes, T.P. (1983) *Networks of Power: Electrification in Western Society, 1880–1930. Johns Hopkins University Press*. doi:10.2307/3104214.

Hughes, T.P. (1994) Beyond the economics of technology. In Granstrand, O. (ed.) *Economics of Technology*. Elsevier.

Kanger, L., Sovacool, B.K. and Noorkõiv, M. (2020) Six policy intervention points for sustainability transitions: A conceptual framework and a systematic literature review. *Research Policy*, 49(7), 104072. https://doi.org/10.1016/j.respol.2020.104072.

Kierkegaard, M.K. (2014) Lights out for traditional bulbs: Lobbyism and government intervention. *International Journal of Private Law*, 7(3), 197–270. doi:10.1504/IJPL.2014.062986.

Klein, H. and Kleinman, D.L. (2002) The social construction of technology: Structural considerations. *Science Technology and Human Values*, 27(1), 28–52. doi:10.1177/016224390202700102.

Köhler, J. *et al.* (2017) *A Research Agenda for the Sustainability Transitions Research Network*. Sustainability Transitions Research Network. doi:10.1016/j.jcp.2005.11.001.

Köhler, J. *et al.* (2019) An agenda for sustainability transitions research: State of the art and future directions. *Environmental Innovation and Societal Transitions*, 31, 1–32. doi:10.1016/j.eist.2019.01.004.

Koretsky, Z. (2021) Phasing out an embedded technology: Insights from banning the incandescent light bulb in Europe. *Energy Research and Social Science*, 82, 102310. https://doi.org/10.1016/j.erss.2021.102310.

Koretsky, Z. (2022) *Unravelling: The Dynamics of Technological Decline*. Maastricht University.

Koretsky, Z. (2023) How do technologies die? Studies of decline in literature on technological change. In Goulet, F. and Vinck, D. (eds) *Doing Without, Doing With Less: The New Horizons of Innovation*. Edward Elgar.

Koretsky, Z. and van Lente, H. (2020) Technology phase-out as unravelling of socio-technical configurations: Cloud seeding case. *Environmental Innovation and Societal Transitions*, 37, 302–317. doi:10.1016/j.eist.2020.10.002.

Koretsky, Z., Zeiss, R. and van Lente, H. (2022) Exploring the dynamics of technology phase-outs through the history of a Soviet computer "Ural" (1955–1990). *Science, Technology & Human Values*. https://doi.org/10.1177/01622439221130139.

Latour, B. (1991) Technology is society made durable. *The Sociological Review*, 38(1), 103–131. doi:10.1111/j.1467-954x.1990.tb03350.x.

Latour, B. (1992) Where are the missing masses? The sociology of a few mundane artifacts. In Bijker, W.E. and Law, J. (eds) *Shaping Technology / Building Society*. MIT Press. doi:10.2307/2074370.

Latour, B. (1999) *Pandora's Hope: Essays on the Reality of Science Studies*. Harvard University Press.

Latour, B. (2005) *Reassembling the Social: An Introduction to Actor-Network Theory*. Oxford University Press.

Latour, B. (2007) Turning around politics: A note on Gerard de Vries' paper. *Social Studies of Science*, 37(5), 811–820. doi:10.1177/0306312707081222.

Law, J. (1987) Technology and heterogeneous engineering: The case of Portuguese expansion. In Bijker, W.E., Hughes, T.P. and Pinch, T. (eds) *The Social Construction of Technological Systems: New Directions in the Sociology and History of Technology: Anniversary Edition*. MIT Press.

Levain, A. *et al.* (2015) Continuous discontinuation – The DDT Ban revisited. In *International Sustainability Transitions Conference "Sustainability Transitions and Wider Transformative Change, Historical Roots and Future Pathways"*.

Levidow, L. and Upham, P. (2017) Socio-technical change linking expectations and representations: Innovating thermal treatment of municipal solid waste. *Science and Public Policy*, 44(2), 211–224. doi:10.1093/scipol/scw054.

Lie, M. and Sørensen, K.H. (1996) *Making Technology Our Own? Domesticating Technology into Everyday Life*. Scandinavian University Press.

Lieu, J. *et al.* (2019) Transition pathways, risks, and uncertainties. In Hanger-Kopp, S., Lieu, J. and Nikas, A. (eds) *Narratives of Low-Carbon Transitions*. Routledge. doi:10.4324/9780429458781-14.

Lindqvist, S. (1994) Changes in the technological landscape: The temporal dimension in the growth and decline of large technological systems. In Granstrand, O. (ed.) *Economics of Technology*. Elsevier.

Loorbach, D., Frantzeskaki, N. and Avelino, F. (2017) Sustainability transitions research: Transforming science and practice for societal change. *Annual Review of Environment and Resources*, 42, 599–626. doi:10.1146/annurev-environ-102014-021340.

Lyle, K. (2021) Interventional STS: A framework for developing workable technologies. *Sociological Research Online*, 26(2), 410–426. doi:10.1177/1360780420915723.

MacKenzie, D. and Spinardi, G. (1995) Tacit knowledge, weapons design, and the uninvention of nuclear weapons. *American Journal of Sociology*, 101(1), 44–99. doi:10.1086/230699.

Mahoney, J. (2000) Path dependence in historical sociology. *Theory and Society*, 29(4), 507–548.

McCarraher, E. (2019) *The Enchantments of Mammon: How Capitalism Became the Religion of Modernity*. Harvard University Press.

Meckling, J. and Nahm, J. (2019) The politics of technology bans: Industrial policy competition and green goals for the auto industry. *Energy Policy*, 126, 470–479. doi:10.1016/j.enpol.2018.11.031.

Merton, R.K. (1949) On sociological theories of the middle range. In *Social Theory and Social Structure*. Free Press.

Michener, W.K. *et al.* (1997) Nongeospatial metadata for the ecological sciences. *Ecological Applications*, 7(1), 330–342. doi:10.2307/2269427.

Minniti, S. (2016) Essay polaroid 2.0 photo-objects and analogue instant photography in the digital age. *Tecnoscienza Italian Journal of Science and Technology Studies*, 7(1), 17–44.

Misa, T.J. (1988) How machines make history, and how historians (and others) help them to do so. *Science, Technology, & Human Values*, 13(3–4), 308–331. doi:10.1177/016224398801303-410.

Mora, L. *et al.* (2021) Assembling sustainable smart city transitions: An interdisciplinary theoretical perspective. *Journal of Urban Technology*, 1–27. doi:10.1080/10630732.2020.1834831.

Murmann, J.P. and Frenken, K. (2006) Toward a systematic framework for research on dominant designs, technological innovations, and industrial change. *Research Policy*, 35(7), 925–952. doi:10.1016/j.respol.2006.04.011.

Oreskes, N. and Conway, E.M. (2010) *Merchants of Doubt: How a Handful of Scientists Obscured the Truth on Issues from Tobacco Smoke to Global Warming*. Bloomsbury Press.

Paterson, M. (2020) Climate change and international political economy: Between collapse and transformation. *Review of International Political Economy*. doi:10.1080/09692290.2020.1830829.

Pel, B. *et al.* (2020) Towards a theory of transformative social innovation: A relational framework and 12 propositions. *Research Policy*, 49(8), 104080. doi:10.1016/j.respol.2020.104080.

Pettigrew, A.M. (1990) Longitudinal field research on change: Theory and practice. *Organization Science*, 1(3), 267–292.

Polanyi, K. (1944) *The Great Transformation: The Political and Economic Origins of Our Time*. Farrar & Rinehart Inc.

Qviström, M. (2012) Network ruins and green structure development: An attempt to trace relational spaces of a railway ruin. *Landscape Research*, 37(3), 257–275. doi:10.1080/01426397.2011.589897.

Rhodes, R.A.W. (1997) *Understanding Governance: Policy Networks, Governance, Reflexivity and Accountability*. Open University Press.

Rip, A. and Kemp, R. (1998) Technological change. In Rayner, S. and Malone, E.L. (eds) *Human Choice and Climate Change*. Battelle Press.

Roberts, J. (2017) Discursive destabilisation of socio-technical regimes: negative storylines and the discursive vulnerability of historical American railroads. *Energy Research & Social Science*, 31, 86–99.

Rogge, K.S. and Johnstone, P. (2017) Exploring the role of phase-out policies for low-carbon energy transitions: The case of the German Energiewende. *Energy Research and Social Science*, 33, 128–137. doi:10.1016/j.erss.2017.10.004.

Rosenbloom, D. and Rinscheid, A. (2020) Deliberate decline: An emerging frontier for the study and practice of decarbonisation. *WIREs Climate Change*. doi:10.1002/wcc.669.

Russell, A.L. and Vinsel, L. (2018) After innovation, turn to maintenance. *Technology and Culture*, 59(1), 1–25. doi:10.1353/tech.2018.0004.

Schopf, H. and Foster, J. (2014) Buried localities: Archaeological exploration of a Toronto dump and wilderness refuge. *Local Environment*, 19(10), 1086–1109. doi:10.1080/13549839.2013.841660.

Sheppard, V. (2020) *Research Methods for the Social Sciences: An Introduction*. BC Campus.

Shove, E. and Pantzar, M. (2005) Fossilisation. *Ethnologia Europaea*, 35(1), 59–62.

Shove, E., Pantzar, M. and Watson, M. (2012) *The Dynamics of Social Practice: Everyday Life and How It Changes*. SAGE.

Simmonds, P.G. *et al.* (1993) Evidence of the phase-out of CFC use in Europe over the period 1987–1990. *Atmospheric Environment Part A, General Topics*. doi:10.1016/0960-1686(93)90125-I.

Sims, B. (2009) *A Sociotechnical Framework for Understanding Infrastructure Breakdown and Repair*. Los Alamos National Lab.

Sims, B. and Henke, C.R. (2012) Repairing credibility: Repositioning nuclear weapons knowledge after the Cold War. *Social Studies of Science*, 42(3), 324–347. doi:10.1177/0306312712437778.

Smith, A., Voß, J.P. and Grin, J. (2010) Innovation studies and sustainability transitions: The allure of the multi-level perspective and its challenges. *Research Policy*, 39(4), 435–448. doi:10.1016/j.respol.2010.01.023.

Soete, L. (2011) *Maastricht Reflections on Innovation*. United Nations University.

Sørensen, K. (2006) Domestication: The enactment of technology. In Berker, T. *et al.* (eds) *Domestication of Media and Technology*. Open University Press.

Sovacool, B.K. and Hess, D.J. (2017) Ordering theories: Typologies and conceptual frameworks for sociotechnical change. *Social Studies of Science*, 47(5), 703–750. doi:10.1177/0306312717709363.

Stegmaier, P. (2023) Conceptual aspects of discontinuation governance: An exploration. In Koretsky, Z. *et al.* (eds) *Technologies in Decline: Socio-Technical Approaches to Discontinuation and Destabilisation*. Routledge.

Stegmaier, P., Kuhlmann, S. and Visser, V.R. (2014) The discontinuation of socio-technical systems as a governance problem. In Borrás, S. and Edler, J. (eds) *The Governance of Socio-Technical Systems: Explaining Change*. Edward Elgar. doi:10.4337/9781784710194.00015.

Stegmaier, P., Visser, V.R. and Kuhlmann, S. (2021) The incandescent light bulb phase-out: Exploring patterns of framing the governance of discontinuing a socio-technical regime. *Energy, Sustainability and Society*, 11(1), 1–22. doi:10.1186/s13705-021-00287-4.

Stegmaier, P. *et al.* (2023) Pathways to discontinuation governance. In Stegmaier, P. *et al.* (eds) *Technologies of Discontinuation: Towards Transformative Innovation Policies*. Edward Elgar.

Stirling, A. (2019) How deep is incumbency? A "configuring fields" approach to redistributing and reorienting power in socio-material change. *Energy Research and Social Science*, 58, 101239. doi:10.1016/j.erss.2019.101239.

Stoler, A.L. (2013a) *Imperial Debris: On Ruins and Ruination*. Duke University Press.

Stoler, A.L. (2013b) *The "Rot Remains": From ruins to ruination, in Imperial Debris: On Ruins and Ruination*. Duke University Press.

Sturm, R. (1993) Nuclear power in Eastern Europe: Learning or forgetting curves? *Energy Economics*, 15(3), 183–189. doi:10.1016/0140-9883(93)90004-B.

Taylor, M. and Taylor, A. (2012) The technology life cycle: Conceptualization and managerial implications. *International Journal of Production Economics*, 140(1), 541–553. doi:10.1016/j.ijpe.2012.07.006.

Turnheim, B. (2012) *The Destabilisation of Existing Regimes in Socio-Technical Transitions: Theoretical Explorations and In-Depth Case Studies of the British Coal Industry (1880–2011)*. University of Sussex. http://sro.sussex.ac.uk/41031/1/Turnheim%2C_Bruno.pdf.

Turnheim, B. (2023) Destabilisation, decline and phase-out in transitions research. In Koretsky, Z. *et al.* (eds) *Technologies in Decline: Socio-Technical Approaches to Discontinuation and Destabilisation*. Routledge.

Turnheim, B. and Geels, F.W. (2012) Regime destabilisation as the flipside of energy transitions: Lessons from the history of the British coal industry (1913–1997). *Energy Policy*, 50, 35–49. doi:10.1016/j.enpol.2012.04.060.

Turnheim, B. and Geels, F.W. (2013) The destabilisation of existing regimes: Confronting a multi-dimensional framework with a case study of the British coal industry (1913–1967). *Research Policy*, 42(10), 1749–1767. doi:10.1016/j.respol.2013.04.009.
Turnheim, B. and Sovacool, B.K. (2020) Exploring the role of failure in socio-technical transitions research. *Environmental Innovation and Societal Transitions*, 37, 267–289. doi:10.1016/j.eist.2020.09.005.
Tushman, M.L. and Rosenkopf, L. (1992) Organizational determinants of technological change: Towards a sociology of technological evolution. *Research in Organizational Behavior*, 14, 311–347.
Unruh, G.C. (2002) Escaping carbon lock-in. *Energy Policy*, 30(4), 317–325. doi:10.1016/S0301-4215(01)00098-2.
Upham, P. *et al.* (2015) Addressing social representations in socio-technical transitions with the case of shale gas. *Environmental Innovation and Societal Transitions*, 16, 120–141. doi:10.1016/j.eist.2015.01.004.
Van de Leemput, D. and van Lente, H. (2023) Caring for decline: The case of 16mm film artworks of Tacita Dean. In Koretsky, Z. *et al.* (eds) *Technologies in Decline: Socio-Technical Approaches to Discontinuation and Destabilisation*. Routledge.
Van Lente, H. and Rip, A. (1998) Expectations in technological developments: An example of prospective structures to be filled in by agency. In *Getting New Technologies Together*. De Gruyter. https://doi.org/10.1515/9783110810721.203.
Van Lente, H., Spitters, C. and Peine, A. (2013) Comparing technological hype cycles: Towards a theory. *Technological Forecasting and Social Change*, 80(8), 1615–1628. doi:10.1016/j.techfore.2012.12.004.
Vinsel, L. and Russell, A.L. (2020) *The Innovation Delusion: How Our Obsession with the New Has Disrupted the Work That Matters Most*. Random House.
Welch, D. and Warde, A. (2015) Theories of practice and sustainable consumption. In Reisch, L.A. and Thøgersen, J. (eds) *Handbook of Research on Sustainable Consumption*. Edward Elgar Publishing. doi:10.4337/9781783471270.00013.
Wyatt, S. (2008) Technological determinism is dead; long live technological determinism. In Hackett, E.J. *et al.* (eds) *The Handbook of Science and Technology Studies*. MIT Press.
Wyatt, S. and Balmer, B. (2007) Home on the range: What and where is the middle in science and technology studies? *Science Technology and Human Values*, 32(6), 619–626. doi:10.1177/0162243907306085.
Yu, D. and Hang, C.C. (2010) A reflective review of disruptive innovation theory. *International Journal of Management Reviews*, 12(4), 435–452. doi:10.1111/j.1468-2370.2009.00272.x.

3

DESTABILISATION, DECLINE AND PHASE-OUT IN TRANSITIONS RESEARCH

Bruno Turnheim

3.1 Introduction

How do fundamental transformations of socio-technical systems occur? What kinds of difficulties are associated with shifting away from undesirable systems of provision? How might such processes be accelerated or oriented?

The socio-technical transitions and innovation literatures ascribe difficulties to set in motion fundamental system transformations to various lock-in mechanisms (Klitkou et al. 2015; Seto et al. 2016; Unruh 2000; Walker 2000), but have so far primarily addressed this question from the perspective of the emergence and development of novelty, examining the systemic obstacles that such "emergent" and positively connoted processes may face. An equally valid way to address this question consists in shifting the gaze away from novelty creation in order to focus on established systems, practices and institutions, and the role of various forms of incumbencies in reproducing existing orders or actively resisting change (Stirling 2019; Turnheim and Sovacool 2020), as well as how these incumbencies may be challenged.

Focusing on *existing* (i.e. already and lastingly stabilised) socio-technical systems enables a closer inspection of the phenomenon of lock-in and inertia, and its incidence on broader patterns of stability and change in socio-technical transitions processes. Starting from the discontinuation of the existing, rather than the emergence of the new, affords alternative lines of reasoning with respect to the core problems of transitions.

According to authors concerned with transitions, such a shift of focus is long overdue:

> Within the fields of innovation studies and transitions theory, processes of emergence and stabilisation are better documented and more widely discussed than those of disappearance, partial continuity and resurrection.
>
> *(Shove 2012)*

DOI: 10.4324/9781003213642-3

> policy mixes favourable to sustainability transitions need to involve both policies aiming for the "creation" of new and for "destroying" (or withdrawing support for) the old.
>
> *(Kivimaa and Kern 2016: 206)*

> current policies are not enough to affect global emissions, or are slow to have a detectable effect, or simply fail to directly address the root cause of the problem: phasing out CO2 emissions from the use of fossil fuels.
>
> *(Peters et al. 2020: 6)*

The destabilisation of existing systems is an emerging research and policy concern related to socio-technical transitions (Bergek et al. 2013; Johnstone and Hielscher 2017; Kivimaa and Kern 2016; Kungl and Geels 2018; Rogge and Johnstone 2017; Turnheim 2012; Turnheim and Geels 2013, 2012). Accelerating socio-technical transitions requires not only the deployment of alternative options, but also breaking away from patterns of inertia and lock-in (Unruh 2002, 2000) that lead existing systems and actors to resist, slow down or prevent transition efforts, i.e. the active phase-out of deeply entrenched systems and related activities (Langhelle et al. 2019; Roberts et al. 2018).

Research about the destabilisation, decline, discontinuation and phase-out of existing systems in the scope of socio-technical transitions has developed significantly in recent years. A number of research orientations can be identified, addressing conceptual, empirical, policy and societal challenges in different ways. This chapter seeks to explore this variety, as a means to reflect on current and future research directions. It contributes to this recent development but also seeks to open up new perspectives.

Indeed, in their review of research on deliberate decline, Rosenbloom and Rinscheid (2020) identify three main strands (destabilisation, phase-out, divestment), but the scope of their exploration leads them to only consider contributions directly relevant to policy and dealing with climate change or carbon-intensive activities. The present contribution seeks a broader engagement with the *process* of socio-technical destabilisation. This leads to considering relevant contributions beyond the scope of transitions studies, climate policy or energy-intensive industries. For instance, the understanding of destabilisation as a process has much to gain from insights concerning harmful substances and socio-technical activities related to food and agriculture or from social science literature beyond transitions studies.

Section 3.2 offers a clarification and disambiguation of related terms: destabilisation, decline, phase-out. Section 3.3 reviews contributions to destabilisation research within transitions studies, structured around six salient themes as entry points for destabilisation research. For each of these themes, I discuss distinct research questions, conceptual elaboration and empirical strategies, as well as ongoing research puzzles. In section 3.4, I return to the themes identified in section 3.3 to formulate six conceptual propositions about destabilisation. The chapter concludes by formulating unanswered questions as avenues for future research endeavours.

3.2 Terminological disambiguation

Before examining the notions deployed around destabilisation in detail, definitional considerations are in order. Indeed, ambiguities exist in research about decline, destabilisation and phase-out of socio-technical systems. These largely relate to boundary conditions and analytical choices, some of which are too often implicitly assumed rather than explicitly formulated. A first necessary step appears to be one of qualifying differences between these concepts and ensuing analytical choices.

Socio-technical *destabilisation* can be understood as a *longitudinal process* by which otherwise relatively stable and coherent socio-technical forms (systems, regimes, institutional arrangements, sets of practices or networks) become exposed to challenges significant enough to threaten their continued existence and their "normal" functioning, triggering strategic responses of core actors within the frame of existing commitments (preservation) and in certain circumstances away from such commitments (transformation). Destabilisation can hence be understood as a *process* involving pressure fronts, strategic responses, and varying commitments to prevailing frames of operation (rules, endowments, etc.). Destabilisation can also be understood as a dynamic *context* for action, involving changing opportunities for navigation and steering from within existing systems (e.g. strategic management of destabilisation contexts by incumbent actors) or from without (e.g. societal contestation or technological alternatives as destabilisation pressures, active destabilisation governance).

System *decline* relates to often a more objectifiable or quantifiable degradation of system performance (e.g., size, economic viability, population, hegemonic power, legitimacy), which can (but rarely does) lead to total decline.[1] Depending on its qualification, decline can hence be understood as a *trend* (e.g., declining performance), a *process* (e.g., system in decline), a possible *outcome* (e.g., decline as consequence of destabilisation) or a *context* (e.g., declining industry as warranting particular kinds of strategies).

Phase-out refers to deliberate (governance) interventions seeking the partial or total discontinuation of a socio-technical form that is deemed undesirable.[2] In practice, phase-outs have largely been restricted to specific products or substances (e.g., DDT, asbestos, mercury, plastic bags, alcohol, class A drugs) and practices (e.g., farm-site slaughter, inner-city driving, indoor smoking),[3] though there are relevant exceptions (e.g., dismantling of tramways, whaling ban). Phase-out is hence a governance *objective*, a form of *intervention*, and a *process* inscribed in a temporal sequence of active discontinuation in phases.

3.3 Core notions deployed within transitions studies

Destabilisation is related to several notions assuming relative centrality within studies of socio-technical transitions: 1) multi-dimensional sources of stability (inertia and lock-in), 2) multi-dimensional sources of change, 3) incumbents as focal actors, 4) processes, pathways and mechanisms, 5) deliberate or purposeful governance, and 6) vulnerability and politics.

3.3.1 Lock-in as structural and *enacted form of stability*

Destabilisation is essentially a process of departure from or challenge of system stability. Within transitions studies, established socio-technical systems are taken to be relatively stable and coherent. System stability is linked to the notion of lock-in and path dependency, which tend to foreclose opportunities for radical innovation and fundamental reconfigurations in favour of more incremental adjustments to existing socio-technical trajectories. For this reason, transitions remain relatively rare phenomena.

On the one hand, *structural* determinants of lock-in are rather well understood from the perspective of economic rationalities (Seto et al. 2016) and cognitive routines (Nelson and Winter 1982), which include sunk investments (e.g., infrastructure, production facilities), increasing returns from economies of scale (Arthur 1989), industry standards, user externalities, network effects and other positive feedback mechanisms reinforcing asymmetrical advantages accruing from cumulative socio-technical development processes. Non-economic dimensions (e.g., institutional and political) are increasingly being considered (Klitkou et al. 2015; Pierson 2000), but are also admittedly more ambiguous due to the "complexity of the goals of politics as well as the loose and diffuse links between actions and outcomes" (Pierson 2000: 260). They include enduring the unusual obduracy of political arrangements, public policies and formal institutions (Pierson 2000), powerful strategic alliances and coalitions with asymmetric access to rule-setting (Roberts et al. 2018), normalised discourses, ideologies and socio-cultural repertoires, and the deep societal embeddedness of user practices and lifestyles. Together, these structural sources of lock-in have the cumulative effect of preventing, limiting or slowing down the development of alternative innovations and socio-technical configurations. Lock-in is an inherently path-dependent process (Arthur 1989; Pierson 2000) resting on self-perpetuating event sequences (Mahoney 2000): past design choices constrain current and future options of system development in ways that favour incremental over radical forms of change and appear as irreversible.

But lock-in is neither permanent nor inevitable: system *un*locking does and has happened. Understanding and characterising structural lock-in requires 1) longitudinal approaches to *how socio-technical configurations have stabilised* along particular trajectories to take on specific forms and shapes, and 2) evaluative-descriptive approaches to *how stable socio-technical configurations actually are* (Klitkou et al. 2015) and *how socio-technical stability may change over time*—notably as alternatives are developed or windows of opportunity for change open up. In practice, the situated analysis of lock-in needs to be cognisant of the variety of relevant dimensions and mechanisms (economic and institutional), their relative importance (e.g. structure depth, interlinkages) and their transient nature.

On the other hand, socio-technical stability is also *enacted* by established actors enjoying central or dominant positions within existing systems that tend to reproduce the conditions of their incumbency (Stirling 2019; Turnheim and Sovacool 2020). Firstly, incumbent actors contribute to the continuity of existing systems, structures and practices through tacit and routinised activity, including maintenance, care and repair, continued investment, service improvements, incremental innovation, or the reproduction of underlying (regime) rules. Such continuation, maintenance and improvement activities may confer sustained relevance and legitimacy to the actors involved and contribute to further entrenching related socio-technical practices, arrangements, infrastructures. They are carried out at various levels, from mundane and invisibilised maintenance work to more strategic (and often framed as remarkable) programmes. Secondly, more active forms of stability enactment include strategic activities aimed at defending and protecting current arrangements and advantages (e.g., favourable policy conditions), notably by incumbent actors (see section 3.3.3) in the face of contestation and other challenges to socio-technical stability (e.g., delegitimation). The repertoire of regime resistance and defence strategies can be approached by distinguishing forms of power available to incumbent actors (see Table 3.1).

It hence appears relevant, as a preliminary step to better understanding destabilisation processes, to develop the means to evaluate system stability and coherence over time. To do this, it is important to identify markers of system stability and coherence, processes of system stabilisation, and the mechanisms—structural and enacted—that contribute to system lock-in and stability. It then becomes possible to evaluate destabilisation as a process involving the waning of stabilisation mechanisms (erosion of structural forms of stability), difficulties to maintain stability in the face of challenges (failure to actively resist pressures) or intended departure from system preservation objectives and rationales (transformation). The interplay between stabilisation and destabilisation is further discussed in section 3.3.4.

3.3.2 Multiple and mutable sources of change

While established socio-technical systems are characterised by lasting stability, they are also exposed to pressures and challenges of varying kinds, which may call into question their normal functioning, expected performance, continued relevance or legitimacy. Such pressures for change may be deflected entirely (e.g., operations resuming after temporary disruption of service), orient incremental improvement and optimisation strategies (e.g., more efficient production modes, substitution of harmful substances), trigger significant adjustments (e.g., new business models,

TABLE 3.1 Types of power harnessed by incumbent actors for resistive purposes

Types of power	*Related incumbent resistance activities*
Instrumental	Mobilising resources (e.g. finance, capabilities, authority, access to decision-making) in immediate interactions with other actors to achieve their goals and interests
Discursive	Mobilising authority and legitimacy to shape what issues are being discussed (agenda-setting) and how they are being discussed (issue framing)
Material	Leveraging technical capabilities and financial resources to promote incremental improvements of existing options over more radical alternatives
Institutional	Reinforcing prevailing political cultures, ideology and structures (e.g. economic liberalism, technocratic styles) to downplay alternative paths and decision rationales

Based on Avelino and Rotmans (2009) and Geels (2014a)

reconfigurations) or eventually make existing systems obsolete. It hence appears important to develop the means to better characterise these sources of pressure, determine and trace their influence over existing systems, namely in order to better qualify *what makes for a destabilising source of change* (as opposed to other sources of change), and *whether specific sources of change may be associated with specific destabilisation patterns*.

Kinds and sources of (destabilising) change can be related to relevant socio-technical dimensions within which systems are embedded, the alignment of which confers stability or the de-alignment of which may generate destabilising conditions. The specific dimensions considered vary according to approaches and are largely a matter for analytical choice. These include usual distinctions between techno-economic dimensions (e.g., significant changes in markets, technologies, infrastructures, scientific knowledge), and socio-political dimensions (e.g., significant changes in ideas, policies, politics or cultures) (Geels 2014b; Turnheim and Geels 2013). Socio-technical approaches make a distinction between three inter-related and partly overlapping ontologies (Geels 2004; Geels and Turnheim 2022): that of technical components and systems (artefacts, technical systems, infrastructures), that of institutions (regulations, conventions, cultural values and beliefs, symbolic meanings, and so on), and that of actors and networks (their actions, motives and interests, forms of organisation and interaction). Accordingly, sources of destabilising change include 1) technical dysfunctions, technological discontinuities or performance erosion, 2) social and political mobilisation, delegitimation, the emergence of new rules or the breakdown of existing rules, and 3) challenges by new actor coalitions, the disbanding of existing coalitions or the accumulation of poor strategic choice. Practice theory distinguishes between three kinds of elements that are tied together in stable practices: materials ("things"), meanings (social and symbolic signification) and competences (forms of understanding and practical knowledgeability)

(Shove et al. 2012). So, the decline, erosion or decay of practices can be understood as the unmaking or breaking of ties between such elements.

Regardless of the chosen frame of reference, a number of observations can be made. First, challenges to stability can emerge along various dimensions: significant change in one dimension may perturbate the relative stability experienced by established systems. Qualitatively distinguishing different sources of change can underpin the identification of ideal destabilisation processes, including technological disruptions and discontinuities, creative destruction (of economic entities), discursive destabilisation (of prevailing frames), delegitimation (of practices, behaviours or specific actors), de-institutionalisation (of rules and conventions), regulatory challenges and so on. Second, destabilisation is likely to result from a combination of sources of change, so it is more fitting to think in terms of *pressure fronts*. Third, destabilisation is not merely the result of threats and challenges but is related to the weakening of stability-conferring ties and alignments within socio-technical configurations. Destabilisation results from the combination of intensifying discontinuities (threats and challenges) and weakening continuities (erosion).

A related distinction concerns the location and distance of sources of change vis-à-vis the definition of an established system's boundaries, i.e., whether sources of change are exogenous or endogenous. Exogenous sources of destabilising change typically relate to challenges outside of existing systems and beyond their immediate environment, such as external threats, uninvited challenges or unforeseen discontinuities (e.g., surprises and shocks). Exogenous pressures tend to be less anticipated and not to be the object of dedicated monitoring or intelligence than their endogenous counterparts. "Landscape changes", such as demographic patterns (e.g., urbanisation), macro-economic trends, geo-political swings, crises and disruptive events (e.g., wars, shocks) are typical exogenous changes. But exogenous changes are not necessarily macro in scale, as with competition from socio-technical alternatives, social movement contestation, or practice and consumption changes, which can start as relatively isolated and follow more gradual emergence patterns yet come to exert significant pressures for change. Endogenous sources of change, by contrast, are more closely linked to established systems and activities or in their immediate vicinity. They tend to be the object of monitoring (through for example performance indicators); regime actors tend to be more knowledgeable about them and so in a better position to anticipate them or perceive them as immediate threats. Endogenous changes include worsening economic performance (e.g., at product, firm or industry level), declining income, slack (which reduces the ability to manage change) or resources, weakening ties between key socio-technical components, changes in political support and coalitions, degraded infrastructures (e.g., material, knowledge), and divergence within organisational fields. Furthermore, endogenous sources of change are likely to materialise as conflicts and contradictions between otherwise aligned elements in configurations. However, the distinction between exogenous and endogenous change is not entirely clear cut: it depends on system boundaries, is a matter for interpretation, and exogenous changes can turn into endogenous pressures for change as they become translated into more concrete concerns and pervade socio-technical configurations.

Lastly, considerations about sources of change in destabilisation processes raise issues about the explanation of change, and the mechanisms by which sources of change can trigger, reinforce or orient destabilisation processes and patterns (see also section 3.3.4). Since destabilisation rarely results from a single pressure or source of change, its analysis calls for representations of the complex causal chains involved: the temporal interaction of sources of change in the formation of *pressure fronts* and related destabilisation patterns.[4] With regard to temporality, a variety of possible destabilisation patterns may be identified according to the intensity, speed, scope and sustained nature of pressure fronts. Suarez and Oliva (2005), focusing on changing industry environments, suggest a distinction between regular change, hyperturbulent change, specific shocks, disruptive change, and avalanche change. Stirling (2014), focused on the interpretation of system vulnerability threats, distinguishes between short-term episodic perturbations (shock interpretations) and long-term enduring pressures (stress interpretations).[5] In practice, destabilisation processes are likely to deviate from such ideal-types with interrupted, cycling or reversing patterns.

With regard to causation, it is useful to distinguish different levels: proximate causal forces, intermediate causal forces and distal causal forces. The shock/stress distinction illustrates the importance of differentiating causes in practice: the 2008 financial crisis may be interpreted as a shock, the primary cause of which is attributable to the "bursting" of a housing market bubble (subprime mortgages) and of a dependent financial bubble in its wake. But beyond this proximate causation, it is also possible to interpret the crisis as resulting from a long-term stress linked to more remote causes: the accumulation of solvency imbalances (between debt obligations and cash flow), the multiplication of financial innovation, lax monetary policy and an increasing disconnect between the financial system and the real economy. Furthermore, the hysteresis-like pattern of many shocks, while influenced by identifiable feedback loops, is largely emergent and contingent in character. Consequently, it is relevant to approach such processes through the analysis of sequences of events, the identification of critical turning points and possible cascading dynamics of change.

So, given that the emergence, evolution and endurance of pressure fronts are an essential source of destabilisation, it is important to develop the means to better characterise, distinguish and evaluate them and their incidence on established systems.

3.3.3 Incumbents as focal actor

Destabilisation is intimately tied to the roles, motives and actions of particular social actors, notably those assuming a *de facto* position of centrality in established socio-technical configurations and those willing to challenge these positions for various motives. Many studies adopt a particular focus on "incumbent actors" and their responses to destabilisation challenges (Andersen and Gulbrandsen 2020; Bergek et al. 2013; Berggren et al. 2015; Bohnsack et al. 2020; Dijk et al. 2016; Hess 2019; Hörisch 2018; Isoaho and Markard 2020; Kungl 2015; Lee and Hess 2019; Lockwood et al. 2020, 2019; Mylan et al. 2019; Raven 2006; Smink et al. 2015; Steen and Weaver 2017; Turnheim and Geels 2013; van Mossel et al. 2018), often portrayed as villains to be dethroned or transformed for system change to occur. While the notion of incumbency denotes particular attributes such as a position of centrality, power, mastery over resources in connection to established system, it is important to move beyond simplistic, monolithic and static portrayals of incumbency attributes and the actors that come to incarnate them (Turnheim and Sovacool 2020), to examine a plurality of forms of incumbency at play and the depth of related socio-technical entanglements (Stirling 2019).

First, "incumbent actors" is often used as shorthand for large firms or multinational corporations, with little consideration for incumbency, centrality and power in other societal spheres (e.g., policy and politics, civil society, knowledge production). Important questions for destabilisation research remain concerning these other forms of incumbency and how they relate to one another, for instance in incumbent coalitions or constellations perpetuating prevailing paradigms.

Second, because incumbency connotes significant vested interests in existing systems, practices and arrangements, it is commonly assumed to exclusively lend to the adoption of resistive postures with respect to radical change, i.e., incumbency as calculated conservatism. In practice, however, the strategic repertoire of incumbents may be broader (e.g., from purely resistive stances to more proactive diversification through various means) and vary significantly from actor to actor. This suggests important questions concerning the range of strategic positions and stances available to and performed by incumbent actors.

Third, incumbent strategies are likely to change over time, as the nature of destabilisation contexts changes (e.g., mounting pressures), as specific opportunities arise or as new frames of reference become available to those actors. This raises questions about the changes in incumbent strategies in various contexts and their determinants.

Fourth, there is a need for greater clarity as to how the notions of "regime" and "incumbency" are related and intersect. Indeed, certain incumbent actors may "disband" from existing regimes and so contribute to their fragmentation (Steen and Weaver 2017), incumbent actors from neighbouring regimes can significantly contribute to niche construction (Berggren et al. 2015; Späth et al. 2016; Turnheim and Geels 2019), and regime rules are reproduced by all sorts of actors—some of which are not considered as particularly powerful, dominant or central to said regime. So, socio-technical regimes and incumbency only partly overlap.

For incumbents, destabilisation ultimately comes with the threat of losing a dominant status, position, sustained relevance or legitimacy (e.g., social licence to operate) but can be met with a broad range of tactical moves and response strategies. Following incumbent actors along destabilisation processes is a relevant analytical entry point, particularly as it enables us to better understand how particular organisations experience, interpret and handle such challenging processes. Correspondingly, incumbent destabilisation processes are being approached from various perspectives deployed at the organisational level, including cognitive and learning approaches, strategic change approaches (e.g., diversification, renewal, ambidexterity), organisational decline, organisational capabilities (Ottosson and Magnusson 2013) or power (Avelino and Rotmans 2009; Geels 2014a). Organisation-centric approaches may further be distinguished according to two dimensions: 1) the depth of challenges and changes (e.g., change in activities, routines, models or core beliefs), and 2) the nature of challenges and changes (e.g., competitive, technological, reputational/legitimacy, socio-political) (see also section 3.3.2).

While incumbents are a relevant entry point to study the destabilisation process, other actors warrant attention and dedicated research, including 1) social groups with no or weaker prior links with existing systems and regimes which may actively contribute to destabilisation (e.g., civil society, new entrants organisations, alternative political alliances) notably due to their interest in challenging positions of power for various motives, 2) vulnerable groups with potentially less agency in destabilisation processes such as workers and local communities (Johnstone and Hielscher 2017) and less visible groups directly affected by decline or phase-out interventions (see also section 3.3.6).

In other words, there is significant scope for conceptual, analytical and empirical elaborations on agency and power in destabilisation processes.

3.3.4 Destabilisation processes, patterns and mechanisms

Destabilisation implies challenges to system stability (see section 3.3.1), is related to different sources of change (see section 3.3.2), and centrally involves actors with strong ties to existing configurations (see section 3.3.3). But at heart, destabilisation is a complex process involving non-linearities, indeterminacies and contingencies. *How to make sense, then, of destabilisation as a process, its generative mechanisms, and variety of possible trajectories and outcomes?*

First, understanding destabilisation as a *process* requires adopting a mode of enquiry that allows describing and explaining *how* and *under which conditions* destabilisation unfolds over time. This implies "understanding both the processes that reproduce durable configurations of social order and those that generate strains or produce events with the capacity to transform social structures" (Clemens 2007: 528), and in what ways the tension between inertial and change forces can evolve. Explaining change in terms of process requires particular attention to (sequences of) events and their enactment:

> Process research is concerned with understanding how things evolve over time and why they evolve in this way..., and process data therefore consist largely

> of stories about what happened and who did what when—that is, events, activities, and choices ordered over time.
>
> *(Langley 1999: 692)*

Certain events in destabilisation sequences may be more influential, while others may be particularly significant because of their location within a sequence or because they activate, orient, accelerate or inhibit key processes. Crises may have an accelerating or orienting effect on destabilisation, while the timely development and legitimation of alternatives may contribute to destabilisation by weakening claims about the need to maintain stability at all costs. Further, while the temporal ordering of relevant destabilising events in sequences and their narrative depiction can produce interesting and relevant stories, they also need to lead to generalisable explanations about the phenomenon at hand and hence come into conversation with theoretical arguments: "process theorization needs to go beyond surface description to penetrate the logic behind observed temporal progressions" (Langley 1999: 694)—following inductive or deductive strategies.

Process tracing offers a useful approach, empirically focused on individual or small-N cases and analytically oriented towards causal inference through the identification of mechanisms as intervening events (Mahoney 2016). Uses of process tracing vary significantly, and can be mobilised both for theory testing or theory development (George and Bennett 2004). In the case of overly complex and overdetermined processes like destabilisation, recourse to "analytical explanation", i.e., a "variety of process-tracing converts a historical narrative into an *analytical* causal explanation couched in explicit theoretical forms" (George and Bennett 2004: 147), appears as a particularly relevant form of process-tracing. The analysis of path dependences, trajectories and branching points along such paths is another useful way forward, provided it does not assume the inevitability of development outcomes (e.g., through consideration of interrupted or resurgent paths) and remains open to counterfactuals.

Second, if process tracing can enable the identification of events-as-mechanisms in individual cases, the comparison of multiple cases can improve knowledge claims and conditional generalisations about a given process, particularly if supported by the development of a typology. Typologies seek the identification of relevant causal mechanisms and pathways that influence the outcomes of a phenomenon by specifying its possible variants as ideal-types within a theoretical space. Again, this may follow an inductive logic (i.e., multiple cases allow the identification of different causal pathways), a deductive logic (i.e., theoretical dimensions inform the space of possible pathways), or an abductive logic (i.e. a combination of both in tentative iterations). The elaboration of a destabilisation typology can hence allow specifying the possible causal pathways that destabilisation processes may follow, with particular attention to conditions and contingent mechanisms. Specifying a variety of potential destabilisation pathways can also guide the selection of appropriate case studies, by maximising the variety of observed types or to search for outliers. Importantly, while single-site studies have accumulated evidence of a variety of destabilisation patterns, a useful next step for the research community would be to comparatively locate these within a coherent repertoire of possible mechanisms.

Third, the temporality of destabilisation warrants particular attention. Similar to socio-technical transitions, destabilisation typically unfolds over multiple decades (Martínez Arranz 2017), although its duration, speed and pace may vary significantly across cases. In ways analogous to the *development* of socio-technical systems, destabilisation may be subject to acceleration phases which may wrongly convey an impression of overall speed, so determining the overall duration of destabilisation largely depends on "when one counts" (Sovacool 2016). Periodisation, whether calling upon ideal-typical phases of development or the identification of crucial events, appears as a useful but not unproblematic way forward. Identifying *when* destabilisation starts and ends is an analytical question. Further, destabilisation, though imprinted in the popular mind by the importance of singular events (e.g., a coal miners' strike, nuclear disaster, or food crisis), is a cumulative process involving layered temporalities. While tempting to focus only on the most dramatic surface events, those provide "the most distorting and unpredictable lens through which to view reality" (Braudel 1970: 148), and hence require attention to overlapping concomitant temporal processes and nested temporalities (e.g. short term, conjectures, longue durée). Under certain conditions, destabilisation processes might appear as particularly rapid, but in most cases they involve long pre-development phases in which contestation builds up without leading to dramatic struggles, because it is kept at bay or overshadowed by the inertia of incremental system developments. Beneath the highly visible surprises, conflicts and struggles are deep structural tensions, latent disagreements between social groups, and more gradual accumulative processes of problem framing, social mobilisation and political contestation, knowledge and innovation development, the articulation of alternative visions of the future, and so on, which contribute to socio-technical instability.[6] In yet other cases, destabilisation may be entirely uneventful (e.g., gradual erosion or drift) when resulting from a lack of maintenance, under-investment or fatigue.

So, destabilisation processes involve a combination of gradual cumulative change and their activation in particular events, which provide focusing devices around which to *a priori* centre empirical efforts, but require investigation of causal chains. Destabilisation is a social process inscribed in multiple temporalities, including long-term social trends, volatile political and public opinion swings, contingent surprises and accidents, medium-term strategic dependencies, abrupt decision reversals and cyclical fluctuations (Braudel 1970; Burke 2005; Sewell Jr 2005), which are combined in particular sequences of events at specific times and places. Understanding destabilisation as a process hence requires explicit analysis of the conditions under which different causal patterns (e.g., distal, intermediate and proximate causes) may accumulate or align. The specification of propositions about such alignments pathways (e.g., their nature, articulation and timing) is key to crafting destabilisation typologies with significant explanatory power, notably as it allows making sense of varied observable temporal patterns such as gradual erosion (Shove 2012), accelerated destabilisation (Andersen and Gulbrandsen 2020), turning points (Abbott 1997), downward spirals or perfect storms (Hambrick and D'Aveni 1988; Kungl and Geels 2018), or interruptions and reversals (Haydu 2010; Sillak and Kanger 2020). Theoretical models of destabilisation may also attend to the cumulative character of the process by distinguishing different typical stages and outcomes, linked to propositions about expected observations: early stages of destabilisation may be characterised by low or divergent degrees of pressure that can be easily denied or deflected by regime actors (Geels 2014a; Lockwood et al. 2019); moving to later stages of destabilisation may require an accumulation and alignment of pressures as well as alternative path creation (Turnheim and Geels 2013). Such stages are likely to also reflect changes in the strategic positioning of actors involved—destabilisation may require increasing coherence of contestation forces and lead to increasing divergence within existing regimes. However, progression through such stages is neither predictable nor inevitable, and likely prone to pushback from collectives under threat. Destabilisation may hence lead to differentiated outcomes, including partial destabilisation, full decline, reorientation or re-creation (Turnheim and Geels 2012), but also continuity and persistence (Newig et al. 2019; Wells and Nieuwenhuis 2012; Winskel 2018), i.e., when destabilisation pressures are effectively deflected.

Fourth, destabilisation is tied to processes of stabilisation in multiple ways. Like stabilisation processes, destabilisation involves ongoing tensions between forces of stability and change, which may ebb and flow, but involves an overall trend away from stability. It can be dialogically related to stabilisation in broader transformation dynamics, particularly in the case of substitution patterns wherein the destabilisation of a given entity enables and requires the stabilisation of another and vice-versa.[7] Destabilisation and stabilisation can also be sequentially related, particularly within the processual logics of punctuated equilibrium (Tushman and Anderson 1986) according to which relatively long periods of stability and incremental change are punctuated by shorter turbulent episodes of radical change in which destabilisation can give way to re-stabilisation, and so on. Given that most transition cases, however, do not follow substitution logics, re-stabilisation is likely to occur at a different structuration level: while the delegitimation of DDT (a powerful and toxic pesticide) led to its subsequent ban (hence fully destabilising the substance's value and use chain), it however led to new forms of legitimation of pesticides concomitant to the regulation of their use (Joly et al. 2022). This example illustrates the importance of analytical considerations about *what* is being destabilised, as well as the potential distinction between destabilisation and "exnovation",[8] notably as they tend to operate at different levels of structure. It also illustrates how destabilisation may lead to the reinforcement of existing structures. Newig et al. (2019), focusing on various types of institutional decline and drawing on Streeck and Thelen (2005a), suggest distinguishing functional change from structural change: 1) adaptation to crises (e.g., technical fixes or enhanced controls) and systemic learning can preserve function and structure, 2) the phase-out of particular substances (e.g., DDT) may alter the structure of underlying institutions but preserve or extend overall function (e.g., pest control) through substitution and adjustments, 3) the repurposing of outdated institutions (e.g., shift from military service to civil service) may be oriented towards new functions without fundamentally altering their structure, and 4) certain institutions may be abolished altogether (structure and function) with or without alternatives. Further, one regime's destabilisation may be tied to another's stabilisation: the stabilisation of a "climate regime" (e.g., the increasing structuration of the problem and solution space related to addressing climate change) is intertwined with the possible destabilisation of a fossil fuels regime.

Though interlinked, stabilisation and destabilisation cannot simply be understood as opposite, reverse or symmetrical processes. Contrasting destabilisation with stabilisation is a useful first step (see Table 3.2), notably as it points towards markers and handles for the evaluation of each process, but it tells us very little about how these unfold. Further, such efforts should be complemented by dedicated conceptual elaboration aimed at explaining and qualifying possible patterns in greater detail.

TABLE 3.2 Contrasting features of socio-technical stabilisation and destabilisation

	Socio-technical stabilisation	*Socio-technical destabilisation*
Processual features	Generative, accumulation, addition	Degenerative, erosion, removal
Structural stability (momentum and inertia)	Increasing: socio-technical embedding in configurations	Decreasing: challenges to parts or entire configurations
Functional stability (purpose and framing)	Oriented towards closure and standardisation: problems meet solutions	Oriented towards opening up: new or unsolved problems, search solutions
Innovation strategies	Focused: Incremental and cumulative system-building (within)	Multiple: radical alternatives (outside) and system transformation (within)
Institutional dynamics	Convergence, relative homogeneity and reproduction	Divergence, disbanding and delegitimation
Politics, controversy and normativity	'evacuated' through relative consensus or hegemony	Tensions and dissensus are central preoccupations and motors of change
Inclusion of relevant social groups	Increasingly selective (outsiders invisibilised)	Increasing visibility of dissenting voices
Infrastructure	Increasingly seamless, reliable, maintained	Increasingly seam-full, failure-prone, eroding

3.3.5 Deliberate, purposeful, intended governance

Within the scope of more pragmatic programmes (Abbott 2004) about destabilisation processes, notably those in search of means of interventions, questions arise concerning whether and how destabilisation can be *purposely* governed (Kivimaa and Kern 2016; Rosenbloom and Rinscheid 2020). Such considerations have been the object of a growing stream of literature under various terminologies, including deliberate destabilisation (Normann 2019; Turnheim and Geels 2012), deliberate decline (Rosenbloom and Rinscheid 2020), phase-out (Andersen and Gulbrandsen 2020; Johnstone and Hielscher 2017; Vögele et al. 2018), exnovation (David 2018; Davidson 2019; Heyen et al. 2017) and innovation through withdrawal (Goulet and Vinck 2017), creative destruction (Kivimaa and Kern 2016) and the governance of discontinuation (Hoffmann et al. 2017; Johnstone and Stirling 2020; Stegmaier et al. 2014).

The governance of socio-technical systems raises several questions. According to Borrás and Edler (2014), key considerations for anyone interested in the governing dynamics of socio-technical change can be appreciated around three pillars: 1) agents and opportunity structures (*Who and what drives change?*), 2) instrumentation (*How is change influenced?*), and 3) legitimacy (*Why is it accepted?*). For Kern (2011), indebted to notions from comparative political economy perspectives (Hall 1993; Hay 2004), ideas, institutions and interests are key dimensions shaping policy and governance processes, which are critical to explaining the tensions between continuity and radical change. Kivimaa and Kern (2016) mobilise three interrelated analytical dimensions of socio-technical change to distinguish policy instruments for "creative destruction" besides "control policies" focusing on changes to 1) regime rules, 2) support for dominant technologies, and 3) networks and actors. Smith et al. (2005: 1507), wary of depictions of policymaking as "coordinating the consensual introduction of elements that are self-evidently required for the smooth operation of a clearly more sustainable innovation system", insist on the importance of considering a variety of contexts and conditions for regime transformation which governance may sustain or alter, but also of attending to the central issues of 1) agency and power, 2) regime structures and membership, 3) uneven and distributed resources, and 4) the performativity of visions and expectations. Building on aforementioned distinctions, the following paragraphs discuss relevant issues arising with the governance of destabilisation.

First, as a social phenomenon, destabilisation is conditioned by the actions and interventions of different social groups—though their agency may be limited and facing important structural determinants. The extent to which destabilisation is purposefully governed varies significantly across cases. Untended destabilisation processes are those involving no *explicit* intention to trigger, slow down, accelerate or orient the difficulties and possible decline experienced by a particular regime. Streeck and Thelen (2005a), in their conceptualisation of institutional change, provide two relevant notions that may apply to cases of untended destabilisation. They define "institutional drift" as processes resulting from the lack of adaptive maintenance in spite of changing external conditions, which may lead to shrinkage, erosion or atrophy of institutions (e.g., degradation of US health care coverage in the face of new risks). They define "institutional exhaustion" as another archetype of gradual breakdown, largely related to failures to anticipate changing conditions whereby institutions deplete the external conditions on which they rely to operate (e.g., revenue balance for social insurance systems). In both cases, what may appear as inaction in the face of destabilisation may more appropriately be understood as a particular form of intervention: neglectful or self-consuming governance. Such cases, however, tend to be viewed in the transitions literature as "unmanaged" or "spontaneous" (and relatively autonomous) forms (Newig et al. 2019), as opposed to more purposeful forms (Smith et al. 2005), which are underpinned by oriented and coordinated efforts. If purposive transitions are those "which have been deliberately intended and pursued from the outset to reflect an explicit set of societal expectations or interests" (Smith et al. 2005: 1502), purposive destabilisation has been defined as a process influenced by "deliberate political steering" (Newig et al. 2019: 17).[9] In-between these two extremes, a variety of destabilisation governance archetypes may be observable. The following paragraphs explore some relevant dimensions.

Second, destabilisation governance may be approached according to the motives and intentions involved. Leaving aside denial, doubt, and resistive stances that may drive action by incumbents, particularly in the early stages (see section 3.3.3), I suggest distinguishing reactive, active and emancipatory motives for destabilisation governance (see Table 3.3). *Reactive* motives are oriented towards mitigating the possible outcomes of destabilisation and decline, as and after it happens. Related actions include slowing down the contraction of declining industries or practices (e.g., protecting declining domestic fisheries or agriculture), reducing the effects of decline though financial assistance (e.g. bailouts) or extended social provisions (e.g., coal miners' fuel allowances), but also dealing with the lasting structural inequalities produced by decline (e.g., regional social policy, priority education zones), infrastructure decommissioning or socio-technical aftercare[10] (Stegmaier et al. 2014). *Active* motives are oriented towards the discontinuation of undesirable systems, with a more forward-looking orientation. Related actions include mobilisation and interventions seeking to trigger the phase-out of products, substances or systems, but also the anticipation of future transformations through increased preparedness (e.g., reskilling strategies, territorial conversion). *Emancipatory* motives are oriented towards opportunities for transforming existing social contracts that destabilisation (or avoided destabilisation) may afford. Related actions include ring-fencing "strategic" activities and sectors, challenging structural forms of power, oppression and neglected interests, delegitimising systems and activities on moral and ethical grounds (e.g., unsustainability, injustice, inequality, uneven access), and empowering alternative pathways of development. Such motives may present scope for complementarity as well as significant points of tension and contradiction.

TABLE 3.3 Motives for destabilisation governance

Type	*Commitment*
Reactive	Mitigating the possible outcomes of destabilisation and decline, as and after it happens
Active	Anticipating and supporting the discontinuation of undesirable systems
Emancipatory	Transforming existing social contracts

Further, motives for destabilisation governance may also be distinguished according to their core normative orientation (i.e., prescriptions about what is "desirable"). Destabilisation motivated by sustainability, justice or low-carbon objectives can be contrasted with destabilisation motivated by no less normative, but different, priorities (e.g., liberalisation transitions of the 1980s or productivity- and scale-oriented transitions in historic agricultural reforms). Similarly, motives for transitions governance may change depending on the position of actors involved vis-à-vis existing regimes and depending on destabilisation contexts. While incumbent actors are more likely to pursue a general orientation towards preservation, continuity and incrementalism, such orientations may change where and when path-insistence is interpreted as less feasible or tenable. Immediate response to acute crises are more likely to be reactive when those are largely unpredicted and unprepared for (e.g., response to Covid-19 in Europe, as opposed to some East Asian responses actively mobilising SARS precedents, related imagery and preparedness), leading to situations more akin to firefighting (Osterholm and Olshaker 2020), but may shift to more active motivations if underlying problems become more widely understood, particularly if combined with emancipatory orientations for transformative change.

Third, while deliberate destabilisation policy is politically difficult (Stegmaier et al. 2014) and relatively new in the context of sustainability and low-carbon objectives, the repertoire of available instruments is broad and extends well beyond policy interventions (Kivimaa and Kern 2016). Phase-out policy, for instance, is currently the object of much experimentation as well as rediscovery of existing instrumentation, as evidenced by the variety of coal phase-out interventions (Spencer et al. 2018) or "policy for incumbency" more generally (Johnstone et al. 2017). Considering the breadth of available instruments requires distinguishing policy interventions from wider governance means and constituencies, as well as how these may be articulated in policy mixes, layering and sequencing, or in changing governance contexts. I suggest a distinction between direct, indirect, experimental and civic interventions (see Table 3.4). *Direct* destabilisation interventions are relatively conventional forms of policy interventions seeking to contribute to phase-out by introducing control policies and altering frame conditions, such as formulating long-term reduction goals (e.g., zero-carbon, zero-pesticides), introducing restrictions (e.g., bans, regulated use), modulations mechanisms (e.g., incentives and disincentives), removing/dismantling undesirable support structures and institutions (e.g., fossil fuel subsidies, R&D funding, decommissioning), reorienting resource flows away from existing regimes, and structural reform. Such interventions are rarely effective on their own, because they face oppositional backlash that tend to weaken and delay action (e.g., litigation, loopholes, exemptions, implementation failure) or lead to reversals, are often implemented as macro-level instruments at a high level of granularity (e.g., carbon emissions trading or regulation), unevenly affect certain actors and communities, and can hence generate all sorts of unintended effects (e.g., exacerbating poverty, affecting livelihoods). *Indirect* destabilisation interventions are those oriented towards addressing (legitimate) oppositions to destabilisation processes, increasing the preparedness of affected or vulnerable groups and overturning structural dependencies. They include compensations for losses and stranded assets (e.g., specific industries, social safety net for affected communities), changes in organisational ownership and control (e.g. nationalisation), reskilling and professional training programmes, regional development and labour adjustment programmes, or infrastructure development to support regional conversions. Such interventions, though providing buffers for the disruptive consequences of decline and increasing adaptive preparedness in certain cases, can have limited or perverse effects when not conditioned to transformative or redistributive outcomes (e.g., bailouts with limited obligations), not tailored to the needs of communities involved (e.g., regeneration programmes driving local residents out, reskilling programmes in the absence of employment opportunities), or not combined with the development of emancipatory opportunities and pathways (e.g. decent jobs, accessible public infrastructures, meaningful lifestyle changes). *Experimental* destabilisation interventions are more novel approaches oriented towards anticipating, triggering and navigating destabilisation as transformative opportunity for introducing lasting systemic change, as well as legitimising such objectives. Given that less practical experience exists with such interventions that do not neatly fit within existing policy roles and responsibilities, they require particular dispositions towards trialling new solutions, real-time evaluation and learning, flexible dispositions concerning rules (e.g., exceptions), tolerance for failure (Kuhlmann et al. 2019), and changes in decision-making procedures and representation to limit the power of incumbents (e.g., more participatory processes, co-production). The scope of experimental destabilisation interventions is particularly broad and systemic, and hence involves the combination of instruments (e.g., in policy mixes), while the perimeter of applications is likely to focus on particularly undesirable or vulnerable sectors and regions. Coal-dependent regions currently appear at the forefront of such experiences, owing to the relative societal and policy purchase that transitions away from coal have recently acquired (Spencer et al. 2018), but similar initiatives are likely to arise in the context of agri-food systems (e.g., zero pesticides), mobility (e.g., car-free cities) or manufacturing (e.g., zero plastics, zero waste). Further, relevant historic exemplars with experimental destabilisation interventions are not restricted to sustainability. Potential exemplars include experiences with tobacco and smoking restrictions, drugs and pharmaceuticals, restrictions on agricultural and fishing practices on ethical grounds (e.g., "humane slaughtering", whaling), polycentric cities (as means to deal with structural trends and challenges arising from concentric, unfair and overspecialised urbanisation patterns) and regional specialisation (e.g., tourism- or knowledge-oriented development in former industrial or agricultural regions). *Civic* destabilisation interventions are those emanating from civil society, social movements and activists. They share a commitment to bringing new problems and issues to the attention of society, mobilising public and political forces around them, and ultimately undermining the continuation of practices and systems seen as undesirable, notably by delegitimising them and promoting alternatives. In terms of strategy, civic interventions may seek to weaken the semantic power and influence of lobby groups and vested interests, develop contesting forms of knowledge and invoke various forms of dissent, but also include material strategies seeking to obstruct or circumvent the normal functioning of socio-technical systems and infrastructures. Interventions vary and may include information campaigns, citizen knowledge and alerts, public protests, peaceful disobedience, boycotts (of products, companies, authorities), consumer opt-out or disconnections, financial divestment (e.g., fossil fuel divestment), the promotion of practices challenging economic paradigms (e.g., degrowth vs capitalist consumerism), as well as civil unrest, production site blockages or asset destruction. Civic interventions are a powerful means of raising attention and mobilising around particularly problematic socio-technical systems and practices.

Fourth, destabilisation governance likely involves a combination of different forms of intervention, different types of actors (policy, civil society, science, industry) and rationales, and their articulation over time. Two notions appear particularly relevant: policy mixes and governance contexts. Approaching destabilisation policy in terms of *policy mixes* enables the identification of complementarities between instruments for a given problem or objective and across policy domains (Rogge and Reichardt 2016), notably in terms of overall coherence, consistency and coordination. Another relevant dimension of policy mixes is their evolution over time (Kivimaa and Kern 2016; Rogge and Reichardt 2016) or issues of

TABLE 3.4 Types of destabilisation governance interventions

Type	*Features*
Direct	Relatively conventional forms of policy interventions seeking to contribute to phase-out by introducing control policies and altering frame conditions
Indirect	Oriented towards addressing oppositions to destabilisation processes, increasing the preparedness of affected or vulnerable groups and overturning structural dependencies
Experimental	Oriented towards anticipating, triggering and navigating destabilisation as transformative opportunity for introducing lasting systemic change, as well as legitimising such objectives
Civic	Emanating from civil society, social movements and activists, they are oriented towards bringing new problems and issues to the attention of society, mobilising public and political forces, and undermining the continuation of practices and systems seen as undesirable

temporal sequencing (Nilsson and Nykvist 2016). While it is too early to say much about the design of destabilisation policy mixes, it is useful to mobilise Streeck and Thelen's (2005a) institutional change typology to qualify possible evolution patterns from current arrangements in terms of their structure- and function-preserving qualities in search of evidence of institutional displacement, layering, drift, conversion or exhaustion. Kivimaa and Kern (2016) suggest that while evidence of policy layering is the most likely pattern, what is really needed for destabilisation policy mixes is a displacement and replacement pattern, notably concerning dominant technologies, key actors and rules.

Approaching interventions in terms of *governance context* broadens the perspective to include non-policy interventions, actors and dynamics, as well enabling conditions for destabilisation governance, i.e., the social, scientific, technical and economic dynamics making political projects of destabilisation possible at a given time and over time. *What are the conditions under which decisive and sustained deliberate destabilisation governance becomes possible, feasible or desirable?* Windows of opportunity for more radical interventions may open up and be seized by change entrepreneurs if problems, proposals and political agendas gain significant traction and become aligned (Kingdon 1984). The dynamic alignment of destabilisation contexts and governance spaces is likely to follow a range of different patterns and is a matter for empirical investigation. For instance, *under which conditions do crises become mobilised as opportunities for the deployment of radical phase-out interventions (as opposed to reinforcement of prevailing arrangements)?*

3.3.6 Vulnerability and politics

This section deals with some of the negative consequences of destabilisation and decline, their linkage to structural problems, and how they might be anticipated and avoided. Indeed, destabilisation raises significant human, social, political and ethical challenges, which may considerably weaken claims of "sustainability"[11] associated with socio-technical transitions if left unattended, and conversely enhance the emancipatory prospects of transition projects if justice, fairness and redistribution are brought centre stage.

Firstly, the destabilisation of socio-technical systems is likely to lead to the decline of certain industries and organisations, the closure of production sites, significant job losses and resulting individual hardships, particularly in communities and regions most dependent on conventional—"unsustainable"—activities. Indeed, by focusing on downfall, closure and abandonment, destabilisation research can usefully support the identification (and anticipation) of potential "losers" of socio-technical transitions. While economic and competitiveness losses in organisations and industries have led to the emergence of "managed decline" arguments and strategies (discussed in sections 3.3.3 and 3.3.5), it is essential to foreground often neglected human and local community perspectives in the process of change[12]—those of "people caught in the cross fire of industrial change" (Cowie and Heathcott 2003: 1). In the context of the British coal phase-out, scholars have pointed towards "the risk of insufficient attention regarding the broader implications of such discontinuity processes around the impacts on local coal communities and future prospects of the workforce" (Johnstone and Hielscher 2017: 457), by highlighting the uneven impacts that closures have on regional economies, workers and deprived communities. Such concerns are manifest around historically declining activities and livelihoods such as farming or heavy industries, notably given important deindustrialisation and delocalisation trends, for which lived experiences of loss and marginalisation abound and tend to persist well beyond the generation directly exposed to such loss (Strangleman 2017). Further, such experiences take on a new form in the context of environmentally motivated transitions, because of their intentional character—ironically, "deliberate destabilisation" risks not involving much local deliberation[13]—and because of the speed of change called for (Newell and Simms 2020), likely to temporally intensify related challenges. Furthermore, while the material, human and social losses of destabilisation tend to be considered in relation to production systems, sites and jobs, it appears relevant to consider the effects of destabilisation in consumption systems and practices too. For instance, phasing-out certain types of foods deemed unhealthy or environmentally problematic may unfairly affect certain communities over others. Similarly, recent social and political backlashes against the introduction of carbon taxes on petrol (e.g., the Gilets Jaunes movement) illustrate the difficulties of imposing restrictions on daily practices without involving concerned users whose livelihoods may be disproportionately dependent on certain modes of consumption (e.g., car-based mobility in rural or peri-urban communities) in the absence of real alternatives (e.g., affordable and accessible public transport). So, destabilisation raises major social and human concerns, which tend to disproportionately affect communities most *dependent* on existing means of production and consumption. These communities are hence most *exposed* to the effects of destabilisation and *vulnerable* to the potential losses associated with such processes but tend to be less visible and represented in related policy debates.

Second, the structural determinants and impacts of destabilisation are tied to particular places, patterns of dependence, vulnerability and opportunity. Concerning social and human dimensions, communities most exposed and vulnerable to the impacts of destabilisation tend to be those with limited resources (income, opportunities, various forms of capital), relatively strong dependence on established systems as means of production and consumption, limited agency or means of representation concerning related strategic decisions, and relatively weak emancipatory prospects (i.e. social groups with little access to alternatives, mobility or relevant infrastructure). So, the human and social impacts of destabilisation are likely to exacerbate *existing* structural inequalities as well as produce *new* forms of disenfranchisement. The place-based character of such structural inequalities has been problematised within human geography perspectives (e.g., geographic political economy, evolutionary economic geography), notably in the context of deindustrialisation in highly specialised regions.

Deindustrialisation studies tell us two fundamental lessons that may be applicable to destabilisation: rather than an ineluctable process, deindustrialisation is related to 1) a rupture of a long-standing "social contract" between worker unions and management, and 2) a process of capital mobility whereby productive activities are displaced to other localities, activities or forms of investment (Cowie and Heathcott 2003). Likewise, destabilisation can be seen as a process linked to mobility and displacement of local activities (which highlights the local, relative and uneven nature of destabilisation and its linkage to the stabilisation of other forms of activity, rather than viewing it as an irremediable and homogenous process), and a process involving struggles and fundamental changes in long-standing arrangements between relevant social actors (which highlights the political and social choices involved in handling destabilisation). So, the injustices related to destabilisation processes are likely tied to uneven power relations and access to capital between social groups, uneven ties and dependencies to places, and related tensions.

Concerning regional economies, sectors and industries, evolutionary economic geography has deployed the notions of regional lock-in (see also section 3.3.1) and relatedness to explain different patterns and outcomes of change in mature industry clusters facing significant challenges (Hassink 2010; Martin 2010). The focus on regional lock-in allows distinguishing adjustment patterns from renewal patterns, and explaining them according to the relative strength of political-institutional resistance to restructuring and to regional economic structure (Hassink 2010). The focus on relatedness (of activities, knowledge and skills in regional economies) enables inferences concerning the adaptiveness or resilience of a regional economy to external shocks: related variety of activities within a regional cluster enables specialisation and growth, but may reduce diversity and adaptability to pressures and shocks (for which unrelated variety may prove more versatile) (Balland et al. 2018; Boschma and Frenken 2011). So, a region's economic structure may determine its vulnerability to future challenges, regional lock-in may "explain the structural economic problems some old industrial areas face, as well as the related persistence of regional economic inequalities in some industrialized countries" (Hassink 2010: 454), and against observations of path dependence and "path insistence", it may be possible to engage more constructively with opportunities for regional path creation, path renewal or path development (Hassink et al. 2019). Blažek et al. (2020) have also suggested a need for greater attention to "negative" regional path developments, i.e., those involving the less appealing yet very likely paths of downgrading, contraction or delocalisation.

Geographical political economy perspectives highlight how the embedding of regional economies in wider political and economic relations generate interdependences and overflows with negative consequences. Regional destabilisation and decline can result in structural poverty and unemployment for place and people "left behind" (Rodríguez-Pose 2018), notably when alternatives are limited—stranded assets *and* stranded communities. Even "positive" regional path developments also have their "dark sides" (MacKinnon et al. 2019), indeed dark and bright sides of regional economic development are related in numerous ways (Phelps et al. 2018). New paths can create new forms of exploitation and inequality, because they lead to poor quality jobs (low value, low pay or precarious), dispossession, displacement or inequalities—notably in peripheries and enclaves within global production networks. Interregional competition can also undermine regional development opportunities in other localities.

Third, there is an inherently political dimension to destabilisation because it raises questions about the representation of the most exposed and vulnerable social groups (i.e., those left behind) and because neglecting to do so can lead to knock-on effects on socio-political stability, collective hopes and the fabric of society.

Policy perspectives on low-carbon transitions tend to emphasise related innovation opportunities and job-creation potential and suggest focusing policy efforts on appropriate jobs and skills development. While sustainability transitions will indeed generate new opportunity pathways for certain sectors, businesses, and forms of employment (e.g., in low-carbon building renovation, renewable energy, or sustainable food production and distribution), they will inevitably also lead to significant job losses in particular sectors and regions (e.g., in extractive, energy-intensive industries or conventional high-input farming). Emerging questions, besides how to re-orient "unsustainable" jobs towards emerging "green" sectors to quantitatively maintain stable employment opportunities, thus concern the scope for 1) avoiding negative and most destructive effects of sustainability transitions on workers and communities historically tied to sectors facing destabilisation, and 2) harnessing transitions as a means to enhance the nature, quality, and decency of employment. Such questions have recently been picked up under the framing of "Just Transitions" (Heffron and Mccauley 2018; Jasanoff 2018; Schwanen 2021; UNRISD 2018), to which trade unions and worker organisations have been actively contributing—though such framing still remains peripheral to mainstream policy concerns (Steward 2015). Workers—unionised or not—though perhaps the largest, most vulnerable and exposed social group to the destructive impacts of destabilisation, remain largely invisibilised in related political debates and policy decisions. A worker perspective is however of central importance if we consider that the destabilisation of industries and organisations can lead to a rupture of the social contract between workers and management, that workers have legitimate concerns about how destabilisation is handled, and that worker intelligence about industrial change contributes to creative solutions. Relevant analytical entry points include the role of workers and trade unions 1) in the production of political discourse and framings of destabilisation, 2) in the crafting of strategic responses to destabilisation, and 3) as mediating forces in destabilisation processes.

The unequal impact of destabilisation on particular communities is also likely to exacerbate political tensions and fuel political discourses capitalising on forms of injustice. Destabilisation and industrial decline can generate material and symbolic grievances, which populist politics mobilises through place-based narratives (Lizotte 2019). The impact of globalisation, automation, and the weakening of trade union representation of manufacturing workers have contributed to the recent political success of right-wing populist (RWP) parties, with "particular appeal amongst [groups commonly referred to as] the 'losers of modernisation' [or] the 'left behind'" (Lockwood 2018: 718). While the electoral appeal of RWPs is by no means restricted to industrial hinterlands, it is in such places that the more evident link with destabilisation can be observed:

> The places that don't matter are becoming tired of being told that they don't matter and are exercising a subtle revenge. They are voting down or threatening to vote down a system they perceive has quelled their potential and driven them down a road in which the future offers no opportunities, no jobs and no hope (Gros 2016; Rodrik 2017). It is as if the declining agricultural areas and rustbelts the world over have had enough of being patronised and have said, rightly or wrongly, that enough is enough: if we are being told that we no longer matter and that we are going down, the whole ship will sink with us.
>
> *(Rodríguez-Pose 2018: 199)*

The rise of global environmental issues as drivers of industry destabilisation have become a new element in this picture, most visible around the climate issue and resulting pressures exerted on high-carbon jobs. Indeed, climate scepticism has become a staple of RWP party discourse, which may be explained through structural arguments related to vulnerability and protection (e.g., hostility to climate policies seen as hitting workers and marginalised communities the hardest) or ideological ones related to anti-establishment and political distrust (e.g., hostility to climate policies seen as the product of a cosmopolitan elite) (Kulin et al. 2021; Lockwood 2018). Regardless, ensuing political discontent and tension around climate policies and "job-killing" arguments are a boon for carbon-intensive industries seeking to lobby more stringent targets and restrictions (Vona 2019).

While this arena currently seems largely captured by populist resentment and "places expressing their fear and outrage about potential futures in which they are economically and culturally irrelevant" (Lizotte 2019: 140), there are also reasons to believe that related conflicts and struggle about socio-environmental justice can spur more hopeful forms of political engagement with destabilisation futures.

3.4 Discussion: Conceptual propositions

This section builds on the six core notions introduced in section 3.2 and seeks to translate them into conceptual propositions. Together, these provide a synthetic reflection of where research debates stand and some pointers for further conceptual developments.

P1: Destabilisation can be seen as a form of challenge, reversal or erosion, of sources of socio-technical stability (lock-ins). Following a distinction between structural and enacted forms of stability, destabilisation involves 1) the breakdown of existing structural patterns, 2) divergence from prevailing action patterns, including a more active mobilisation of resistive power.

P2: Destabilisation is likely to result from a combination of multiple and mutable sources of change. Sources of change may vary in terms of dimensions, intensities, scope or kinds of causation. The intensity of individual and combined sources of change is likely to change over time and, if sustained, may lead to escalating destabilisation pressures. Destabilisation patterns may be distinguished according to the configuration of pressure fronts and their evolution over time.

P3: Incumbent actors are central actors of destabilisation with major stakes in and resources for maintaining prevailing configurations. Incumbency is related to important power and resource asymmetries that tend to reinforce stability tacitly (through reproduction of rules, practices and advantages) and actively (strategic moves to resist destabilisation). However, incumbency is also a plural and potentially transient kind of attachment. Other frontline actors of destabilisation processes include those invested in challenging forms of incumbency and vulnerable groups.

P4: Destabilisation is a non-linear, indeterminate and contingent process calling for the analysis of causal mechanisms and the conditions of their activation. A process approach to destabilisation should consider how sequences of events and particular conditions can activate causal mechanisms. The development of a destabilisation typology, specifying causal pathways, is a useful way forward, particularly if it is combined with a comparative approach oriented towards the exploration of a variety of pathways empirically. The temporality of destabilisation processes needs to be deconstructed, namely to disentangle punctual crises from more continuous stresses and to make sense of a variety of temporal profiles. Critically exploring the linkages between destabilisation and stabilisation is a useful avenue for theoretical development.

P5: The deliberate destabilisation of undesirable socio-technical regimes appears as an emerging, yet politically thorny, horizon for governance. It raises important questions related to 1) limits to agency and coordination vis-à-vis destabilisation processes and their governance; 2) a broad array of (potentially countervailing) motives for destabilisation governance with important implications for what is to be governed and how; 3) a variety of governance instruments extending beyond the conventional remit of innovation policy; 4) the need for thinking through combinations of policy instruments (policy mixes, policy sequencing, policy layering) and the types of governance contexts that may constitute destabilisation as a legitimate object of governance.

P6: Destabilisation and phase-out have human, social and political implications, which are exacerbated by uneven patterns of dependence, exposure and vulnerabilities. While such implications are likely to become more visible as destabilisation pressures materialise, they need to be foregrounded and anticipated. Place-based losses are likely to exacerbate existing structural inequalities as well as produce new forms of disenfranchisement. Destabilisation research has much to gain from engaging with place-based accounts of vulnerability, dependence and opportunity. Destabilisation is inherently political, raising issues of political representation and expressions of political discontent.

3.5 Conclusion: Emerging and unanswered problems

To conclude, I would like to evoke several unresolved issues and emerging puzzles, as we take a sidestep to engage with the flipsides, the dark sides, the losses and less hopeful aspects of socio-technical transitions. These are what I see as crucial next steps for destabilisation research: my wish list for a collective and distributed research programme.

Elaborating typologies of destabilisation pathways. Typologies of transitions pathways have assumed a central role in transitions studies, as a heuristic to make sense of developments in historical cases and to shed light on transitions in the making. Taking destabilisation seriously as a process opens the way for similar typological development. Typologies of destabilisation pathways should be oriented towards the exploration of a variety of possible processes and outcomes, enable pattern recognition in cases, and specify key degenerative mechanisms. This chapter has offered a number of relevant dimensions that such typological work may foreground (see section 3.3.4).

Harnessing and expanding empirical variety. Destabilisation research needs to supplement rich single-site case studies with ways to draw comparatively on the growing number of existing cases. To some extent, such work is underway, but it may largely benefit from the specification of destabilisation pathways. Furthermore, it appears important that the comparison of destabilisation patterns extends transversally across sectors, contexts and time periods. A shift towards more systematic comparisons also raises issues about the appropriate unit of analysis and implies trade-offs with deeper engagement with the various scales and sites of destabilisation. So, systematic comparison should not be a substitute for single case studies, but rather go alongside a deep engagement with non-standard cases that can challenge existing concepts and frames.

Dealing critically with shocks and temporality. Shocks and crises play a central role in imaginaries, discourse and research related to destabilisation. Destabilisation and breakdown are often ascribed or confined to exceptional circumstances, disruptions and external shocks challenging the otherwise normal operation of systems. Sections 3.3.2 and 3.3.4 have however demonstrated how destabilisation involves multiple and mutable sources of change that may combine in a variety of patterns, ranging from the dramatic to the more gradual or seemingly uneventful. It therefore appears important to situate crises in longer developmental sequences, by attending to what precedes and what may follow crises and the particular circumstances that may lead crises to activate or enable fundamental reconfigurations (crisis/transformation) and those that may lead crises to reinforce prevailing logics and dynamics (crisis/preservation). Section 3.3.5 has further underlined a very fruitful distinction between the discontinuation of governance (in exceptional times) and the governance of discontinuation (Stegmaier et al. 2014), which is an invitation to critically reflect on tensions between preservation and transformation rationales in the face of crises.

Dealing with uncertainty, anticipation and imagination. Destabilisation is an inherently uncertain and unpredictable process. That being said, taking the symmetry argument seriously requires engaging with the types of futures that a destabilisation focus may help envision. An obvious motive is that destabilisation scenarios may be a powerful tool to anticipate, and so possibly avoid, some of the direr consequences of destabilisation evoked in section 3.3.6. For this purpose, destabilisation research needs to engage more with modelling and other types of scenario techniques. Destabilisation futures are not necessarily hopeful and may be inhabited by more monsters than their Promethean counterparts, but there are possibly also joyous and enchanting aspects to future destabilisation, withdrawals and reductions waiting to be uncovered and engaged with. Living with less, without or in the ruins may also have its bright sides—or simply be necessary. Serious engagement with destabilisation imaginaries—dark and bright—appears as a crucial endeavour.

Dealing with vulnerability, justice and politics. Destabilisation comes with significant and lasting negative consequences on industries, regions, communities and individuals. To be sure, destabilisation is becoming a very current issue in the context of sustainability transitions, because an innovation-only approach is proving its limits to enable system-wide reconfigurations and because as sustainability innovations overflow out of their niches they are generating new forms of unintended consequences, backlash or resistance from incumbent actors. Transitions studies has always highlighted the role of struggles, but this new phase for transitions (Markard 2018; Turnheim et al. 2018) may be characterised by qualitatively and quantitatively different processes, notably ones in which struggles and politics take centre stage. By proposing a conceptual side-step, destabilisation research affords new ways of thinking about forms of dependence to existing systems, vulnerabilities to change, and the kinds of politics that emerge from situations of impending loss and marginalisation. The just transitions framing is one way forward (Jasanoff 2018), as is thinking through the emancipatory prospects of transitions (Stirling 2015), or thinking more seriously about the social determinants and implications of loss (Elliott 2018).

Dealing with expertise and knowledge. The way it has been framed here, destabilisation research involves an epistemological wager. Shifting the gaze to the existing and its challenges is already delivering significant insights as evidenced by the richness of existing conceptual elaborations (see section 3.2) and promises to bring STS traditions to new ground and debates. As destabilisation research and practice moves forward, it will no doubt spring up new matters of fact (what is destabilisation?) and matters of concern (how should destabilisation be handled?). One yet unaddressed question concerns what this new focus on destabilisation will do to knowledge communities and forms of expertise. Destabilisation bears promises of re-invigorating certain forms of knowledge and expertise around new questions as well as possibly destabilising epistemic communities currently less well equipped to think through its problems.

Notes

1 Indeed, in practice, there are very few historical examples of total decline of material or socio-technical systems, which instead tend to lose their centrality and significance as prime engines of socio-technical evolution.

2 Undesirability can be predicated upon various motivations, including environmental, aesthetic, cultural, economic arguments, and always a matter of perspective.

3 It is worth noting that practice bans usually take the form of restrictions on the modalities and locations of allowed use. The notion of perimeter (of use, of ban) hence appears relevant.
4 It is useful to distinguish between different forms of causation. Given the path dependent and contingent nature of socio-technical processes, *process theory* approaches are likely to be more useful than *variance-based* approaches, because they attend to the multiple and mutable causes involved in developmental process and event chains rather than attributing a specific outcome or event (e.g. the closure of a firm, the toppling of a political regime) to a set of variables via immutable forms of causation. Consequently, process tracing, event-sequence analysis and narrative analysis are relevant strategies to uncover the processes, contexts and conditions that can explain destabilisation and its outcomes (Smith et al. 2005; Yazar et al. 2020).
5 Stirling's analysis underscores the importance of interpretive schemes and heuristics as determining the style of action mobilised, with for example shock-interpretations tending towards conservative responses and stress-interpretations towards more radical changes in the underlying conditions.
6 In other words, and following a distinction between causal mechanisms according to the amount of energy (or pressure) they require posited by Bunge (1997), destabilisation may involve a combination of Type I causation (involving energy transfer) and Type II causation (wherein a very small cause may *trigger* a disproportionate effect). Type II causation hinges upon 1) triggering mechanisms or events, and 2) latent system instability.
7 Such metamorphic analogies can be found in the work of Joseph Schumpeter on the relationship of "gales of creative destruction" wherein industrial mutations involve the breakdown of industrial structures to create new ones—a perspective that has been heavily criticised for its pro-innovation bias (Joly 2019).
8 "Exnovation" (David 2017) or "innovation through withdrawal" (Goulet and Vinck 2017), initially developed as a principle for managing innovation within firms (Kimberly 1981) whereby organisations should devote more attention to divesting themselves of or discarding old innovations to make space for new ones, has become increasingly popular in debates about the deliberate phase-out of undesirable innovations (Davidson 2019; Heyen et al. 2017). While the focus tends to be on the removal of particular technologies or practices, rather than underlying systems and industries, it is pragmatically oriented towards the identification of particularly problematic, vulnerable or changeable elements with the intent to induce wider changes (Newig et al. 2019).
9 Stegmaier et al. (2014), focusing on *discontinuation as a purposeful action sui generis*, suggest another useful distinction between *enacted discontinuation*, which consists in setting change in motion, and *emergent discontinuation*, which consists in seizing prevailing developments of change.
10 Indeed, "abandoned socio-technical systems do not vanish completely and some continued governance effort is necessary long after their exit" (Stegmaier et al. 2014: 121–122). This raises the specific problem of "governing socio-technical aftercare", seeking to "control the loose ends of 'undead' regime and system parts" following decisions to phase-out, which may include dealing "legally, politically and technically" with remaining stocks and artefacts.
11 Notably by restricting them to environmental benefits, or by losing sight of distributive outcomes.
12 On this particular point, see Andy Stirling's advocacy of a "worm-eye" view on transition, focused on human experience and practice, as opposed to the more common "eagle-eye" view, focused on top-down management and governance (Stirling 2019).
13 I am indebted to Marc Barbier for intransigently pointing this out.

References

Abbott, A. (1997) On the concept of turning point. *Comparative Social Research*, 16, 85–105.
Abbott, A. (2004) *Methods of Discovery: Heuristics for the Social Sciences*. W.W. Norton & Co.

Allan, B.B. (2017) Producing the climate: States, scientists, and the constitution of global governance objects. *International Organization*, 71, 131–162. https://doi.org/10.1017/S0020818316000321.

Andersen, A.D. and Gulbrandsen, M. (2020) The innovation and industry dynamics of technology phase-out in sustainability transitions: Insights from diversifying petroleum technology suppliers in Norway. *Energy Research and Social Science*, 64, 101447. https://doi.org/10.1016/j.erss.2020.101447.

Arthur, W.B. (1989) Competing technologies, increasing returns, and lock-in by historical events. *The Economic Journal*, 99, 116–131.

Avelino, F. and Rotmans, J. (2009) Power in transition: An interdisciplinary framework to study power in relation to structural change. *European Journal of Social Theory*, 12, 543–569. https://doi.org/10.1177/1368431009349830.

Balland, P.A., Boschma, R., Crespo, J. and Rigby, D.L. (2018) Smart specialization policy in the European Union: Relatedness, knowledge complexity and regional diversification. *Regional Studies*. https://doi.org/10.1080/00343404.2018.1437900.

Bergek, A., Berggren, C., Magnusson, T. and Hobday, M. (2013) Technological discontinuities and the challenge for incumbent firms: Destruction, disruption or creative accumulation? *Research Policy*, 42, 1210–1224.

Berggren, C., Magnusson, T. and Sushandoyo, D. (2015) Transition pathways revisited: Established firms as multi-level actors in the heavy vehicle industry. *Research Policy*, 44, 1017–1028. https://doi.org/10.1016/j.respol.2014.11.009.

Blažek, J., Květoň, V., Baumgartinger-Seiringer, S. and Trippl, M. (2020) The dark side of regional industrial path development: towards a typology of trajectories of decline. *European Planning Studies*, 28, 1455–1473. https://doi.org/10.1080/09654313.2019.1685466.

Bohnsack, R., Kolk, A., Pinkse, J. and Bidmon, C. (2020) Driving the electric bandwagon: The dynamics of incumbents' sustainable product innovation. *Business Strategy and the Environment*, 29, 727–743.

Borrás, S. and Edler, J. (eds) (2014) *The Governance of Socio-Technical Systems: Explaining Change*. Edward Elgar Publishing.

Boschma, R. and Frenken, K. (2011) Technological relatedness, related variety and economic geography. In Cooke, P., Asheim, B., Boschma, R., Martin, R., Schwartz, D. and Tödtling, F. (eds), *Handbook of Regional Innovation and Growth*. Edward Elgar.

Braudel, F. (1970) History and the social sciences: The long term. *Information*, 9, 144–174.

Bunge, M. (1997) Mechanism and explanation. *Philosophy of the Social Sciences*, 27, 410–465.

Burke, P. (2005) *History and Social Theory*. Polity Press.

Clemens, E.S. (2007) Toward a historicized sociology: Theorizing events, processes, and emergence. *Annual Review of Sociology*, 33, 527–549. https://doi.org/doi:10.1146/annurev.soc.33.040406.131700.

Cowie, J. and Heathcott, J. (2003) The meanings of deindustrialization. In Cowie, J. and Heathcott, J. (eds) *Beyond the Ruins: The Meanings of Deindustrialization*. ILR Press.

David, M. (2017) Moving beyond the heuristic of creative destruction: Targeting exnovation with policy mixes for energy transitions. *Energy Research and Social Science*, 33, 138–146. https://doi.org/10.1016/j.erss.2017.09.023.

David, M. (2018) The role of organized publics in articulating the exnovation of fossil-fuel technologies for intra- and intergenerational energy justice in energy transitions. *Applied Energy*, 228, 339–350. https://doi.org/10.1016/j.apenergy.2018.06.080.

Davidson, D.J. (2019) Exnovating for a renewable energy transition. *Nature Energy*, 4, 254–256. https://doi.org/10.1038/s41560-019-0369-3.

Dijk, M., Wells, P. and Kemp, R. (2016) Will the momentum of the electric car last? Testing an hypothesis on disruptive innovation. *Technological Forecasting and Social Change*, 105, 77–88. https://doi.org/10.1016/j.techfore.2016.01.013.

Elliott, R. (2018) The sociology of climate change as a sociology of loss. *Archives Europeennes de Sociologie*, 59, 301–337. https://doi.org/10.1017/S0003975618000152.

Geels, F.W. (2004) From sectoral systems of innovation to socio-technical systems: Insights about dynamics and change from sociology and institutional theory. *Research Policy*, 33, 897–920. https://doi.org/10.1016/j.respol.2004.01.015.

Geels, F.W. (2014a) Regime resistance against low-carbon transitions: Introducing politics and power into the multi-level perspective. *Theory, Culture & Society*, 31, 21–40. https://doi.org/10.1177/0263276414531627.

Geels, F.W. (2014b) Reconceptualising the co-evolution of firms-in-industries and their environments: Developing an inter-disciplinary Triple Embeddedness Framework. *Research Policy*, 43, 261–277.

Geels, F.W. and Turnheim, B. (2022). *The Great Reconfiguration: A Socio-Technical Analysis of Low-Carbon Transitions in UK Electricity, Heat, and Mobility Systems*. Cambridge University Press.

George, A.L. and Bennett, A. (2004) *Case Studies and Theory Development in the Social Sciences*. MIT Press.

Goulet, F. and Vinck, D. (2017) Moving towards innovation through withdrawal: The neglect of destruction. In Godin, B. and Vinck, D. (eds), *Critical Studies of Innovation: Alternative Approaches to the Pro-Innovation Bias*. Edward Elgar. https://doi.org/10.4337/9781785367229.00014.

Hall, P.A. (1993) Policy paradigms, social learning, and the state: The case of economic policymaking in Britain. *Comparative Politics*, 25(3), 275–296.

Hambrick, D.C. and D'Aveni, R.A. (1988) Large corporate failures as downward spirals. *Administrative Science Quarterly*, 33, 1–23.

Hassink, R. (2010) Locked in decline? On the role of regional lock-ins in old industrial areas. Boschma, R. and Martin, R. (eds) *The Handbook of Evolutionary Economic Geography*. Edward Elgar. https://doi.org/10.4337/9781849806497.00031.

Hassink, R., Isaksen, A. and Trippl, M. (2019) Towards a comprehensive understanding of new regional industrial path development. *Regional Studies*, 53, 1636–1645. https://doi.org/10.1080/00343404.2019.1566704.

Hay, C. (2004) Review: Ideas, interests and institutions in the comparative political economy of great transformations. *Review of International Political Economy*, 11, 204–226. https://doi.org/10.1080/0969229042000179811.

Haydu, J. (2010) Reversals of fortune: Path dependency, problem solving, and temporal cases. *Theory and Society*, 39, 25–48. https://doi.org/10.1007/s11186-009-9098-0.

Heffron, R.J. and Mccauley, D. (2018) Geoforum: What is the "just transition"? *Geoforum*, 88, 74–77. https://doi.org/10.1016/j.geoforum.2017.11.016.

Hess, D.J. (2019) Incumbent-led transitions and civil society: Autonomous vehicle policy and consumer organizations in the United States. *Technological Forecasting and Social Change*, 151, 119825. https://doi.org/10.1016/j.techfore.2019.119825.

Heyen, D.A., Hermwille, L. and Wehnert, T. (2017) Out of the comfort zone! Governing the exnovation of unsustainable technologies and practices innovations. *GAIA*, 26, 326–331.

Hoffmann, S., Weyer, J. and Longen, J. (2017) Discontinuation of the automobility regime? An integrated approach to multi-level governance. *Transportation Research Part A: Policy and Practice*, 103, 391–408. https://doi.org/10.1016/j.tra.2017.06.016.

Hörisch, J. (2018) How business actors can contribute to sustainability transitions: A case study on the ongoing animal welfare transition in the German egg industry. *Journal of Cleaner Production*, 201, 1155–1165. https://doi.org/10.1016/j.jclepro.2018.08.031.

Isoaho, K. and Markard, J. (2020) The politics of technology decline: Discursive struggles over coal phase-out in the UK. *Review of Policy Research*, 37, 342–368. https://doi.org/10.1111/ropr.12370.

Jasanoff, S. (2018) Just transitions: A humble approach to global energy futures. *Energy Research and Social Science*, 35, 11–14. https://doi.org/10.1016/j.erss.2017.11.025.

Johnstone, P. and Hielscher, S. (2017) Phasing out coal, sustaining coal communities? Living with technological decline in sustainability pathways. *Extractive Industries and Society*, 4, 457–461. https://doi.org/10.1016/j.exis.2017.06.002.

Johnstone, P. and Stirling, A. (2020) Comparing nuclear trajectories in Germany and the United Kingdom: From regimes to democracies in sociotechnical transitions and discontinuities. *Energy Research and Social Science*, 59, 101245. https://doi.org/10.1016/j.erss.2019.101245.

Johnstone, P., Stirling, A. and Sovacool, B. (2017) Policy mixes for incumbency: Exploring the destructive recreation of renewable energy, shale gas "fracking", and nuclear power in the United Kingdom. *Energy Research and Social Science*, 33, 147–162. https://doi.org/10.1016/j.erss.2017.09.005.

Joly, P.-B. (2019) Reimagining innovation. In Lechevalier, S. (ed.) *Innovation Beyond Technology: Science for Society and Interdisciplinary Approaches*. Springer.

Joly, P.-B., Barbier, M. and Turnheim, B. (2022) Gouverner l'arrêt des grands systèmes sociotechniques. In Goulet, F. and Vinck, D. (eds) *Faire sans, Faire avec Moins: Les Nouveaux Horizons de L'innovation*. Presses des Mines.

Kern, F. (2011) Ideas, institutions, and interests: Explaining policy divergence in fostering "system innovations" towards sustainability. *Environment and Planning C: Government and Policy*, 29, 1116–1134. https://doi.org/10.1068/c1142.

Kimberly, J. (1981) Managerial innovation. In Nystrom, P. (ed.) *Handbook of Organizational Design*. Oxford University Press.

Kingdon, J.W. (1984) *Agendas, Alternatives, and Public Policies*. Little Brown and Company.

Kivimaa, P. and Kern, F. (2016) Creative destruction or mere niche support? Innovation policy mixes for sustainability transitions. *Research Policy*, 45, 205–217.

Klitkou, A., Bolwig, S., Hansen, T. and Wessberg, N. (2015) The role of lock-in mechanisms in transition processes: The case of energy for road transport. *Environmental Innovation and Societal Transitions*, 16, 22–37. https://doi.org/10.1016/j.eist.2015.07.005.

Kuhlmann, S., Stegmaier, P. and Konrad, K. (2019) The tentative governance of emerging science and technology: A conceptual introduction. *Research Policy*, 48, 1091–1097. https://doi.org/10.1016/j.respol.2019.01.006.

Kulin, J., Johansson Sevä, I. and Dunlap, R.E. (2021) Nationalist ideology, rightwing populism, and public views about climate change in Europe. *Environmental Politics*. https://doi.org/10.1080/09644016.2021.1898879.

Kungl, G. (2015) Stewards or sticklers for change? Incumbent energy providers and the politics of the German energy transition. *Energy Research and Social Science*, 8, 13–23. https://doi.org/10.1016/j.erss.2015.04.009.

Kungl, G. and Geels, F.W. (2018) Sequence and alignment of external pressures in industry destabilisation: Understanding the downfall of incumbent utilities in the German energy transition (1998–2015). *Environmental Innovation and Societal Transitions*, 26, 78–100. https://doi.org/10.1016/j.eist.2017.05.003.

Langhelle, O., Meadowcroft, J. and Rosenbloom, D. (2019) Politics and technology: Deploying the state to accelerate socio-technical transitions for sustainability. In

Meadowcroft, J., Banister, D., Holden, E., Langhelle, O. and Linnerud, K. (eds) *What Next for Sustainable Development? Our Common Future at Thirty*. Edward Elgar Publishing.

Langley, A. (1999) Strategies for theorizing from process data. *Academy of Management Review*, 24, 691–710.

Lee, D. and Hess, D.J. (2019) Incumbent resistance and the solar transition: Changing opportunity structures and framing strategies. *Environmental Innovation and Societal Transitions*. https://doi.org/10.1016/j.eist.2019.05.005.

Lizotte, C. (2019) Where are the people? Refocusing political geography on populism. *Political Geography*, 71, 139–141. https://doi.org/10.1016/j.polgeo.2018.12.007.

Lockwood, M. (2018) Right-wing populism and the climate change agenda: Exploring the linkages. *Environmental Politics*, 27, 712–732. https://doi.org/10.1080/09644016.2018.1458411.

Lockwood, M., Mitchell, C. and Hoggett, R. (2019) Unpacking "regime resistance" in low-carbon transitions: The case of the British capacity market. *Energy Research and Social Science*, 58, 101278. https://doi.org/10.1016/j.erss.2019.101278.

Lockwood, M., Mitchell, C. and Hoggett, R. (2020) Incumbent lobbying as a barrier to forward-looking regulation: The case of demand-side response in the GB capacity market for electricity. *Energy Policy*, 140, 111426.

MacKinnon, D., Dawley, S., Pike, A. and Cumbers, A. (2019) Rethinking path creation: A geographical political economy approach. *Economic Geography*, 95, 113–135. https://doi.org/10.1080/00130095.2018.1498294.

Mahoney, J. (2000) Path dependence in historical sociology. *Theory and Society*, 29, 507–548. https://doi.org/10.1023/A:1007113830879.

Mahoney, J. (2016) Mechanisms, Bayesianism, and process tracing. *New Political Economy*, 21, 493–499. https://doi.org/10.1080/13563467.2016.1201803.

Markard, J. (2018) The next phase of the energy transition and its implications for research and policy. *Nature Energy*. https://doi.org/10.1038/s41560-018-0171-7.

Martin, R. (2010) Roepke lecture in economic geography – Rethinking regional path dependence: Beyond lock-in to evolution. *Economic Geography*, 86, 1–27. https://doi.org/10.1111/j.1944-8287.2009.01056.x.

Martínez Arranz, A. (2017) Lessons from the past for sustainability transitions? A meta-analysis of socio-technical studies. *Global Environmental Change*, 44, 125–143. https://doi.org/10.1016/j.gloenvcha.2017.03.007.

Mylan, J., Morris, C., Beech, E. and Geels, F.W. (2019) Rage against the regime: Niche-regime interactions in the societal embedding of plant-based milk. *Environmental Innovation and Societal Transitions*, 31, 233–247. https://doi.org/10.1016/j.eist.2018.11.001.

Nelson, R.R. and Winter, S.G. (1982) *An Evolutionary Theory of Economic Change*. The Belknap Press of Harvard University Press.

Newell, P. and Simms, A. (2020) How did we do that? Histories and political economies of rapid and just transitions. *New Political Economy*. https://doi.org/10.1080/13563467.2020.1810216.

Newig, J., Derwort, P. and Jager, N.W. (2019) Sustainability through institutional failure and decline? Archetypes of productive pathways. *Ecology and Society*, 24, 18–31.

Nilsson, M. and Nykvist, B. (2016) Governing the electric vehicle transition – Near term interventions to support a green energy economy. *Applied Energy*, 179, 1360–1371. https://doi.org/10.1016/j.apenergy.2016.03.056.

Normann, H.E. (2019) Conditions for the deliberate destabilisation of established industries: Lessons from U.S. tobacco control policy and the closure of Dutch coal mines. *Environmental Innovation and Societal Transitions*, 33, 102–114. https://doi.org/10.1016/j.eist.2019.03.007.

Osterholm, M. and Olshaker, M. (2020) *Chronicle of a Pandemic Foretold: Learning from the COVID-19 Failure – Before the Next Pandemic Arrives*. Foreign Affairs.

Ottosson, M. and Magnusson, T. (2013) Socio-technical regimes and heterogeneous capabilities: The Swedish pulp and paper industry's response to energy policies. *Technology Analysis and Strategic Management*, 25, 355–368. https://doi.org/10.1080/09537325.2013.774349.

Peters, G.P., Andrew, R.M., Canadell, J.G., Friedlingstein, P., Jackson, R.B., Korsbakken, J. I., Quéré, C. Le and Peregon, A. (2020) Carbon dioxide emissions continue to grow amidst slowly emerging climate policies. *Nature Climate Change*, 10, 3–6.

Phelps, N.A., Atienza, M. and Arias, M. (2018) An invitation to the dark side of economic geography. *Environment and Planning A*, 50, 236–244. https://doi.org/10.1177/0308518X17739007.

Pierson, P. (2000) Increasing returns, path dependence, and the study of politics. *The American Political Science Review*, 94, 251–267.

Raven, R.P.J.M. (2006) Towards alternative trajectories? Reconfigurations in the Dutch electricity regime. *Research Policy*, 35, 581–595. https://doi.org/10.1016/j.respol.2006.02.001.

Roberts, C., Geels, F.W., Lockwood, M., Newell, P., Schmitz, H., Turnheim, B. and Jordan, A. (2018) The politics of accelerating low-carbon transitions: Towards a new research agenda. *Energy Research and Social Science*, 44, 304–311. https://doi.org/10.1016/j.erss.2018.06.001.

Rodríguez-Pose, A. (2018) Commentary: The revenge of the places that don't matter (and what to do about it). *Cambridge Journal of Regions, Economy and Society*, 11, 189–209. https://doi.org/10.1093/cjres/rsx024.

Rogge, K.S. and Johnstone, P. (2017) Exploring the role of phase-out policies for low-carbon energy transitions: The case of the German Energiewende. *Energy Research and Social Science*, 33, 128–137. https://doi.org/10.1016/j.erss.2017.10.004.

Rogge, K.S. and Reichardt, K. (2016) Policy mixes for sustainability transitions: An extended concept and framework for analysis. *Research Policy*, 45, 1620–1635.

Rosenbloom, D. and Rinscheid, A. (2020) Deliberate decline: An emerging frontier for the study and practice of decarbonization. *Wiley Interdisciplinary Reviews: Climate Change*. https://doi.org/10.1002/wcc.669.

Schwanen, T. (2021) Achieving just transitions to low-carbon urban mobility. *Nature Energy*. https://doi.org/10.1038/s41560-021-00856-z.

Seto, K.C., Davis, S.J., Mitchell, R., Stokes, E.C., Unruh, G. and Ürge-Vorsatz, D. (2016) Carbon lock-in: Types, causes, and policy implications. *Annual Review of Environment and Resources*. https://doi.org/10.1146/annurev-environ-110615-085934.

Sewell Jr, W.H. (2005) *Logics of History: Social Theory and Social Transformation*. University of Chicago Press.

Shove, E. (2012) The shadowy side of innovation: Unmaking and sustainability. *Technology Analysis and Strategic Management*, 24, 363–375. https://doi.org/10.1080/09537325.2012.663961.

Shove, E., Pantzar, M. and Watson, M. (2012) *The Dynamics of Social Practice: Everyday Life and How it Changes*. SAGE.

Sillak, S. and Kanger, L. (2020) Global pressures vs. local embeddedness: the de- and restabilization of the Estonian oil shale industry in response to climate change (1995–2016). *Environmental Innovation and Societal Transitions*, 34, 96–115. https://doi.org/10.1016/j.eist.2019.12.003.

Smink, M.M., Hekkert, M.P. and Negro, S.O. (2015) Keeping sustainable innovation on a leash? Exploring incumbents' institutional strategies. *Business Strategy and the Environment*, 24, 86–101. https://doi.org/10.1002/bse.1808.

Smith, A. and Stirling, A. (2007) Moving outside or inside? Objectification and reflexivity in the governance of socio-technical systems. *Journal of Environmental Policy and Planning*, 9, 351–373. https://doi.org/10.1080/15239080701622873.

Smith, A., Stirling, A. and Berkhout, F. (2005) The governance of sustainable socio-technical transitions. *Research Policy*, 34, 1491–1510. https://doi.org/10.1016/j.respol.2005.07.005.

Sovacool, B.K. (2016) How long will it take? Conceptualizing the temporal dynamics of energy transitions. *Energy Research & Social Science*, 13, 202–215.

Späth, P., Rohracher, H. and Von Radecki, A. (2016) Incumbent actors as niche agents: The German car industry and the taming of the "Stuttgart e-mobility region". *Sustainability*, 8. https://doi.org/10.3390/su8030252.

Spencer, T., Colombier, M., Sartor, O., Garg, A., Tiwari, V., Burton, J., Caetano, T., Green, F., Teng, F. and Wiseman, J. (2018) The 1.5°C target and coal sector transition: At the limits of societal feasibility. *Climate Policy*, 18, 335–351. https://doi.org/10.1080/14693062.2017.1386540.

Steen, M. and Weaver, T. (2017) Incumbents' diversification and cross-sectorial energy industry dynamics. *Research Policy*, 46, 1071–1086. https://doi.org/10.1016/j.respol.2017.04.001.

Stegmaier, P., Kuhlmann, S. and Visser, V.R. (2014) The discontinuation of socio-technical systems as a governance problem. In Borrás, S. and Edler, J. (eds) *The Governance of Socio-Technical Systems: Explaining Change*. Edward Elgar. doi:10.4337/9781784710194.00015.

Steward, F. (2015) *Policies and Practices to Promote Work Enhancing Pathways in the Transition to a Low Carbon Economy – Europe*. ACW International Policy Working Group.

Stirling, A. (2014) From sustainability, through diversity to transformation: Towards more reflexive governance of vulnerability. In Hommels, A., Mesman, J. and Bijker, W.E. (eds) *Vulnerability in Technological Cultures: New Directions in Research and Governance*. MIT Press.

Stirling, A. (2015) Emancipating transformations: from controlling "the transition" to culturing plural radical processes. In Scoones, I., Leach, M. and Newell, P. (eds) *The Politics of Green Transformations*. Routledge.

Stirling, A. (2019) How deep is incumbency? A "configuring fields" approach to redistributing and reorienting power in socio-material change. *Energy Research & Social Science*, 58, 101239. https://doi.org/10.1016/j.erss.2019.101239.

Strangleman, T. (2017) Deindustrialisation and the historical sociological imagination: Making sense of work and industrial change. *Sociology*, 51, 466–482. https://doi.org/10.1177/0038038515622906.

Streeck, W. and Thelen, K. (eds) (2005a) *Beyond Continuity: Institutional Change in Advanced Political Economies*. Oxford University Press.

Streeck, W. and Thelen, K. (2005b) Introduction: Institutional change in advanced political economies. In Streeck, W. and Thelen, K. (eds) *Beyond Continuity: Institutional Change in Advanced Political Economies*. Oxford University Press.

Suarez, F.F. and Oliva, R. (2005) Environmental change and organizational transformation. *Industrial and Corporate Change*, 14, 1017–1041. https://doi.org/10.1093/icc/dth078.

Turnheim, B. (2012) *The Destabilisation of Existing Regimes in Socio-Technical Transitions: Theoretical Explorations and In-Depth Case Studies of the British Coal Industry (1880–2011)*. University of Sussex.

Turnheim, B. and Geels, F.W. (2012) Regime destabilisation as the flipside of energy transitions: Lessons from the history of the British coal industry (1913–1997). *Energy Policy*, 50, 35–49. https://doi.org/10.1016/j.enpol.2012.04.060.

Turnheim, B. and Geels, F.W. (2013) The destabilisation of existing regimes: Confronting a multi-dimensional framework with a case study of the British coal industry (1913–1967). *Research Policy*, 42, 1749–1767. https://doi.org/10.1016/j.respol.2013.04.009.

Turnheim, B. and Geels, F.W. (2019) Incumbent actors, guided search paths, and landmark projects in infra-system transitions: Re-thinking Strategic Niche Management with a case study of French tramway diffusion (1971–2016). *Research Policy*, 48, 1412–1428. https://doi.org/10.1016/j.respol.2019.02.002.

Turnheim, B. and Sovacool, B.K. (2020) Forever stuck in old ways? Pluralising incumbencies in sustainability transitions. *Environmental Innovation and Societal Transitions*, 35, 180–184. https://doi.org/10.1016/j.eist.2019.10.012.

Turnheim, B., Wesseling, J., Truffer, B., Rohracher, H., Carvalho, L. and Binder, C. (2018) Challenges ahead: Understanding, assessing, anticipating and governing foreseeable societal tensions to support accelerated low-carbon transitions in Europe. In Foulds, C. and Robinson, R. (eds) *Advancing Energy Policy: Lessons on the Integration of Social Sciences and Humanities*. Palgrave Pilot. https://doi.org/10.1007/978-3-319-99097-2_10.

Tushman, M.L. and Anderson, P. (1986) Technological discontinuities and organizational environments. *Administrative Science Quarterly*, 31, 439–465.

UNRISD (2018) *Mapping Just Transition(s) to a Low-Carbon World*. The Just Transition Research Collaborative.

Unruh, G.C. (2000) Understanding carbon lock-in. *Energy Policy*, 28, 817–830. https://doi.org/10.1016/S0301-4215(01)00098–00092.

Unruh, G.C. (2002) Escaping carbon lock-in. *Energy Policy*, 30, 317–325. https://doi.org/10.1016/S0301-4215(01)00098–00092.

van Mossel, A., van Rijnsoever, F.J. and Hekkert, M.P. (2018) Navigators through the storm: A review of organization theories and the behavior of incumbent firms during transitions. *Environmental Innovation and Societal Transitions*, 26, 44–63. https://doi.org/10.1016/j.eist.2017.07.001.

Vögele, S., Kunz, P., Rübbelke, D. and Stahlke, T. (2018) Transformation pathways of phasing out coal-fired power plants in Germany. *Energy, Sustainability and Society*, 8. https://doi.org/10.1186/s13705-018-0166-z.

Vona, F. (2019) Job losses and political acceptability of climate policies: why the "job-killing" argument is so persistent and how to overturn it. *Climate Policy*, 19, 524–532. https://doi.org/10.1080/14693062.2018.1532871.

Walker, W. (2000) Entrapment in large technology systems: Institutional commitments and power relations. *Research Policy*, 29, 833–846.

Wells, P. and Nieuwenhuis, P. (2012) Transition failure: Understanding continuity in the automotive industry. *Technological Forecasting and Social Change*, 79, 1681–1692. https://doi.org/10.1016/j.techfore.2012.06.008.

Winskel, M. (2018) Beyond the disruption narrative: Varieties and ambiguities of energy system change. *Energy Research and Social Science*, 37, 232–237. https://doi.org/10.1016/j.erss.2017.10.046.

Yazar, M., Hestad, D., Mangalagiu, D., Ma, Y., Thornton, T.F., Saysel, A.K. and Zhu, D. (2020) Enabling environments for regime destabilization towards sustainable urban transitions in megacities: comparing Shanghai and Istanbul. *Climatic Change*. https://doi.org/10.1007/s10584-020-02726-1.

4

CONCEPTUAL ASPECTS OF DISCONTINUATION GOVERNANCE

An exploration

Peter Stegmaier

4.1 Introduction: Focus and phenomenon

The fact that socio-technical regimes do not last forever, but decline, is receiving increasing attention in research on the transition of socio-technical systems and innovation dynamics. There is a great need for this attention, because for a long time these research directions have not or only marginally dealt with how existing systems and their supporting regimes (as well as parts of them) destabilise or how innovations break down before they are fully established. The *complex interrelationships* are not easy to reconstruct. Often, an obvious dimension of this gets lost in the background or is difficult to consider on top of it: the active destabilisation or promotion of destabilisation, the intended exit or termination of support for existing regimes, pursued by a wide variety of actors, alone or together, in the public and political spheres, but just as well in the private and corporate spheres.

Not even a reasonably agreed *terminology* has emerged.[1] Therefore, it must be stated here which terms are mobilised to designate what. "Discontinuation" means the rather actively pursued exit from a socio-technical regime, in contrast to "destabilisation" as rather passively developing (Turnheim 2023). The notion of "decline" itself is used as a very general term to describe the *circumstance* of progressive destabilisation or discontinuation. The term "phase-out" is employed when it is literally a matter of gradual development. Exit and abandonment are used as relatively non-specific terms to describe the opposite of maintenance, stabilisation, or continuation. Since this paper is about the political, administrative, and managed governance work on an exit from a socio-technical regime, I will mostly speak of "discontinuation" as a more or less purposeful pursuit of a goal.[2] "Discontinuity" then rather refers to the circumstance of a targeted or achieved end of such a regime. Finally, I choose the concept of the socio-technical regime from the multi-level perspective (Geels 2019) as a reference point for the

DOI: 10.4324/9781003213642-4

governance efforts: as the object-related and theoretical context of choice for the considerations.

This chapter is about the purposeful discontinuation of a socio-technical regime, no matter what stage or path of development, maturity, or age. When speaking of socio-technical regimes, I may be talking about narrower or broader domains, the immediate usage environment of a device (e.g., light bulb) as well as the associated value chain (light bulb manufacturing) or embedding in other regimes (household, power supply, lamp design, and many more). I assume that some form of concerted, purposeful action or series of actions, using means of shaping the public or private, will occur if a socio-technical system is to be ended. This can involve large-scale technology as well as small devices, entire infrastructures as well as production lines or product ranges. Discontinuation affects technology as well as the science, politics, economy, everyday practice, or law that supports it. It affects knowledge and ignorance, forgetting and preserving, strategies and routines, individual and collective action. Not everything that happens during discontinuation is induced through governance, but it is a research focus to better understand whether, in which ways, and to which extent governance plays a role. I am in fact interested in *configurations that shape the situations bearing the possibility (if recognised and used) to deliberately change/discontinue a socio-technical regime through governance.*

We still know too little about discontinuation and destabilisation to make a further-reaching exclusionary theory selection. The phenomenon itself cannot be subsumed under existing concepts, patterns, and models—existing research was too often blind to the interplay between discontinuity, socio-technical phenomena, and governance, and therefore the optics for grasping it in an adequate way need to be invented. Since we are talking about a rather new and emergent research field, I suggest, firstly, to use a *mix of perspectives* and, secondly, to understand these ways of seeing as *heuristics*—conceptual propositions to see phenomena associated with discontinuation from a governance-related viewpoint.

In the following, I will develop arguments about how existing knowledge about discontinuation can be used to argue for certain governance approaches that offer the chance to get to the bottom of discontinuation governance-in-action (section 4.2). I will then introduce a number of relevant existing concepts that can help us to become aware of discontinuation governance, if only as a precursor to case-specific and cross-case concepts that should emerge from the research (section 4.3). I conclude with a series of challenges that discontinuation governance research is currently facing.

4.2 Governance as key focus

If we want to put the "governance" of discontinuation on the agenda, we should clarify what can be meant by "governance". Given the plethora of questions relevant to discontinuation itself, this conceptualisation cannot be exhaustive at this point. Given the newness of the research field, it would also not be appropriate to make a definitional determination. Instead, it is wise to suggest heuristic aspects that might be relevant to consider. Ultimately, it will be a task of any research on discontinuation governance to develop an appropriate concept of governance itself or to specify alternatives such as "policy", "politics", "administration", and so on that fit the respective object of research. However, it would be a pity if no effort was made to specifically shape the concept of governance or if it was only done implicitly. Then this aspect would perhaps be black boxed instead of being made fruitful.

The term governance is used to indicate that the notion of hierarchical "steering" or "government action" as the tackling of collective problems falls short of capturing political processes and situations in complex societies, political systems, and states. The questions that regularly arise are: who is doing what, what is it about, in which structures is the action embedded, which structures is it trying to meet or escape and which means and forms of action are used. A general starting point for research with a focus on governance of discontinuation is that by governance we address a process of mutual shaping a political, market, technoscientific, or any other social order. The assumption, then, is that governance-making and governance structures in some political ways have as their aim and purpose the stabilisation, maintenance/repair, and/or destabilisation of a given order. It is political when someone asserts him- or herself alone or with others and achieves a somewhat binding order that is meant to help address collective problems. The various social contexts of these socio-technical orders are also relevant.

This applies to any order that a community, an organisation, or a group establishes for itself, more or less on the basis of procedures and rules. This can happen not only by direct control, but also by influencing (or coordinating, negotiating) relationships between more or less independent actors against the background of their resources, arenas, rules, instruments, and in rather heterarchical or rather hierarchical contexts. Does governance mean a more tentative or directive approach (Kuhlmann et al. 2019)? We would want to see which relationships characterise a governance and how, which orientations and motivations are at work and/or expressed, and how (far) individual actions are coordinated. This we would need to understand also in temporal contexts, so what was before the current situation and in which direction it moves. Here it is helpful to look at the development in its patterns and manifestations as well as individual situations in the course of this development. The not-so-frequent nor well-known discontinuation governance especially must be searched for how an opportunity actually arises that discontinuation becomes central. Which failed attempts, abortions, preparations, changes of plans, power structures and discourse struggles first opened up the chance to distance oneself from something familiar, "normal" and accepted for a long time? Do stable orders break down from delegitimisation, disinterest, non-action or destruction, withdrawal or redefinition of knowledge and norms, perception, or interpretations? Does a regime destabilise from dysfunctionality, erosion, exhaustion, weakness, or fragile statics, from the inability to be adapted or updated (Joly et al. 2022)? In addition to process and situation orientation, the analysis of problem orientation is also cultivated (Kohoutek et al. 2013). It examines how discontinuation governance takes place (as process), in which constellation it asserts itself and ends a socio-technical regime (situation), and finally along which problem perceptions and agendas it finds its authoritative orientation.

It doesn't suffice only to mention many aspects that need to be looked at more closely typically, but also to point out symmetries. This means that when the focus is on governance-making, one should not ignore the structural circumstances; or, conversely, that one should not overlook the specific action of influence or triggering action in the face of all the structures and abstract developments; and that one should put events and justifications, knowledge- and value-related perceptions, explanations, and justifications in relation to each other. One can read volumes about the usual policies and governance arrangements, routines, and improvisations. But what is discontinuation—how do you do it, why, what for, what triggers it, hinders it, or ends it? How far is discontinuation presenting itself as a governance innovation?

In order to study how discontinuation is actively addressed by governance, it needs to be seen how much public and private or mixed elements are de-aligned; how much corporate (private) governance and public policy (alone or together) are defining the subject of discontinuation (e.g., of actors in a household or in public transport). It must be clarified for each individual case in which planetary, global, regional, national to local interconnections discontinuation takes place. How are specific and meta-governance frameworks related, are their developments interdependent or do they contradict each other? Is a meta-governance framework built up first before a specific regime is brought down?

4.2.1 Discontinuation as a social construct

When governance tackles something, we observe that it is typically because those involved have agreed that there is a *problem* that needs to be addressed (Colebatch 2006; Colebatch et al. 2010; Dewey 1927). The fact that there is a problem for a collective, what it is for them, how to recognise it, what it has to do with other or previous problems, how one usually reacts to it, and so on are all questions of shared perception, interpretation, and negotiation of what to do with it (Kingdon 1984; Hoppe 2010) in the light of norms and theories, patterns of consciousness and frames as our (shared) interpretations of how situations are organised. Conflicts also arise as to what one sees, whose view is correct or whether/how one usually reacts to it with governance. Normative questions also arise whether one should see something one way or another, what one should do with it, what is the right (factually appropriate) or best (politically, morally, ethically, etc.) reaction to it.[3] If one considers these processes of the social production of occasions to engage in governance, it becomes clear why one would do well to explore how active or non-active discontinuation occurs, how it comes about that discontinuation governance does not always receive the same attention, although it may already be a means of choice again and again on a smaller scale. It is therefore also worthwhile to see how discontinuation governance can experience an upswing when, for example, as is currently the case, the most diverse upheavals are being discussed and discontinuation is seen as a way to go, as in energy issues, technology development, environmental protection, or industrial policy (or for other areas, see, e.g., Sato 2002; Bauer et al. 2012; Princen et al. 2015).

If both the way of acting, which can be understood as discontinuation governance, and the occasion to do so, must be continuously produced and maintained, i.e., are not determined as innate or natural law conditions, then we can see them as social constructs. Then we can say discontinuation governance is communicatively constructed, be it at the beginning or as it is sustained during social negotiations (Eberle 2019: 265). Negotiation is thereby understood as a generic category referring to the negotiated order as a result of all kinds of interactions, not only as one form of shaping social order among others (cf. Strauss 1978, 1993: 57). Following Berger and Luckmann (1966), we can see discontinuation governance and its result, discontinuity, as the co-construction of a discontinuous reality: as objective facticity *and* subjective meaning. This means asking how it is possible that human activity should produce or discontinue a world of things and vice versa: the *co-construction of continuous or discontinuous reality* (ibid.: 27). This means that something is encountered as a social matter that cannot easily be ignored or escaped, *and* that this has meaning for an actor. For instance, in 2009, at the start of the phase-out, the incandescent light bulb (ILB) was an object that followed norms that were increasingly perceived as delegitimate, an object that industry wanted to get rid of, and the ILB users knew how to interpret this situation, e.g., by stockpiling the ILB for use beyond the end of production because users liked the light quality, or, conversely, they shared the perception that ILBs needed to be replaced for energy-saving reasons and hurried to exchange the old ones for the new ones.

At the same time, it is clear that precisely discontinuation and destabilisation entail a change and thus a partial dissolution of an existing social order (in our case better: socio-technical order) that has hitherto been preserved *grosso modo*. The *fluidity of order*, order as something mobile, plastic, turns out to be sometimes lossy while analysing and governing discontinuation and destabilisation: not everything remains forever. What just belonged can lose legitimacy or even become obsolete.

4.2.2 The agency–structure nexus

There are many reasons to assume that discontinuity is not purely a question of *active* governance action. Socio-technical regimes are too large and complex to always be brought down with simple actions. Where active action is taken with this aim in mind, *passive* tendencies may also come into play, intended developments may go hand in hand with unintended ones, and planned developments may go hand in hand with unplanned ones. Active attempts may fail, take a long time with an uncertain outcome, or have a partial rather than comprehensive impact. Existing destabilising tendencies may first be discovered, waited for, and eventually exploited, but not triggered completely on their own. Before active action can take effect, a great deal of preparatory work may have to be done, such as discovering and picking up on existing trends, exploiting other processes for discontinuation, and laying a variety of foundations so that discontinuation can become conceivable, communicable, acceptable, realisable, and justifiable. Times and windows of opportunity to become active can be wrongly chosen or less sustainable than thought.

At the same time, however, one cannot claim that discontinuities occur without any *intervention or active influence, impetus, or intention*. Socio-technical regimes may be doomed to decline because they are not maintained and developed, or even neglected, or because they overlook problems that over time build up to existential threats. Complex systemic relationships can lead to discontinuation and destabilisation just as well as any other condition. Destabilisation may not lead to the end (Turnheim 2023); this may need active influence (cf. Le Quéré et al. 2019). Decline could ultimately be accepted—otherwise one could try to resist it and keep something alive, pour financial resources into it and declare war on its ending. From this point of view, the failure to intervene and resist—the failure to act, the veto by doing nothing—can also be an active measure[4] (cf. Weber 1978; Streeck and Thelen 2005). Turnheim (2023: 56) calls this "inaction" a "particular form of intervention: neglectful or self-consuming governance".

But one can see how decisions are made that lead to discontinuity, at least in the areas subject to this decision-making power. When NASA pulled out of the space shuttle programme, realistically speaking, it will not have been the decision of the NASA leadership alone, nor will it have been made entirely of its own free will. But a formal act of termination certainly took place. When Siemens sold its nuclear division, this did not mean that nuclear technology ended everywhere, but that it ended at Siemens—together with the management and knowledge structures, the personnel and the value chain that was once built up and maintained. Thus, it seems advisable to search for those *combinations* of active action and existing developments, of new and existing processes and efforts that make the governance of discontinuity tangible. It will be a matter of identifying situational as well as processual, structural, and mobile aspects of the governance of discontinuity.

In practice, there are hardly any cases in which the *public* or *private* can manage entirely on their own. On the contrary, we regularly see that these areas are somehow intertwined—for example, when Siemens sold its nuclear division as a company, this was a corporate matter closely linked to one country's national nuclear policy; when the government closes nuclear power plants, this affects the energy industry, among many other things; households must be prepared for changes in electricity quality and prices, and must figure out whether or not to support a more sustainable energy policy.

The underlying problem is, in brief, that people act, resulting in structures that in turn influence action (Berger and Luckmann 1966; Giddens 1984). It is therefore important to consider how discontinuation governance as action reacts to existing structures or questions old ones, dissolves them, or even creates new ones. *Action and structures intertwine*. Structures of discontinuing action are individual pieces or types of actions and actors, concatenations, clusters, habitualisations, institutionalisations, socialisations, knowledge structures and distributions, and so on. If a discontinuation has already taken place, any further discontinuation already refers to prior knowledge, patterns of action, even instruments and other institutionalised (repertoires of) routines equipped with rules and knowledge of rules (Cairney 2012; Colebatch 2006). Or there is recourse to modes of governance that do not come directly from discontinuation; work is done analogously to other governance. Combinations of exceptional and analogue modes of governance can be found.

From a governance analysis perspective, it is also important to remember that not all problems are *perceived* or *addressed* by decision-makers, partly because it would be too burdensome to address them or break them down into manageable parts. Thus, starting to exit can seem like an oversized hurdle because it has never been done before (if that is indeed the case) or could trigger consequences (side-effects, collateral damages) that encompass more than just the discontinuation issue.

4.2.3 The things–meanings nexus

Discontinuation governance targets a spectrum from perceptions, frames, knowledge structures on the one hand, to built environment, architecture, and infrastructure on the other hand. What has been introduced as a social construction of discontinuity is realised both in the breaking up of thinking patterns and dismantling of what has been built in stone and steel. It is about breaking down previously valid knowledge and old certainties as well as dismantling production plants, buildings, and the structural realisation of infrastructure in hardware and information technology structures such as software.

What has been said so far about governance can also be applied to the terms policy and politics (or whatever distinctions the respective contextual language offers). It is not so important which of the terms one uses, but for what and how. Much of what is said about policy today comes very close to the concept of governance, especially the departure from top-down and governmental focus (Colebatch 2009). One can also see the work on problem conceptions, policy alternatives and decision-making politics as dimensions of governance (Kingdon 1984). Governance is also realised in material and digital structures.

4.3 Existing conceptualisations

There are several concepts that have been around for a while to help understand decline in governance, policy, or administration studies. These are the two prominently represented terms—"termination" and "dismantling"—but alongside them the lesser-noted should also be mentioned. In the following, some typical perspectives in research on science, technology, innovation, and policy will be identified. It will be discussed which aspects might be helpful for our collection of building blocks for a conceptualisation of discontinuation governance. It starts with concepts from the field of policy and governance studies and ends with references from science and technology studies. In between, I highlight the emergence of discontinuation aspects in socio-technical transition studies as well as in studies on socio-technical-economic, innovation, and more recent discontinuation governance studies.

4.3.1 *Discontinuation concepts in studies of governance, policy, administration*

Although there is some attention for questions of discontinuation, work on "termination", "dismantling", "reversal", "removal", or "retrenchment" never really became a core topic in policy studies research and textbooks. The abandonment or dismantling of socio-technical systems, deeply embedded in society and the economy, has not been studied yet from that angle.[5] Our focus is somewhat different from these existing literatures, however, for two main reasons: first, the focus isn't on a particular actor or social group (firm, industry, or an entire policy branch such as welfare), but on technologies and socio-technical systems and the regimes that carry them that involve and interact with many actors. More importantly, we are interested in a *particular kind of discontinuation*, which is driven or *motivated* by the desire on the part of relevant governance actors *to address one or more perceived particular social issues* (e.g., protection of the EU lighting industries and climate change, if you think of the incandescent light bulb phase-out within the EU ecodesign framework; safety and environmental effects with regards to the ban of DDT by the Stockholm Convention; safety and risk issues in the case of the German nuclear phase-out). The perception of adverse effects of a specific socio-technical system is at the core of the mobilisation process. It very often requires production of knowledge. Second, our emphasis is on understanding by whom and how "issues" are being *articulated*, framed, and put on political agendas in related actor arenas—in other words, how "problem"-driven political agendas manage to question and *challenge incumbent governance arrangements* within and around a socio-technical system, and possibly may result in *deliberate discontinuation governance efforts*, which may be more or less successful over time.

Policy termination

In the policy studies literature, there is some attention for the governance of discontinuation of particular policies and programs, often referring to the term "*policy termination*" (cf. Bauer 2009). Policy termination may either result from a changed formulation or perception of a policy problem, or from a changed formulation or perception of a policy solution (van de Graaf and Hoppe 1996: 211–227; Stegmaier et al. 2014: 114; on perception change regarding dam removal, see Clark 2009: 403). This can easily be translated into "governance termination" as the termination of governance that contributes to maintaining a socio-technical system, if we want to opt for the broader notion of "governance" that encompasses policy as one stream next to problem recognition and politics, as well as the focus on the broader collectivity of those involved (cf. Kingdon 1984).

Studies of policy termination offer some addition to our insights and hints for further research. Bardach conceives of the *politics of policy termination* "as a special case of the policy adoption process" which "is exceedingly difficult" (Bardach 1976: 123; cf. Turnhout 2009). He compares different cases of programmes, policies, and organisational entities with less or more technoscientific orientation, including the US participation in the Vietnam War. He distinguishes the "explosion type endings" occurring "through a change in policy effected by a single authoritative decision" from the "very long whimper" type coming "not from a single policy-level decision but from a long-term decline in the resources by which a given policy is sustained" (ibid.: 125), the latter gradual development is also called "decrementalism" (Lambright and Sapolsky 1976). Termination meets "coalitions of proponents and opponents" in contest over terminating a policy (Bardach 1976: 126), dividing the first into "Oppositionists", "Economizers", and "Reformers", the second described as the guardians of the status quo in administration, political system, and organised interest. A first obstacle for easily deciding to pursue termination seems to be that policies are often built on a nimbus of having an infinite future (ibid.: 128). Second, opponents to termination, Bardach suggests, mobilise useful interrelationships between political and moral order (ibid.: 127), as expressed "in the general moral repugnance… towards the deliberate disruption of arrangements which people have learned to rely on for a significant portion of their livelihoods or careers" (ibid.: 128). Third, Bardach suggests policymakers are "reluctant to admit—or seem to admit—past mistakes" (ibid.: 129). Fourth, pro-termination coalitions seem to hesitate to damage an existing regime (ibid.: 129). Finally, some regimes seem to reward novelty and innovation rather than termination—whereby the latter is apparently not so easily regarded as the actual innovative approach. Termination is facilitated, Bardach suggests, by "change in administration", "[d]elegitimation of the ideological matrix in which the policy is embedded", "turbulences" which shake optimistic expectations, "cushioning the blow" in case of policy failure or heavy crisis, and the experience that strategic "designing policies for eventual termination would facilitate their transformation or even their complete destruction when the time was ripe" (Bardach 1976: 130).

Frantz (2002) reconstructed for a case how a government has used resources and political argument to make a policy termination successful: how goals were articulated, how problems were defined, and how a variety of solutions to the problem were used. Frantz discovered that it is not, as often assumed, the government and administration that hold the key to attempts at termination at all. She shows how much "the skilful use of political argument and resources empowered the opponents of closure in their battle against the outwardly more powerful forces of the government… how the government succeeded only when it made better use of its political resources" (Frantz 2002: 25).

In another study, Frantz (1997) carved out how the high costs of policy termination come about. She starts from the observation that often "policy terminations are promoted as cost saving moves when, in fact, there are often considerable short term costs" (Frantz 1997: 2097), both monetary and psychological (cf. Behn 1978; deLeon 1978). Frantz, based on a study of termination in a public health system, categorises costs to prevent damage to community, constituents, and staff (Frantz

1997: 2111). In a strongly economically dominated governance-making environment, it would be extremely important for discontinuation governance research to know 1) what costs are incurred, 2) how they are calculated, and 3) how they compare with (cost) alternatives to discontinuation solutions (cf. also Joly et al. 2022: 45). Choices, framings (cf. Liersch and Stegmaier 2022), concerns, justice perceptions (cf. Johnstone and Hilescher 2017), resistance against discontinuation (cf. deLeon 1978), among others, could thus be better contextualised. In this context, it is also relevant to see whether and which measures are taken to cushion the impact of discontinuation (cf. Hogwood and Peters 1982), which is probably one key aspect of a special governance to discontinue another governance as opposed to governance that is dedicated to coordinating and deciding about the process. Again for health policy, Sato (2002) and Sato and Narita (2003) reconstructed the abolition of the leprosy prevention law in Japan in the 1990s through leadership. Sato (2002) introduced the figure of the "skillful terminator" who set the issue on the agenda, negotiated, achieved consensus, and got the thus delegitimised policy abolished. This I would call a "discontinuation entrepreneur" (a variant of a policy entrepreneur or institutional entrepreneur, who seeks policy initiative and change as an antidote to persistence or inertia in a field under pressure to change). Interestingly, the role of science remained instrumental and limited (Sato and Frantz 2005). How expertise is used to initiate or justify discontinuation is a relevant question for research.

Policy dismantling

Policy dismantling stands for "a distinctive form of policy change, which involves the cutting, reduction, diminution or complete removal of existing policies" (Bauer and Knill 2012: 31) with the two dimensions of policy intensity and policy density. Bauer et al. (2012), on social and environmental policy for instance, emphasise the quantifiable shrinking of spending for a policy branch, which as a consequence of this leads to, for example, a dry out of welfare state measures or a weakening of air pollution control. "Policy dismantling" is usually used as an opposite term to "policy expansion" (cf. Knill et al. 2018), whereby Gravey and Jordan (2019) also distinguish "limited expansion" and "limited dismantling". Both authors also present a dismantling strategy typology, which helps to better grasp active dismantling action: "no dismantling decision", "passive dismantling strategies", "active dismantling decision", and "active dismantling strategies" (cf. Bauer and Knill 2014). This can be a key typology also for discontinuation governance-in-action.

There is also attention in policy literature for the dismantling of public administration institutions when a state lacks a system of integrative political institutions (Vogelsang-Coombs and Keller 2013). Gravey and Jordan (2019) are shifting the dismantling focus from a mostly national to an international level. Policy dismantling touches upon technoscientific questions when addressing environmental (Lenschow et al. 2020) and energy policy (Bürgin 2018; Barnett et al. 2020; Prontera 2021). In addition to this advanced concept of dismantling, we often find only metaphorical use, which at least suggests a sense of the active and deliberate de-aligning and shrinking of state and welfare institutions, but which is not problematised further in its mechanics and practice (Pike et al. 2018; McCarthy 2017).

It has been correctly observed that discontinuation leads to the shrinking or even dissolution of an existing governance. However, this insight should not lead to overlooking the sometimes considerable effort required to invent and operate a governance of discontinuation. At the latest when a discontinuation falters, one should realise that active attempts can also get bogged down, incorrectly dosed or addressed, or even fail. The problem with this is twofold: firstly, it is not easy to identify what results in a discontinuation in the overall character, what it is based on, at which often confusingly distributed points it starts and ends. Second, it is not easy to accurately measure the intensity, extensiveness, and power of discontinuation efforts in relation to shrinkage or contraction. This is quite similar to the question of how and how far dis-inscription by using an "anti-programme" to disinscribe the original programme (Akrich and Latour 1992) actually works.

Although the observation of shrinking and expanding policies is a valuable one, it doesn't account for the policy that is expanded to shrink another policy; exactly what this chapter focuses on with "active, purposeful discontinuation governance". In other words, discontinuation is not mere retreat and downsizing, it is the construction of new forms of governance to support the discontinuation of existing orders. Some dismantling research supports this view, such as the observations—that also help to develop the "dismantling" notion further—about how "two major types of knowledge-boundary dismantling work" ("full boundary dismantling work", "boundary perforating") helped to shift the locus of innovation in NASA as a knowledge-intensive organisation as well as the identity of professionals (Lifshitz-Assaf 2018). Whether one sees it positively or negatively from the point of view of the professionals, this study indicates that something else is emerging, partly because something existing has been dismantled. This deserves attention for dismantling and discontinuation research: what changes and emerges as a result of the abandonment?

Policy reversal

The notion of *policy reversal* is used as a concept signifying the opposite of "policy adoption" and is defined as the "undoing of past policy" (Lowry 2005: 395), of, for instance, regulatory policies through deregulation, or criminal, social, taxation policy through its repeal, perhaps not being "mere innovations or marginal modifications of existing policies, but rather replacements of accepted policy goals with opposite goals" (Lowry 2005: 395). This is a problematic definition, as it doesn't include the aspect that something must happen to the "accepted goals" before they might be considered for reversal, such as an erosion of acceptance and legitimation, as well as the redefinition of what the problem and its policy answer is. It also doesn't account for the extra effort required to find a *governance to discontinue* another governance. The way the notion is used also assumes a symmetry between adoption and reversal, while experience teaches that adoption and reversal (or implementation and discontinuation) usually involve rather different attitudes, justifications, objectives, crafts, and competencies, among others.

In environmental management and policy there is a small, slowly growing body of reflections on the decision-making and the entire policy process of *dam removal* (Lejon et al. 2009; Clark 2009; Johnson and Graber 2002). Two decades ago, the "science of dam removal" in the US was called "undeveloped and most agencies faced with dam removal lack a coherent purpose for removing dams" (Doyle and Stanley 2003: 453). This means, there is a very late-coming learning and institutionalisation process going on for the governance of the abandonment of a very particular assemblage of socio-technical infrastructure, that of river dams for energy production, that has been developed with the technological and economic modernisation of countries.

Still, policy scholarship scarcely uses dam removal politics as an example for studying *policy reversal* (Lowry 2005). The latter suggests investigating whether the reversal of policy, from building to removing dams, creates "a new type of politics… different patterns of policy diffusion and adoption" (Lowry 2005: 395). This resonates with our focus here on a governance that emerges or is used to end another governance. With regards to American states, Lowry finds that "these patterns of political behavior for policy reversals are roughly comparable to those for policy adoption" (p. 395). However, he also sees some differences in that reversals may come from national rather than from regional governance levels, they are more gradual than for new governance, and states will adopt reversals in view of fiscal health and relevant interest groups rather than of innovation. Policy reversal tends to profit from "the entrenchment of status quo interests,… the willingness of the state to pursue innovations in related areas. Measures of fiscal capacity, urbanization, ideology, and institutional capacity, however, will not be significant determinants" (Lowry 2005: 402). Several policy-oriented studies on dam removal use Kingdon's streams model (1984) to further investigate the evolution of the idea of dam breaching into a viable policy alternative until the streams of problem perceptions, alternatives, and opportunities converge (Haeuber and Mitchener 1998; Clark 2010; Lowry 2005).

The discussion about the nexus between discontinuation and failure was excluded. On the one hand, there is the risk of placing the two perspectives too quickly and unclearly next to or within each other, which can at least quickly lead to a judgement that what is being terminated must have failed after all. On the other hand, the question should be raised as to how the relationship between discontinuation and failure is de facto negotiated and whether, in the process, judgements of failure fulfil a function or whether discontinuation is also carried out with other perceptions and justifications.

4.3.2 Discontinuation in studies of socio-technical regimes

Existing transition theory focuses on system (regime, innovation) dynamics, but *active governance-making* has long been black boxed (Geels and Schot 2007; Rip 2012). Theorising governance (or policy, politics) is usually blind to *socio-technical* aspects of change (Cerna 2013; Stachowiak 2013). Theories of socio-technical transition or change either look at the transformation patterns with technology in focus (Geels and Schot 2007; Dolata 2013) or at the governance side without any specific socio-technical focus (Borrás and Edler 2014). However, questions of governance are constitutive to understanding (from a practitioner's point of view, also to implementing) socio-technical system transitions. Questions need thus to be raised over who governs, whose framings count and whose objectives are prioritised (Smith and Stirling 2010), which problem perceptions and power relations are used to give direction to discontinuation governance, which knowledges and normativities are transformative (Stirling 2014; Stegmaier et al. 2014), and which instruments and legitimations (Borrás and Edler 2014) are mobilised.

The *governance practice-oriented* approaches (cf. Köhler et al. 2019) that have socio-technical regimes or technological social contexts on their radar can be understood as reflexive attempts at shaping new technologies, or transforming socio-technical regimes, such as "constructive technology assessment" (CTA; Rip 2018; Rip et al. 1995; cf. Aukes et al. 2023), "strategic niche management" (SNM; Rip and Kemp 1998; Geels and Raven 2006; Schot and Geels 2008), "ethical, legal and societal issues (ELSI) studies" (Forsberg 2014; Myskja et al. 2014; Rip 2009; Zwart and Nelis 2009; Yesley 2008), "transition management" (TM; Kern and Smith 2008; Kemp and Loorbach 2006; Smith and Kern 2009), or "transformative innovation policy" (TIP; Schot and Steinmueller 2018a, 2018b). They have all developed quite detailed understandings of how to influence and govern socio-technical change if one wants to achieve something specific in the process. CTA and SNM strive for a more socially acceptable technology development through shared learning and other interaction. ELSI and RRI, the same, as well as a normatively responsible technology development. TM and TIP above all aim at system change towards more sustainable societies. As a by-product, there are also indications of the conditions of the circumstances of such efforts, which advance the understanding of governance with regard to socio-technical dynamics itself; however, this is not the focus. What is missing here are indications of specific ways of dealing with discontinuation and destabilisation. Even if such developments do occur, they are not the main goal, or cannot be communicated in this way, so as not to drive the promise of rather painless change *ad absurdum*.

Discontinuation can be interpreted as a kind of intended regime change. In the light of the technological substitution pathway described by Geels and Schot (2007: 410) discontinuation can be thought of as being the case when a technology drops off the present socio-technical regime as the result of (or at least associated with) a specific moment of shock in the broader political-cultural landscape (also cf. Pierson 1994: 266). This may indeed hold true for the abandonment of nuclear energy right after the 1986 Chernobyl and 2011 Fukushima-Daiichi disasters. The shock hypothesis does, however, obviously not apply to such abandoned socio-technical systems and technologies as the incandescent light bulb (ILB), DDT, or stuttering efforts to phase-out of the internal combustion car engine. Rather it seems that more diffuse, less abrupt changes in the landscape and the offer of alternative technologies from niches (like energy-saving lamps and related technologies, less dangerous pesticides, and alternative car engine technologies) can be associated with boosting discontinuation in these areas, in combination with policies and political initiatives pertinent for changes.

Some studies emphasise policy-driven discontinuation in terms of regime destabilisation through collective action as "script for performative dramas: pressure mechanisms, obstruction mechanisms and overflow mechanisms" (Baigorrotegui 2019) and national decarbonisation pathways in the context of specific energy policies and cultures (Stephenson et al. 2021). User focus in policy-focused transitions suggests that to achieve decarbonisation, "users… need to be involved in niche construction, as well as in regime destabilisation" forming "a recursive and interactive relationship" between users and policy (Martiskainen et al. 2021: 137).

The research on socio-technical systems warns us not to underestimate, given the complexity, how difficult it is to imagine a central or omnipotent body from which one could simply control such a system, govern it authoritatively. Exceptions, such as the 2011 decision in Germany to phase out nuclear energy after the Fukushima-Daiichi reactor disaster, show how powerful politics can be, but then only for a specific national framework in which nuclear energy production and power plant infrastructure are abandoned. Mostly, it will come down to reconstructing the multiple threads, overlaps and fractures in the assemblage of a section of a system along which attempts are made to redirect a system.

Essential here are the considerations of Joly et al. (2022) on the circumstances under which governance can work in the direction of destabilisation and discontinuation. They address less the actual negotiations, arrangements, windows of opportunity, and actions. They suggest paying attention to three complementary dimensions: 1) social mobilisation, 2) objectification and publication of hidden costs, and 3) construction of credible public policies (ibid.: 45). This can only be a start, because discontinuation governance is not limited to this. Besides the adaptation of normal rules, routines, and procedures, quite specific instruments and regulations, strategies and coalitions are forged, agendas are set and existing ones are changed (Geels and Turnheim 2022), to name a few aspects—for instance, the threads that need to be pulled together until a "policy of 'controlled rundown'" (Turnheim and Geels 2012), which directly and indirectly removes the preconditions for continuation, can be implemented and maintained. In the light of a multi-levelled perspective on socio-technical systems, it would be important to locate these processes and follow their dynamics across the levels and along the pathways. It would be particularly fruitful to apply this perspective also to governance dynamics, for example to show how discontinuation governance emerges, matures, and grows out of niches—and how the discontinued technology returns into a niche, continues, and triggers a need for aftercare governance in marginal remnants after its decline.

4.3.3 Discontinuation in studies of socio-technical-economic regimes

Discontinuation has an economic dimension. Discontinuing a socio-technical regime or a part of it involves not only public governance, civic movements, but also *industrial regime, corporate management, and consumer choice* (cf. Geels 2014; Girod et al. 2014). The latter are often active agents, not just passive respondents to external pressure. However, in both roles they sometimes use the option of not continuing (e.g., giving up on less sustainable practices), divesting (e.g., large pension funds or insurers exiting from investments in fossils), leaving a sector (Siemens giving up its nuclear branch in 2011), or dealing with the fact that an industry, market, or business model is in decline (e.g., Sabatier et al. 2012). With a view to governance, we do not only include state actors, but all elements that negotiate, build, maintain, or dissolve the socio-technical order (in this case) with or against each other (Stegmaier et al. 2014; Hoppe 2010; Benz 2006). However, it should not be expected that product and service discontinuation will be the same for all kinds of industry or business (Crowley 2017). So, here we look at how the socio-technical-economic regime deals with developments that (could) lead to discontinuity (in this volume, with a focus on the linkages from economic to socio-technical governance and sustainability). We find research that already captures this: for example, how civic movements and economic interests divide on the question of expanding or dismantling fossil energy production (Curran 2020; Stephenson et al. 2021) and both within formal negotiation formats (e.g., the German "coal commission") and informal protest platforms (e.g., the struggle for the Hambach forest and open-cast lignite mining, with forest occupations and demonstrations, police clearing of the forest and judicial condemnation of the clearing as unlawful; Liersch and Stegmaier 2022).

In other cases, the economic dimension will also manifest itself as a policy that accompanies, promotes, sets the framework for, or also limits the economy. *Economic policy* or the governance of the economy always includes dealing with the discontinuation of innovation, the divestment of economic activities, not least the expiry of support measures and subsidies, and the closure of companies and the dismissal of workers. Seen in this light, there is a wealth of potential topics based on which a discontinuation governance of the economic could be developed. The field will only be roughly outlined here; for reasons of space, it cannot be treated exhaustively. This topic is mentioned here as a reminder not to neglect the economic dimension in connection with policymaking and governance arrangements. In any case, public policy and private business activities are not only seen in their interrelationship as external to each other, but are also treated as a linked, decidedly public–private partnership. Thus, both economic activities that tend to be independent and those that are actively linked to public political-administrative activities must also be included in discontinuation governance considerations, as well as the structures.

In political economic thought, for instance, on a macro level, historical discontinuity is presented as major *disorder* (in the form of World War II) that has created "the conditions for the creation of a new period of order in the international political economy", "a new global economic order" (Biersteker 1993: 16) replacing the previous imperialist, socialist, and nation-state-focused liberal orders, as well as the "international economic *dis*order of the inter-war years" (ibid.: 14). As part of that discontinuation process is the emerging process of decolonisation—the "breakup of the old colonial empires" (ibid.: 15). Interestingly, Biersteker also reflects the accompanying de-alignment and re-alignment of theoretical systems either intellectually paving the way for or being yielded by changes of this scale (ibid.: 9, 21). Similarly, Hall (1993) has suggested an investigation of the linkages between changing policy paradigms, periodic discontinuities in policymaking, and directions of research programmes.

Decline in political economic thought

Those who use the concept of "decline" today must keep in mind that it has had a bad reputation, at least among British economic historians, because one side has used it to sing a swan song to the welfare state Britain (Chalmers 1985; Tomlinson 2000; Fry 2005), while the other side has found the diagnosis of decline unsubstantiated. On the meta-level, there is literature that discusses decline for its use in struggle about theory and interpreting the direction of British politics (Hall 1993; Bernstein 2004). "Decline" is seen as a fighting concept, and some of the proponents of describing the decline of Great Britain are accused of unsound scholarly work (Middleton 2006), holding against it studies that promise to prove growth (Matthews et al. 1982). So, one can do without the use of the decline term, hope that readers have bad memories—or redefine "decline" and explicitly separate it from the old debate. As far as I can see, research on phase-out, socio-technical destabilisation, and the "death" of technology has used the notion rather innocently, without explicitly distinguishing itself from the economic-historical debate.

From a political economic standpoint, decline is thus associated with an ailing state (Mokosch 2019), with collapsing senescent industries (that have lobbied so successfully against adjustment to challenges that they collapse all the more; Brainard and Verdier 1997; cf. Hillman 1982), with transition and recovery (Stuart and Panayotopoulos 1999). Transitions in less prominent sectors, such as recording studios (Leyshon 2009) or free music software (Harkins and Prior 2021), are sometimes discussed in connection to decline, as are how central actors in the computer industry decline in power as a consequence of socio-technical change (Dolata 2009). The literature on urban decline reminds us that by no means is the study of discontinuation and destabilisation confined to national-level analyses, but urban social worlds (Friedrichs 1993) can offer manifold and complex frames of analysis, such as on strategic options for dealing with population losses (van Leuven and Hill 2021). Finance policy is another field often neglected in discontinuation and destabilisation research.

Divestment

Divestment shows how political business is. Divestment is a term used in particular in connection with the exit from fossil investments. It is about social movements and campaigning, business strategy and policy framing business behaviour towards climate warming. Informed by the rather recent field of research in global environmental politics on the political economy of commodity trade, Neville (2020: 4), for instance, investigates social movements and strategies of "divestment from fossil fuels as climate action, considering the unintended or spillover consequences of reinvestment in other industries" (ibid.). Protagonists are NGOs focusing upon climate change and justice (Ayling and Gunningham 2017: 136). Neville observes that because of the perception that fossil fuels destabilise the climate, divestment considers "fossil fuel markets as risky for investors" (2020: 4).

8w?>According to the economic logic of divestment, the external costs are internalised, which leads to investors moving to other areas and recognising that sustainable management leads to savings and new opportunities (cf. Halstead et al. 2019). Consequently, Neville sees the resulting dismantling of the fossil fuels regime, which in turn may lead to a transformation of related regimes, the economy, and society. Neville suggests that "[d]ivestment efforts delegitimize the fossil fuel sector, countering its political and cultural power" (2020: 4; cf. Bergman 2018). The questions that follow for further discontinuation governance research are to what extent these causal attributions are empirically confirmed in each case, which corporate and public governance frameworks are created or used for this, and how the systemic destabilisation and the actively pursued regime discontinuation are weighted.

From a financial political viewpoint, it should also be asked how and how far the readjustment of "political economic practices of financialization, capitalization, and assetization" (Neville 2020: 8) through divestment leads to again growth or degrowth, and whether and how this changes these practices. What is *divestive discontinuation* in practice and as a regime arrangement? Other research considers the role of international institutions, such as OECD, OPEC, WTO, UNFCCC, and IEA, in governing the transition away from fossil fuels (van Asselt 2014), non-state governance in climate policy (Ayling and Gunningham 2017), and open confrontation at extraction sites over exiting from investments or keeping them going (Curran 2020; Liersch and Stegmaier 2022).

Exit

In *management and organisation theory*, there is literature on the decline of particular organisations (Cameron et al. 1987; Caves et al. 1984; Mone et al. 1998; Lamberg and Pajunen 2005), which tends to focus on unsuccessful adaptation (e.g., because of an inability to overcome inertia or "wrong" strategies) to changing environments. Ghezzi (2013) distinguishes enterprise-driven (and firm-specific) from environment-driven discontinuities. In much of this literature, the decline of an organisation is seen as a long-term process that goes through different phases. Decline is seen not only as an inevitable result of external impulses that can hardly be ignored, "but also as an endogenous and strategic reaction to performance problems" (Turnheim and Geels 2012: 37). Weitzel and Jonsson (1989) suggest distinguishing the five stages of "blinded", "inaction", "faulty action", "crisis", and "dissolution". They argue that the lack of an appropriate response leads to the decline of organisations. The literature on organisational decline also emphasises that there is not one decline path, but several possible pathways depending on the nature and speed of external pressures and endogenous responses. Turnheim and Geels (2012, 2013) have actively drawn on this literature and theories to make sense of destabilisation and decline at the level of industries and inter-organisational fields.

Recent management studies tend to view exit as a natural part of a company's overall strategy (cf. Cefis and Marsili 2007; Graebner and Eisenhardt 2004; Villalonga and McGahan 2005; Mohamad et al. 2015). From this perspective, exit is not necessarily synonymous with failure since the decision to exit can also mean an increase in efficiency or the realisation of profits (Cefis and Marsili 2007: 1; Dupleix and D'Annunzio 2018).

Obsolescence

Obsolescence reminds us of the fact that discontinuity is caught between planned and unplanned, desired and undesired. In the area of consumer products, the loss of functionality, which leads to the replacement or disposal of the product, has a positive character from the point of view of manufacturers and sellers, who have an interest in products not lasting forever and not becoming hits on the second-hand market, so that new products need to be purchased. Studies could show to what extent obsolescence is an expression of the throwaway culture[6] that is obsessed with disposing of products (while they could just as well be maintained, repaired, mothballed, used as a source of spare parts, museumised, etc.). In other areas, the end of usability is more associated with a loss of investment and market share. Legislators and consumers may each have opposing stances on the industry view. Of course, there are consumers who like to replace old clothes with new and fashionable ones; old electronic items with new ones with updated features.[7] But at the same time, consumers may be more interested in ensuring that what they buy doesn't break down too quickly, because otherwise they lose more money than they have or want to spend (Kuppelwieser et al. 2019). The legislator stands between them if, on the one hand, they want to promote production and trade (and skim off taxes), i.e., they do not necessarily promote the longevity of products, but, on the other hand, the legislator may also want to limit the unbridled throwaway mentality on sustainability considerations in some areas (Ober et al. 2017).

However, this attribution of obsolescence is not only based on how far it is planned, but also on replaceability with new things and dysfunctionality in the face of changed circumstances (Mellal 2020). It is carried out for products as well as for production plants and processes, organisations, and bodies of knowledge (Warmington 1974). Mellal also points to a form of obsolescence that is still often overlooked, namely the situation where an alternative would be available but it does not move forward even though it could. The question is why we continue with the old; what political conditions are set so that the new alternative is not chosen and the old technology is not terminated; and how much discontinuation governance knows about such hidden alternatives based on existing but unused obsolescence.

How to deal with the fact that knowledge becomes obsolete (Margulles and Raia 1967) when a socio-technical regime comes to an end (and something else takes its place) is also a question discussed as a problem of obsolescence, but only very little for discontinuation and destabilisation. Linked to this is the even more complicated situation when this occurs, but remnants of the old regime remain and need to be taken care of. In this case, the old knowledge is largely fading away, but it still must be preserved in relevant parts in order to be available for special uses, maintenance obligations, waste products, or late effects.

We will find here a variety of interconnections between business and public, economic, and political occasions and practices, structures and instruments that aim at or at least entail discontinuation. Thus, this area calls for in-depth observation, analysis, and discussion of where 1) obsolescence is employed or abandoned, desired, or not tolerated, 2) how this discontinuation is engineered or subverted in practical terms, and 3) how this is related to political and normative ideas about how it could be justified to let some things break faster or less quickly, to accelerate or stop the decline.

4.3.4 Discontinuation in studies of innovation regimes

Studies of innovation regimes and sustainability policies have also discovered discontinuation. They treat it not only with exit and destabilisation terms, but also with the distinctive concepts of "exnovation" and "outnovation". The difference between the notions of ex- and outnovation seems to be that while exnovation is used to address the life-cycle end as the end of an innovation, outnovation addresses the letting go of a part of a larger socio-technical system in a move that can in itself be quite innovative in doing things differently than before. Exnovation occurs rather passively, outnovation deliberately.

Exnovation

The notion of "exnovation" was originally used in terms of innovation through *ex*ternal impulses,[8] as a quasi-branded consultancy term. Recently, it has also been used in environmental sociology to contrast the introduction of the new against the discontinuation of the old (Kropp 2014; David 2014; Gross and Mautz 2015), and the social and governance nature of the process is emphasised that abolishes technologies thereby making space for new ones (Paech 2013; Sveiby et al. 2012). David and Gross (2019) emphasise that "exnovation" addresses the "'natural' flip-side of innovation". From the perspective of the firm, "exnovation" was apparently first introduced by Kimberly (1981)

> who described innovation as a series of processes which in combination define an "innovation life-cycle". The final process of the model is exnovation, where the organisation must discard existing practice associated with a previously implemented innovation, thereby allowing the adoption of a new innovation, where the life cycle starts again.
>
> *(Patterson et al. 2009: 27)*

A fruitful aspect implied is that innovation can become obsolete. The question for governance and management is whether there is enough capacity and capability to end one innovation or maintenance path and turn to another. Problematic with the "exnovation" notion is that authors often jump directly from in- to exnovation, for instance, when explaining that "exnovation implies active rejection of an innovation that has been invested in previously" (Gross and Mautz 2015: 3; Kimberly 1981: 91), no mention of the fact that what was an innovation might have been the norm for quite some time before it becomes questioned to such an extent that its discontinuation might become visible on the governance horizon. Furthermore, although seen as a governance task, there is hardly any specification of an "exnovation governance" (be it public or corporate governance).

Outnovation

The notion of "outnovation" has been used by Levain et al. (2015) to consider discontinuation as a specific type of innovation, in terms of an "inverse innovation" or "innovation through withdrawal" as suggested by Goulet and Vinck (2012). Outnovation means a change, which proceeds through the withdrawal or subtraction of a part—or the whole—of a socio-technical regime (Pellissier 2021: 172). Pellissier suggests the notion of "outnovation" is an act of innovation policy that removes a substance or product from the regime, as was the case for DDT: innovation by substitution (although in other cases it could also lead to other pathways). Levain et al. (2015) and especially Pellissier emphasise the active part of registering and banning a pesticide—as public action. Pellissier shows that the use of pesticides is not free but must be carried out in accordance with rules; that registration does not distinguish between products that are considered toxic and those that are not, but selects those whose use seems to be compatible with the registration rules. A tension can thus be seen within the registration system, which, on the one hand, allows products under certain conditions, but, on the other hand, can prohibit them (Pellissier 2021: 172).

In this way, Pellissier also raises questions about the continuity of regimes. She explains that bans on pesticides enrich the set of instruments of controlled use, while the regime does not change radically, but may even be reinforced (ibid: 195–196). However, Pellissier also points out that the prohibition of a substance or product for which no alternatives by substitution exist, or do not yet exist, could lead to a different situation. Prohibition without substitution challenges the dialectic between innovation and prohibition (ibid.: 196). It will be an important task of research on decline, discontinuation, and destabilisation to find contrasting cases and to identify the constellations and processes and then learn to distinguish between them.

In all these empirical questions, the concept of "outnovation" is synonymous with that of discontinuation. Authors who use the notion of discontinuation, however, do not emphasise that it is about the countermovement to an innovation, but refer to any termination, whether of an incomplete, recently completed, tentatively established, or firmly established regime or one that is increasingly perceived as weak and undesirable.

4.3.5 Discontinuation governance studies: Pathways and configurations

As result of their comparative studies of discontinuation governance, Stegmaier et al. (2023) view "discontinuation" as a property of a trajectory[9] in which the constituting relations become misaligned to such an extent that the trajectory's distinctive character is lost, as one possible result of various permutations of distributed agency, emergence, contingency, or deliberate governance.[10] Discontinuation governance entails two complementary aspects, in which groups of governance actors undertake towards and eventually within a window of opportunity a spectrum of actions (including use of policy instruments) intended directly to effect both the discontinuation of a trajectory itself (governance of discontinuation) and of governance practices that help stabilise it (discontinuation of governance). The governance of discontinuation is here understood as governing the discontinuation of a particular governance problem as the result of a changed framing (formulation, perception) of a socio-technical regime (cf. van der Graaf and Hoppe 1996). The discontinuation of governance practices, in turn, is seen as the discontinuing of a particular way of solving a policy or a governance problem as the result of a changed framing (formulation, perception) of a problem or solution.

Their research found that a trajectory can become the addressee of discontinuing (besides and most in contrast to "building" and/or "maintaining") governance. Since the empirical reality of governance targets is complex, often composed of nested entities, and the targeted trajectories are rather "moving targets", the "targeted trajectory" (e.g., DDT, incandescent light bulbs, or nuclear energy production) is distinguished from the "wider trajectory" (pesticides, electric lighting/energy efficiency for energy-using products, and the nuclear industrial-technical-military complex). When a phase-out or ban occurs, some characteristic patterns can be seen: specific configurations that evolve over time (pathways). Both are important: the configuration as a particular structural state, and the emergent character of it when on the way. While these patterns could also be used to describe dynamics that emerge, they are intended here as markers of strategic directions of governance action: 1) how these modes of influence on a socio-technical target object are used in a targeted way, or 2) how a bundle of measures leads recognisably towards this direction. In contrast, the phase model of destabilisation enactment (Turnheim and Geels 2012), for example, can be understood as a reactive or response scheme, in which it is largely a matter of reacting appropriately and counteracting destabilisation, but if this does not help and the de-alignment trend intensifies, at least still accompanying or managing the decline. The discontinuation governance pathways, also in their specific sequence in each case, follow more the logic of a proactive scheme which is based on the assumption that discontinuation is intended and should therefore be fostered.

When a "window of opportunity" opens, different patterns can be observed:

- Ending pathway: Two forms of governance-induced ending, often associated with the notions of "*phase-out*" and "*ban*", which are the incremental or the abrupt discontinuation of a trajectory.

- Weakening pathway: Three preparatory modes of discontinuation—*control* (producing intelligence, limiting use by critical observation/social control effect), *restriction* (scope of usage), and *reduction* (scope of production)—which together can lead to final discontinuation in terms of ban or phase-out, but which can also "only" lead to a development that persists, but in a somewhat constrained, retrenched, or limited way.

These pathways cover the core of each discontinuation governance phenomenon: the way to ultimate closure, at once or in steps. Another set of pathways captures discontinuity in relation to continuity after a window of opportunity has opened and has been used or not used:

- Expiration pathway: Two possible progressions, either what is discontinued gets *replaced* (substituted)—here, the relation of the discontinued trajectory to the one that takes its place is part of the discontinuation governance; or *abandonment* occurs, which means that discontinuation governance is independent of anything that could take the place of the old and that the gap it leaves is not filled directly.
- Continuance pathway: When no discontinuation occurs at all, in cases in which efforts of discontinuation governance are induced and a destabilisation of the trajectory could be observed, but this doesn't lead to discontinuation (not yet, not within foreseeable time, or due to flaws in the process or advocacy coalitions breaking apart before success).

These types describe rough alternatives. In the individual cases and with even more detailed historical considerations, it will probably be found that some trajectories do not completely disappear for a long time. What is left is often a remnant of usage and knowledge, infrastructure, and function, for a transitional period, until all its functions are replaced by new ones (think of plastic production, which cannot be done without petroleum products, if the goal is fossil-free production). At other times, it may be a matter of decision, for example if DDT can still be used for very limited purposes or if lightbulbs are intended for special uses. In short, it seems as if almost nothing disappears completely at first (cf. Stegmaier 2023).

4.3.6 How (far) STS grasps discontinuation and its governance

Abandonment of technologies and socio-technical systems occurs not infrequently. However, similar to governance, administration, and policy studies, science and technology studies is not known for a broad consideration of discontinuation and destabilisation. The emergence of novel technoscience in the form of scientific, technological, medical, engineering knowledge, artefacts, and practices is usually prioritised over all forms of decline. However, it is Kuhn (1962) who had already posed the key question in his studies on scientific revolutions:

> What is the process by which a new candidate for paradigm replaces its predecessor? Any new interpretation of nature, whether a discovery or a theory, emerges first in the mind of one or a few individuals... How are they able to, what must they do, to convert the entire profession or the relevant professional subgroups to their way of seeing science and the world? What causes the group to abandon one tradition of normal research in favor of another?
>
> *(Kuhn 1962: 144)*

The empirical cases are legion. However, it is crucial to see how socio-technical systems, technological regimes, or technologies are (or have been) disappearing or are being ended. Research on discontinuation and destabilisation has not yet made a real breakthrough. The authors of the studies of the Ural computer (Koretsky, Zeiss and van Lente 2022), unravelling (Koretsky 2023) and 16 mm film (van de Leemput and van Lente 2023) set out to change this. And they have precursors who have developed important clues, as we see below.

Trajectories of erosion, decay, and fossilisation

Shove and Walker conceptualise the "trajectories of erosion, decay, and fossilisation" (Shove and Walker 2007: 767) as parallel movements to innovation. This prominent and extensive focus on decline in STS turns out to be at the same time a particularly problematic way of prioritising progress. It overlooks the fact that, firstly, innovations can also be aborted; secondly, that sometimes innovations are deliberately intended to function as bridging technologies (halogen lamps as a temporary replacement for conventional incandescent lamps, as well as energy-saving and LED lamps as long as they were not yet fully functional; hybrid drives as an intermediate step to electric cars; gas as an intermediate step to all-round energy supply with renewables while nuclear and coal are discontinued); and thirdly, that innovations of governance are in some cases newly developed in order to be able to undermine or bring down a strongly established existing regime. Moreover, Shove and Walker limit themselves to such downward trajectories as evolutionary rather than managed or controlled processes; to decline as the result of an inherently uncontrollable process (Pantzar and Shove 2010: 459). Shove portrays emergence and novelty as bright, whereas the unmaking, erosion, and decay as dark—the "shadowy side of innovation" (2012). Shove reproduces the "conceptual and empirical emphasis on novelty" (2012: 364) that she criticises.

Unviability

However, there are in STS a few important studies addressing the issue of ending directly. Latour presents the drama of the end of the automated train system Aramis in France, which he compares to the Concorde supersonic aeroplane in side-remarks (Latour 2002/1992). In each of the cases, their respective technological-political dimensions are profoundly linked. One of the socio-technical regimes was abandoned before its introduction, the other only after three decades of operation, during which it was always controversial. The termination of political support for the Aramis project led to the death of the socio-technical system as a whole. When it had almost matured sufficiently, the political will to use it was still lacking—and it never came. To fully mature (for example, the complex software) would still have required a not easily determinable time and an enormous additional budget. Some parts that were too complicated had already been simplified during the development process, while others, such as the control software, were not. Latour makes an important observation here: that when projects become too complex, individual parts can be left out without terminating everything (discontinuation of unviable components). If this is not done in sufficient numbers,[11] the whole system could become unwieldy, too complex, and possibly be up for disposal. What was not managed was both technological sophistication and the excessive expectations of the stakeholders—a tension between technology maturation and project management that is difficult to resolve and points to the close link between governance and technology: the specific concoction of the socio-technical regime. In the end, it's hard to decide whether Aramis was feasible, either technologically or regarding its governance. It is a question of perspective whether one sees Aramis as a collapse of innovation or the end of a short lifetime. The fundamental patterns do not seem very different.

Concorde, the civil supersonic aircraft, was a technically functioning system, albeit with quirks, which was only terminated after a fatal accident at Paris Charles de Gaulle airport after 27 years of operation. This is remarkable because the end had already been threatened during its development: the costs of production and operation were high, acceptance of the enormous noise pollution among the population in the neighbourhoods of airports was low, and potential customers did not want to take the acceptance risk (Japan cancelled an order because of this). In 1973, the US even banned civil supersonic flight. Concorde was a politically charged intergovernmental project of the French and British governments, from which the British partners did not withdraw only because they did not want to incur the wrath of the French over the necessary breach of contract. With Concorde, then, we register a tense mixture of political will (France) and political unwillingness (UK), technological maturity (basic functionality) and technological inadequacy (noise). Moreover, the political-regulatory framework changed due to bans on civil supersonic flight. It can be said that Concorde was kept alive artificially for some time (with great effort, with many popular rejections), while Aramis did not even see the light of day, but was stopped before birth.

De-inscription and non-use

For cases in which technologies are already in use, script analysis may offer another lead, e.g., when Akrich and Latour (1992) refer to "de-inscription" as one form of rejecting or renegotiating what is prescribed by designers' programmes for users. While designers assume projected users with typical forms of use, interests, competences, and taste, and inscribe this vision like a "programme" in an artefact (Latour and Woolgar 1979; Akrich 1992), actual users may not comply with the programme of action and the delegation of roles. Users may then elaborate an alternative "anti-programme" and use it to de-inscribe the original programme thereby reshaping the artefact. The artefact as such will not (necessarily) be abandoned. Through the concept of the gender script, it is proposed to capture all the work that involves not only the inscription but also the de-inscription of representations of masculinity and femininity in technical artefacts (Oudshoorn and Pinch 2003). In many studies there are some elements embedded that relate to governance or policy. Discontinuation comes in when users refuse to use or change the inscribed, intended programmes for use. In that respect, the slowly growing scholarship on non-use and no-longer-use after long use, or after test use before full adoption, always has the potential to be re-read for a discontinuation governance focus. Here, non-users are seen as active agents in the destabilisation of technologies during or after introduction, before or after longer use (cf. Oudshoorn and Pinch 2003; Wyatt 2003; Melby and Toussaint 2016; Weiner and Will 2016).

Conceptual discontinuity

The ANT approach with its symmetrisation of human and non-human entities among many other categories as a methodological programme and objective breaks with conventional presuppositions about the separation of nature, technology, and society. Instead, hybrids such as the ozone hole are considered, composed of CFCs, refrigerators, industrialists, politicians, chemists, and meteorologists (Latour 1993). In this way, ANT not only lays the groundwork for methodological and theoretical rethinking, but also political potential and, importantly in this paper, the inclusion of politics and governance, administration, and regulation in STS analyses. The diagnosed crisis of the moderns also becomes the level of consideration for the crisis of large-scale, unsustainable business-as-usual modernisation and other challenges that politics and governance are already forced to consider together in practice. In the same way, feminist STS have offered challenges to rediscover the customary order of gender relations and to re-inform and re-address socio-political issues of notions of gender, assigned function and performance in science and technology. Breaking down gender stereotypes goes hand in hand with breaking down the practices and rules that assign an unequal role to female members of society. Gender research suggests discontinuation needs. Later, new biotechnological and medical STS research showed how nature can become both a construction and demolition site for outdated concepts of the body, gender, science models and the boundaries of technology (vis-à-vis bodies and brains). Postcolonial research showed how American-European or Western-Northern science and technology, in conjunction with politics and policy, have provincialised the horizon and what might be de-universalised in response. With this rough overview, I only want to hint at the great and broad potential of STS to address in much greater depth numerous questions relevant to the political shaping of society, also from the perspective of the leveraged, deconstructed, delegitimised, and discarded thought models and socio-technical regimes.

Doing with less or without

Goulet and Vinck note that responses to the problems facing modern industrialised and technologised societies involve reducing or even abandoning certain technologies, substances, or other artefacts essential to ways of life and production; and that this is even happening at an increased rate. They suggest a critique of the sociology of translation in that it tends to overemphasise the making of new connections and the breaking of existing associations (attachments). They focus on mechanisms of dissociation and detachment. To speak of mechanisms of dissociation and detachment suggests a view that sees not whole substantial terminations or exits, but the making of cut-offs, disconnections, reductions, dissolutions, and distances from each other of elements that have hitherto been seen as belonging together. In this way, parts remain, while others no longer have a function, or at best do so in a different place and in a different way. The aim is to better understand the detachment processes at work in most innovations (Goulet and Vinck 2012). In doing so, Goulet and Vinck refer to an entire discipline of innovation studies from Schumpeter to organisational studies, transition studies, regime dynamics to Callon, which consider innovation without a detailed look at withdrawal and destruction. The focus is on complementing innovation research with the withdrawal perspective. Key forms found are banning through delegitimation and prohibition, as well as disintermediation as "removal of an actor or an intermediary object… considered harmful" (Goulet and Vinck 2017: 108).

Maintenance and care as forms of doing less with new things are only recently receiving more attention (see van de Leemput and van Lente 2023), yet almost never in terms of the ending of care-taking or maintenance endeavours, such as the category of "aftercare" for the care-taking of some remaining components of a socio-technical regime until they are also finished, as introduced by Stegmaier et al. (2014) to address typical policymaking practices in the discontinuation context.

4.4 Challenges for research about discontinuation governance

From the preceding considerations, we identify the following key challenges for describing and operating discontinuation governance. Why both? Because, on the one hand, we still have a lot to learn about discontinuation work and circumstances and, on the other hand, because practitioners are also regularly faced with the question of identifying challenges and opportunities for action.

4.4.1 Extinction: Discontinuation in studies of bio-physical / socio-technical-ecological regimes

A largely neglected field in studies of decline, discontinuation, and destabilisation is the reduction of biodiversity and extinction of species as a side effect or declared goal of management and governance actions. This is also about the overexploitation of nature through clearing, settlement, and reclamation, through the destruction of habitats and populations living there and entire species besides alteration beyond recognition or destruction of landscape and habitat through resource extraction. Examples like deforestation, poaching, and extinctive hunting of predatory animals extend far back into human history. Overfishing and habitat degradation are more recent effects of industrialisation and human overpopulation. As unintended side effects, religiously legitimised subjugation of nature, or targeted profit-seeking, many of these examples involve human action, not infrequently legitimised by permissive laws or tolerated by lax law enforcement. Socio-technical systems are related to this in various ways, whether as means or ends. Socio-technical and economic systems carry the potential of a discontinuation practice with them as part of their basic equipment, whenever they are based on mining, harvesting, or fishing. Not everything leads immediately to extinction, but problems such as overexploited fish stocks suggest that food production means risking extinction of stocks. Other systems have the simple goal of material and profit exploitation until nothing is left: oil exploration, rare earths, fishing as an industry in itself, for instance. Side effects occur when a socio-technical economic system exploits natural resources and thereby undermines other livelihoods. This area goes beyond socio-technical change and is sometimes referred to as socio-technical-ecological change (Smith and Stirling 2010; Folke et al. 2005). Extinction as a form of discontinuation is yet to be discovered. This context also includes reflections on the "finite earth" view, which straddles The Limits of Growth report (Meadows et al. 1972) and Chakrabarty's analysis of the planetary biological and geomorphological role of humans with the opportunity to destroy the livelihoods of themselves and other life forms through interventions they cannot control (Charkrabarty 2021).

4.4.2 When discontinuation governance is more than just public policy

Another challenge relates to the difficulty of discovering governance in areas that are not commonly associated with or do not have clear links to "governing". For this reason, I have mentioned the above terms from the field of management and economics. Why should discontinuation governance only occur in governmental, administrative contexts and not also in private-public contexts (private companies providing or using civil infrastructure) or be based on informal rather than formal institutions and organisations (such as civic engagement) or public-private partnerships?

4.4.3 Values, political-societal missions, normative attitudes, and power

There is a broad socio-legal dimension to discontinuation that requires far more attention. Markard et al. (2023) remind us that "When technologies decline, there will be struggles between actors that seek to disrupt legitimacy (delegitimation) and those that try to maintain or re-establish legitimacy (legitimation)". The struggles are indeed manifold and broader than this. Legitimacy is a means to achieve something in the governance arena—for or against discontinuation. As we have seen in case studies (Liersch and Stegmaier 2022; Oudelaar 2015), government agencies sometimes act outside the legal bounds (eviction of a protest camp on an illegal legal basis), and citizens' groups or activists also operate in grey areas of legality (occupation of land owned or granted for exploitation by open-cast mining companies and power plant operators). In doing so, they try to delegitimise the other and to present and maintain their own position as legal and legitimate. Due to the slow pace of justice, research will face a long haul, especially when going through the instances and in relation to current political, managerial, and activist actions. We see the nesting of normativity at various levels, procedural innovation, and areas of law, especially in European and federal contexts. Little attention is paid to the global level of contractual discontinuation governance under international law (e.g., the Minamata Convention banning and phasing out mercury in consumer products; cf. Bulten 2016). At the same time, there are often extra-judicial, discursive, mediatised, and little formalised negotiations of legitimacy, which is why one must do much more than investigate legal texts and court decisions.

4.4.4 Size, scale, and magnitude

Comparing studies of decline, discontinuation, and destabilisation, one may get the impression that preference is given to larger socio-technical systems (or production lines, product ranges) associated with big infrastructures or widespread substances. Size, scale, and magnitude seem to make them weightier examples. However, one should not underestimate the interconnectedness of smaller objects and their governance, such as a "simple" light bulb (Stegmaier et al. 2021). On closer examination, the incandescent light bulb relates to industrial and environmental policy, the change in what is regarded as illuminants and as lighting culture, with energy issues, jobs, technological innovations, and competences in semiconductors, for example. The Ural computer can be read as an example of the ability to operate large-scale computing systems at the transition to accommodating large computing power in smaller devices, while the Ural project also failed due to the development and availability of parts and software code, among other things (Koretsky 2023). Instead, research into such unfamiliar realms, as suggested by Latour (1996), benefits from avoiding the measurable and looking for connections and their meaning as they are made or cut. The same is true for how narrow or wide a domain is, or for the nestedness of the immediate usage environment, value networks, and embedding regimes.

4.5 Conclusion

Discontinuation is not the opposite of innovation, but of continuation. Innovation in discontinuation involves 1) novel socio-technical systems that offer the chance to substitute the incumbent, and 2) innovative governance strategies and formats dedicated to the specific discontinuation goal. A governance to discontinue a governance can thus be found as well as a governance to be discontinued. It is about the connections between both in context.

Nor should we be under the misconception that discontinuation governance is only applied to climate-damaging or ecologically undesirable socio-technical systems. On the contrary, it also affects politically and economically undesirable systems, and sometimes also puts an end to the expansion of sustainable energy production (as can be seen in the example of the actively undermined and ruined photovoltaic and wind power industry in Germany after the nuclear phase-out, where instead the focus was on supposedly cheap gas and oil, but at the same time obstacles were erected to renewables; cf. IRENA and ILO (2021: 45) on job losses since 2011).

There is a considerable need for research, because for a long time the counter-movement to the emergence of new socio-technical systems and their regimes has remained underexposed (Koretsky et al. 2023). There is a need to better understand practices, rules, and the conditions of the possibility that a discontinuation effort is undertaken and successful. For practitioners in politics, business, civil society and the media, science and technology, it has always been important to somehow manage to get rid of and end things that are no longer useful or desired. It seems that this focus has often been dealt with rather implicitly and informally—apart, of course, from the big debates and struggles over nuclear phase-out or nuclear capping, discrimination, or the absence of the rule of law. Increasingly, however, we are seeing open political, economic, and general social struggles about the direction of development of socio-technical regimes, for example regarding the crises of growth, climate, biodiversity, food, air quality or the exploitation of natural resources. Exiting fossils and terminating policies is becoming quite a trendy approach to socio-technical and societal transition. One speaks directly about what and how something can or should be stopped: activists against a fossil-fuelled world, governments and managements in fear of missing the boat on technological development, or that in crises, the supply of energy, raw materials, and supplier products fails. Incidentally, this also applies to the opponents of climate protection policy, who in turn want to put an end to the transformation efforts.

Our research helps to open eyes to the challenges of making something that has been practised and maintained for a long time stay the same or change it so much that it is no longer the same as it was before. It is about stopping in order to be able to continue.[12]

Notes

1 Throughout this volume, we present our different and various justified ways of speaking and conceptualising in order to make a series of explicit contributions to this. This is also the case here.

2 In this general sense, but without the theoretical frame, Kern and Howlett (2009) have used the verb "discontinued" and the noun "discontinuation" in the context of energy policy and transition management. In general policy studies literature, Berry referred to the "discontinuation" of health policy programmes and their "rapid dismantling" (Berry 1974: 354). Lindenberg (1989: 364) addressed the "discontinuation" of public services. We introduced it with the sketch of a research programme on discontinuation governance of socio-technical systems a few years later (Stegmaier et al. 2012a, 2012b).

3 This refers to the negotiation of what is the case and what should be done about it, and how: such as perceptions of problems, of situations/factors etc. that lead to problems for those who perceive them, the struggle about who defines with/against whom with which

explanation and justification, who has a more plausible or legitimate view on a matter, or what the governance answer could or should look like. This is also about how views on problems and their handling are communicated and translated in the policy realm and beyond, what objects are at stake, how they can serve as "boundary objects" (Star 2010) for processing from different sides without consensus or are not suitable for this.

4 Weber includes this in his definition of social action "which includes both failure to act and passive acquiescence" (1978: 131). Streeck and Thelen (2005: 29) speak of "institutional exhaustion" as a specific kind of institutional change leading to "a process in which behaviors invoked or allowed under existing rules operate to undermine these", which aligns with Weber's "failure to act and passive acquiescence".

5 It would be worth a study of its own to find out why this side of the coin has been so neglected.

6 See https://blogs.griffith.edu.au/social-marketing-griffith/2020/04/14/my-2020-resolution-being-the-change-i-want-to-see/ for an iconic article published in *LIFE* magazine in August 1955 featuring the notion "throwaway living".

7 Joly et al. (2022) explain that the debate on planned obsolescence may point to the drift of a system that values permanent change and novelty for novelty's sake.

8 Cf. http://4managers.de/management/themen/exnovation/.

9 We focus on "streams" and "trajectories" to emphasise the processual and interactive character of socio-technical and governance phenomena and to avoid ex ante assumptions about levels and hierarchies.

10 This and the following definitions have already been used in this form in conference presentations (Stegmaier and Kuhlmann 2016; Stegmaier 2017).

11 Or too much, so that, in other cases, not enough remains from the envisaged system.

12 If this feels strangely wrong, to stop something, this should also be investigated: how it is that continuation feels so much better than discontinuation. One reason could be that there is not always a sense of urgency for change and termination, for stopping to do things that we are used to doing.

References

Akrich, M. (1992) The de-scription of technical objects. In Bijker, W.E. and Law, J. (eds), *Shaping Technology/Building Society*. MIT Press.

Akrich, M. and Latour, B. (1992) A summary of a convenient vocabulary for the semiotics of human and nonhuman assemblies. In Bijker, W.E. and Law, J. (eds), *Shaping Technology/Building Society*. MIT Press.

Asselt, H.V. (2014) *Governing the Transition away from Fossil Fuels: The Role of International Institutions*. Stockholm Environment Institute Working Paper No. 2014–2007. Stockholm Environment Institute – Oxford Centre.

Aukes, E., Stegmaier, P., and Schleyer, C. (2022) Guiding the guides: Doing 'Constructive Innovation Assessment' as part of innovating forest ecosystem service governance. *Ecosystem Services*, 58, 101482. https://doi.org/10.1016/j.ecoser.2022.101482.

Ayling, J. and Gunningham, N. (2017) Non-state governance and climate policy: The fossil fuel divestment movement. *Climate Policy*, 17(2), 131–149.

Baigorrotegui, G. (2019) Destabilization of energy regimes and liminal transition through collective action in Chile. *Energy Research & Social Science*, 55, 198–207.

Bardach, E. (1976) Policy termination as a political process. *Policy Sciences*, 7(2), 123–131. www.jstor.org/stable/4531635.

Barnett, B., Wellstead, A.M. and Howlett, M. (2020) The evolution of Wisconsin's woody biofuel policy: Policy layering and dismantling through dilution. *Energy Research & Social Science*, 67, 101514.

Bauer, M. W. (2009) The policy termination approach: critique and conceptual perspectives. Lehrstuhl Politik und Verwaltung, Working Paper series. Humboldt University.

Bauer, M.W. and Knill, C. (2012) Understanding policy dismantling: An analytic framework. In Bauer, M.W., Jordan, A., Green-Pedersen, C. and Hèritier, A. (eds) *Dismantling Public Policy: Preferences, Strategies, and Effects*. Oxford University Press.

Bauer, M.W. and Knill, C. (2014) A conceptual framework for the comparative analysis of policy change: Measurement, explanation and strategies of policy dismantling. *Journal of Comparative Policy Analysis: Research and Practice*, 16(1), 28–44.

Bauer, M.W., *et al.* (eds) (2012) *Dismantling Public Policy: Preferences, Strategies, and Effects*. Oxford University Press.

Behn, R.D. (1978) How to terminate a public policy: A dozen hints for the would-be policy terminator. *Policy Analysis*, 4(3), 393–414.

Benz, A. (2006) Governance in connected arenas – political science analysis of coordination and control in complex control systems. In Jansen, D. (ed.) *New Forms of Governance in Research Organizations. From Disciplinary Theories towards Interfaces and Integration*. Springer.

Berger, P.L. and Luckmann, T. (1966) *The Social Construction of Reality: A Treatise in the Sociology of Knowledge*. Doubleday.

Bergman, N. (2018) Impacts of the fossil fuel divestment movement: Effects on finance, policy and public discourse. *Sustainability*, 10(7), 2529.

Bernstein, G.L. (2004) *The Myth of Decline: The Rise of Britain since 1945*. Pimlico.

Berry, D.E. (1974) The transfer of planning theories to health planning practice. *Policy Sciences* 5, 343–361.

Biersteker, T.J. (1993) Evolving perspectives on international political economy: Twentieth-century contexts and discontinuities. *International Political Science Review*, 14(1), 7–33.

Borrás, S. and Edler, J. (2014) The governance of change in socio-technical and innovation systems: Some pillars for a conceptual framework. In Borrás, S. and Edler, J. (eds) *The Governance of Systems Change*. Elgar.

Brainard, S.L. and Verdier, T. (1997) The political economy of declining industries: Senescent industry collapse revisited. *Journal of International Economics*, 42(1–2), 221–237.

Bulten, M. (2016) *Understanding Global Discontinuation Governance: An Explorative Case Study on the Minamata Convention on Mercury* (Master Thesis, Enschede, University of Twente).

Bürgin, A. (2018) The impact of Juncker's reorganization of the European Commission on the internal policy-making process: Evidence from the Energy Union project. *Public Administration*, 98(2), 378–391.

Cairney, P. (2012) *Understanding Public Policy: Theories and Issues*. Palgrave Macmillan.

Cameron, K.S., Kim, M.U. and Whetten, D.A. (1987) Organizational effects of decline and turbulence. *Administrative Science Quarterly*, 32(2), 222–240. https://doi.org/10.2307/2393127.

Caves, R.E., Fortunato, M. and Ghemawat, P. (1984) The decline of dominant firms, 1905–1929. *The Quarterly Journal of Economics*, 99(3), 523–546.

Cefis, E. and Marsili, O. (2007) *Going, Going, Gone: Innovation and Exit in Manufacturing Firms*. ERIM Report Series Research in Management, ERIM Research Program: "Organizing for Performance", ERS-2007–2015-ORG, 1–20. https://doi.org/http://hdl.handle.net/1765/9732.

Cerna, L. (2013) *The Nature of Policy Change and Implementation: A Review of Different Theoretical Approaches*. OECD.

Chalmers, M. (1985) *Paying for Defence: Military Spending and British Decline*. Pluto Press.

Chakrabarty, D. (2021) *The Climate of History in a Planetary Age*. University of Chicago Press.

Clark, B.T. (2009) River restoration in the American West: Assessing variation in the outcomes of policy change. *Society and Natural Resources*, 22, 401–416. https://doi.org/10.1080/08941920801914528.

Clark, B.T. (2010) Agenda setting and issue dynamics revisited: Dam removal on the lower Snake River. Paper prepared for the Annual Meeting of the Western Political Science Association, San Francisco, CA, April. http://papers.ssrn.com/sol3/papers.cfm?abstract_id=1580368.

Colebatch, H. (ed.) (2006) *The Work of Policy: An International Survey*. Lexington Books.

Colebatch, H. (2009) *Policy*. Open University Press.

Colebatch, H., *et al.* (eds) (2010) *Working for Policy*. Amsterdam University Press.

Crowley, F. (2017) Product and service innovation and discontinuation in manufacturing and service firms in Europe. *European Journal of Innovation Management*, 20(2), 250–268.

Curran, G. (2020) Divestment, energy incumbency and the global political economy of energy transition: The case of Adani's Carmichael mine in Australia. *Climate Policy*, 20(8), Special Issue: "Curbing Fossil Fuel Supply to Achieve Climate Goals", 949–962.

David, M. (2014) *Exnovation-Governance im Nachhaltigkeitskontext: Annäherung an eine Typologie*. https://regierungsforschung.de/exnovation-governance-im-nachhaltigkeitskontext-annaeherung-an-eine-typologie/.

David, M. and Gross, M. (2019) Futurizing politics and the sustainability of real-world experiments: what role for innovation and exnovation in the German energy transition? *Sustainability Science*, 14, 991–1000.

deLeon, P. (1978) A theory of policy termination. In May, J.V. and Wildavsky, A.B. (eds) *The Policy Cycle*. Sage.

Dewey, J. (1927) *The Public and its Problems*. Holt.

Dolata, U. (2009) Technological innovations and sectoral change. Transformative capacity, adaptability, patterns of change: An analytical framework. *Research Policy*, 38, 1066–1076.

Dolata, U. (2013) *The Transformative Capacity of New Technologies: A Theory of Sociotechnical Change*. Routledge.

Doyle, M.W. and Stanley, E.H. (2003) Toward policies and decision-making for dam removal. *Environmental Management*, 31(4), 453–465.

Dupleix, M.D. and D'Annunzio, C. (2018) El éxito del fracaso: Casos de discontinuidad de iniciativas empresariales jóvenes en el sector de software y servicios informáticos. *Estudios Gerenciales*, 34(148), 262–278.

Eberle, T.S. (2019) Variations of constructivism. In Pfadenhauer, M. and Knoblauch, H. (eds) *Social Constructivism as Paradigm? The Legacy of the Social Construction of Reality*. Routledge.

Folke, C., *et al.* (2005) Adaptive governance of social-ecological systems. *Annual Review of Environment and Resources*, 30, 441–473.

Forsberg, E.-M. (2014) Institutionalising ELSA in the moment of breakdown? *Life Sciences, Society and Policy*, 10(1), 1–16.

Frantz, J.E. (1997) The high costs of policy termination. *International Journal of Public Administration*, 20(12), 2097–2119.

Frantz, J.E. (2002) Political resources for policy terminators. *Policy Studies Journal*, 30(1), 11–28.

Friedrichs, J. (1993) A theory of urban decline: economy, demography and political elites. *Urban Studies*, 30(6), 907–917.

Fry, G.K. (2005) *The Politics of Decline: An Interpretation of British Politics from the 1940s to the 1970s*. Palgrave.

Geels, F.W. (2014) Reconceptualising the co-evolution of firms-in-industries and their environments: Developing an inter-disciplinary Triple Embeddedness Framework. *Research Policy*, 43, 261–277.

Geels, F.W. (2019) Socio-technical transitions to sustainability: A review of criticisms and elaborations of the Multi-Level Perspective. *Current Opinion in Environmental Sustainability*, 39, 187–201. https://doi.org/10.1016/j.cosust.2019.06.009.

Geels, F.W. and Raven, R. (2006) Non-linearity and expectations in niche-development trajectories: Ups and downs in Dutch biogas development (1973–2003). *Technology Analysis and Strategic Management*, 18(3–4), 375–392.

Geels, F.W. and Schot, J. (2007) Typology of socio-technical transition pathways. *Research Policy*, 36(3), 399–417.

Geels, F.W. and Turnheim, B. (2022) *The Great Reconfiguration: A Socio-Technical Analysis of Low-Carbon Transitions in UK Electricity, Heat, and Mobility Systems*. Cambridge University Press.

Ghezzi, A. (2013) Revisiting business strategy under discontinuity. *Management Decision*, 51 (7), 1326–1358.

Giddens, A. (1984) *The Constitution of Society: Outline of the Theory of Structuration*. Polity.

Girod, B., *et al.* (2014) Climate policy through changing consumption choices: Options and obstacles for reducing greenhouse gas emissions. *Global Environmental Change*, 25, 5–15.

Goulet, F. and Vinck, D. (2012) L'innovation par retrait: Contribution à une sociologie du détachement. *Revue Française de Sociologie*, 2(532), 195–224.

Goulet, F. and Vinck, D. (2017) Moving towards innovation through withdrawal: The neglect of destruction. In Godin, B. and Vinck, D. (eds), *Critical Studies of Innovation: Alternative Approaches to the Pro-Innovation Bias*. Edward Elgar Publishing.

Graebner, M.E. and Eisenhardt, K.M. (2004) The seller's side of the story: Acquisition as courtship and governance as syndicate in entrepreneurial firms. *Administrative Science Quarterly*, 49(3), 366–403.

Gravey, V. and Jordan, A.J. (2019) Policy dismantling at EU level: Reaching the limits of "an ever-closer ecological union"? *Public Administration*, 98(2), 349–362.

Gross, M. and Mautz, R. (2015) *Renewable Energies*. Routledge.

Haeuber, R.A. and Mitchener, W.K. (1998) Policy implications of recent natural and managed floods. *BioScience*, 48(9), 765–772.

Hall, P.A. (1993) Policy paradigms, social learning and the state: The case of economic policymaking in Britain. *Comparative Politics*, 25(3): 275–296.

Halstead, M., *et al.* (2019) The importance of fossil fuel divestment and competitive procurement for financing Europe's energy transition. *Journal of Sustainable Finance & Investment*, 9(4), 349–355.

Harkins, P. and Prior, N. (2021) (Dis)locating democratization: Music technologies in practice. *Popular Music and Society*, 45(1), 84–103.

Hillman, A.L. (1982) Declining industries and political-support protectionist motives. *The American Economic Review*, 72(5), 1180–1187.

Hogwood, B.W. and Peters, B.G. (1982) The dynamics of policy change: Policy succession. *Policy Sciences*, 14(3), 225–245.

Hoppe, R. (2010) *The Governance of Problems: Puzzling, Powering and Participation*. Policy Press.

Hyysalo, S., Jensen, T.E. and Oudshoorn, N. (eds) (2016) *The New Production of Users: Changing Innovation Collectives and Involvement Strategies*. Routledge.

IRENA and ILO (2021) *Renewable Energy and Jobs – Annual Review 2021*. International Renewable Energy Agency, International Labour Organization.

Johnson, S.E. and Graber, B.E. (2002) Enlisting the social sciences in decisions about dam removal. *BioScience*, 52(8), 731–738.

Johnston, P. and Hielscher, S. (2017) Phasing out coal, sustaining coal communities? Living with technological decline in sustainability pathways. *The Extractive Industries and Society*, 4, 457–461.

Joly, P.-B., *et al.* (2022) Gouverner l'arrêt des grands systèmes sociotechniques. In Goulet, F. and Vinck, D. (eds) *Faire sans, Faire avec Moins: Les Nouveaux Horizons de L'innovation*. Presses des Mines.

Kemp, R. and Loorbach, D. (2006) Transition management: A reflexive governance approach. In Voss, J.-P., Bauknecht, D. and Kemp, R. (eds) *Reflexive Governance for Sustainable Development*. Elgar.
Kern, F. and Howlett, M. (2009) Implementing transition management as policy reforms: A case study of the Dutch energy sector. *Policy Sciences*, 42, 391.
Kern, F. and Smith, A. (2008) Restructuring energy systems for sustainability? Energy transition policy in the Netherlands. *Energy Policy*, 36, 4093–4103.
Kimberly, J.R. (1981) Managerial innovation. In Nystrom, P.C. and Starbuck, W.H. (eds) *Handbook of Organizational Design*, Vol. 1. Oxford University Press.
Kingdon, J.W. (1984) *Agendas, Alternatives, and Public Policies*. Little, Brown.
Knill, C., Steinebach, Y. and Fernández-i-Marín, X. (2018) Hypocrisy as a crisis response? Assessing changes in talk, decisions, and actions of the European Commission in EU environmental policy. *Public Administration*, 98(2), 363–377.
Köhler, J., Geels, F.W., Kern, F., Markard, J., Onsongo, E., Wieczorek, A., Alkemade, F., Avelino, F., Bergek, A., Boons, F. and Fünfschilling, L. (2019) An agenda for sustainability transitions research: State of the art and future directions. *Environmental Innovation and Societal Transitions*, 31, 1–32.
Kohoutek, J., *et al.* (2013) Conceptualizing policy work as activity and field of research. *Central European Journal of Public Policy*, 7(1), 28–59.
Koretsky, Z. (2023) Dynamics of technological decline as socio-material unravelling. In Koretsky, Z. *et al.* (eds) *Technologies in Decline: Socio-Technical Approaches to Discontinuation and Destabilisation*. Routledge.
Koretsky, Z., van Lente, H., Turnheim, P. and Stegmaier, P. (2023) Introduction: The relevance of technologies in decline. In Koretsky, Z. *et al.* (eds) *Technologies in Decline: Socio-Technical Approaches to Discontinuation and Destabilisation*. Routledge.
Koretsky, Z., Zeiss, R. and Van Lente, H. (2022) Exploring the dynamics of technology phase-outs through the history of a Soviet computer "Ural" (1955–1990). *Science, Technology & Human Values*. https://doi.org/10.1177/01622439221130139.
Kropp, C. (2014) *Exnovation – Nachhaltige Innovationen als Prozesse der Abschaffung*. Unpublished manuscript.
Kuhlmann, S., *et al.* (2019) The tentative governance of emerging science and technology – A conceptual introduction. *Research Policy*, 5, 1091–1097.
Kuhn, T.S. (1962) *The Structure of Scientific Revolutions*. University of Chicago Press.
Kuppelwieser, V.G., *et al.* (2019) Consumer responses to planned obsolescence. *Journal of Retailing and Consumer Services*, 47, 157–165.
Lamberg, J.-A. and Pajunen, K. (2005) Beyond the metaphor: The morphology of organizational decline and turnaround. *Human Relations*, 58(8), 947–980.
Lambright, W.H. and Sapolsky, H.M. (1976) Terminating federal research and development programs. *Policy Sciences*, 7(2), 199–213. Latour, B. (1993) *We Have Never Been Modern*. Harvard University Press.
Latour, B. (1996) On actor-network theory: A few clarifications. *Soziale Welt*, 47(4), 369–381.
Latour, B. (2002) *Aramis, or the Love of Technology*. Harvard University Press.
Latour, B. and Woolgar, S. (1979) *Laboratory Life: The Social Construction of Scientific Facts*. Sage.
Le Quéré, C., *et al.* (2019) Drivers of declining CO2 emissions in 18 developed economies. *Nature Climate Change*, 9, 213–217.
Lejon, A.G.C., Renöfält, B.M. and Nilsson, C. (2009) Conflicts associated with dam removal in Sweden. *Ecology and Society*, 14(2), 4–22.
Lenschow, A., Burns, C. and Zito, A. (2020) Dismantling, disintegration or continuing stealthy integration in European Union environmental policy? *Public Administration*, 98(2), 340–348.

Levain, A., Joly, P.-B., Barbier, M., Cardon, V., Dedieu, F. and Pellissier, F. (2015) Continuous discontinuation – The DDT Ban revisited. International Sustainability Transitions Conference: Sustainability transitions and wider transformative change, historical roots and future pathways.

Leyshon, A. (2009) The software slump: Digital music, the democratisation of technology, and the decline of the recording studio sector within the musical economy. *Environment and Planning A*, 41, 1309–1331. https://doi.org/doi: 10.1068/a40352.

Liersch, C.C. and Stegmaier, P. (2022) Keeping the forest above to phase out the coal below: The discursive politics and contested meaning of the Hambach Forest. *Energy Research & Social Science*, 89, 1–19.

Lifshitz-Assaf, H. (2018). Dismantling knowledge boundaries at NASA: The critical role of professional identity in open innovation. *Administrative Science Quarterly*, 63(4), 746–782.

Lindenberg, M. (1989) Making economic adjustment work: The politics of policy implementation. *Policy Sciences*, 22, 359–394.

Lowry, W.R. (2005) Policy reversal and changing politics: State governments and dam removals. *State Politics and Policy Quarterly*, 5(4), 394–419.

Margulles, N. and Raia, A.P. (1967) Scientists, engineers, and technological obsolescence. *California Management Review*, 10(2), 43–48.

Markard, J., Isoaho, K. and Widdel, L. (2023) Discourses around decline: Comparing the debates on coal phase-out in the UK, Germany and Finland. In Koretsky, Z. *et al.* (eds) *Technologies in Decline: Socio-Technical Approaches to Discontinuation and Destabilisation*. Routledge.

Martiskainen, M., *et al.* (2021) User innovation, niche construction and regime destabilization in heat pump transitions. *Environmental Innovation and Societal Transitions*, 39, 119–140.

Matthews, R.C.O., Feinstein, C.H. and Odling-Smee, J. (1982) *British Economic Growth 1856–1973: The Post-War Period in Historical Perspective*. Oxford University Press.

McCarthy, M.A. (2017) *Dismantling Solidarity: Capitalist Politics and American Pensions since the New Deal*. Cornell University Press.

Meadows, D. H., Meadows, D. L., Randers, J., & Behrens, W. W. (1972) *The Limits to Growth. A Report for the Club of Rome's Project on the Predicament of Mankind*. Universe.

Melby, L. and Toussaint, P. (2016) "We walk straight past the screens": The power of the non-user of a hospital information system. In Hyysalo, S., Jensen, T.E. and Oudshoorn, N. (eds) *The New Production of Users: Changing Innovation Collectives and Involvement Strategies*. Routledge.

Mellal, M.A. (2020) Obsolescnece – A review of the literature. *Technology in Society*, 63, 101347.

Middleton, R. (2006) Review article: The political economy of decline. *Journal of Contemporary History*, 41(3), 573–586.

Mohamad, A., *et al.* (2015) Business discontinuity among small and medium enterprises. *Advanced Science Letters*, 21(6), 1763–1766.

Mokosch, G. (2019) Breaking the spiral: Institutions and Italy's economic malaise. *The International Spectator: Italian Journal of International Affairs*, 54(3), 153–155.

Mone, M.A., McKinley, W. and Barker, V.L. (1998) Organizational decline and innovation: A contingency framework. *The Academy of Management Review*, 23(1), 115–132.

Myskja, B.K., Nydal, R., Myhr, A.I. (2014) We have never been ELSI researchers – there is no need for a post-ELSI shift. *Life Sciences, Society and Policy*, 10(9), 1–17.

Neville, K.J. (2020) Shadows of divestment: The complications of diverting fossil fuel finance. *Global Environmental Politics*, 20(2), 3–11.

Ober, E., *et al.* (2017) Planned obsolescence: The government's choice?PLATE conference, Delft University of Technology, 8–10 November.

Oudelaar, N. (2015) *The Influence of Non-Governmental Organisations on European Union Policy Processes – An Explorative Study on the Role of Non-Governmental Organisations during the Policy Process that Went Prior to the Incandescent Light Bulb Phase-Out* (Master Thesis, Enschede, University of Twente).

Oudshoorn, N. and Pinch, T. (2003) Introduction: How users and non-users matter. In Oudshoorn, N. and Pinch, T. (eds) *How Users Matter: The Co-Construction of Users and Technology*. MIT Press.

Paech, N. (2013) Economic growth and sustainable development. In Angrick, M. *et al.* (eds) *Faktor X, Re-source – Designing the Recycling Society*. Springer.

Pantzar, M. and Shove, E. (2010) Understanding innovation in practice: A discussion of the production and re-production of NordicWalking. *Technology Analysis & Strategic Management*, 22(4), 447–461.

Patterson, F., Kerrin, M., and Gatto-Roissard, G. (2009) *Characteristics & Behaviours of Innovative People in Organizations*. City University.

Pellissier, F. (2021) *Tuer les pestes pour protéger les cultures. Sociohistoire de l'administration des "pesticides" en France* (Doctoral dissertation, Université Gustave Eiffel).

Pierson, P. (1994) *Dismantling the Welfare State?*Cambridge University Press.

Pike, A., Coombes, M., O'Brien, P. and Tomaney, J. (2018) Austerity states, institutional dismantling and the governance of sub-national economic development: The demise of the regional development agencies in England. *Territory, Politics, Governance*, 6(1), 118–144.

Princen, T., *et al.* (eds) (2015) *Ending the Fossil Fuel Era*. MIT Press.

Prontera, A. (2021) The dismantling of renewable energy policy in Italy. *Environmental Politics*, 30(7), 1196–1216.

Rip, A. (2009) Futures of ELSA. *EMBO Reports*, 10(7), 666–670. https://doi.org/10.1038/embor.2009.149

Rip, A. (2012) The context of innovation journeys. *Creativity and Innovation Management*, 21 (2), 158–170.

Rip, A. (2018) *Futures of Science and Technology in Society*. Springer VS.

Rip, A. and Kemp, R. (1998) Technological change. In Rayner, S. and Malone, E.L. (eds) *Human Choice and Climate Change 2*. Battelle Press.

Rip, A., Misa, T.J. and Schot, J. (eds) (1995) *Managing Technology in Society: The Approach of Constructive Technology Assessment*. Pinter.

Sabatier, V., *et al.* (2012) When technological discontinuities and disruptive business models challenge dominant industry logics: Insights from the drugs industry. *Technological Forecasting & Social Change*, 79, 949–962.

Sato, H. (2002) Abolition of leprosy isolation policy in Japan: Policy termination through leadership. *Policy Studies Journal*, 30(1), 29–46.

Sato, H. and Frantz, J.E. (2005) Termination of the leprosy isolation policy in the US and Japan: Science, policy changes, and the garbage can model. *BMC International Health and Human Rights*, 5(3), 1–16.

Sato, H. and Narita, I. (2003) Politics of leprosy segregation in Japan: The emergence, transformation and abolition of the patient segregation policy. *Social Science & Medicine*, 56, 2529–2539.

Schot, J. (1992) Technology in decline: A search for useful concepts. *British Journal for the History of Science*, 25(1), 5–26.

Schot, J. and Geels, F. (2008) Strategic niche management and sustainable innovation journeys: Theory, findings, research agenda, and policy. *Technology Analysis and Strategic Management*, 20(5), 537–554.

Schot, J. and Steinmueller, W.E. (2018a) Three frames for innovation policy: R&D, systems of innovation and transformative change. *Research Policy*, 47, 1554–1567. https://doi.org/doi.org/10.1016/j.respol.2018.08.011.

Schot, J. and Steinmueller, W.E. (2018b) New directions for innovation studies: Missions and transformations. *Research Policy*, 47, 1583–1584.

Shove, E. and Walker, G. (2007) Commentary: CAUTION! Transitions ahead: Politics, practice, and sustainable transition management. *Environment and Planning A*, 39, 763–770.

Shove, E. (2012) The shadowy side of innovation: unmaking and sustainability. *Technology Analysis & Strategic Management*, 24(4), 363–375.

Smith, A. and Kern, F. (2009) The transitions storyline in Dutch environmental policy. *Environmental Politics*, 18(1), 78–98.

Smith, A. and Stirling, A. (2010) The politics of social-ecological resilience and sustainable socio-technical transitions. *Ecology and Society*, 15(1), 11.

Stachowiak, S. (2013) Pathways for change: 10 theories to inform advocacy and policy change efforts. ORS Impact. www.syrialearning.org/system/files/content/resource/files/main/2013-10-ors-10-theories-to-inform-advocacy-and-policy-change-efforts.pdf

Star, S.L. (2010) This is not a boundary-object. *Revue d'anthropologie des connaissances*, 41(1), 18–35.

Stegmaier, P. (2023) Aftercare, or doing less with discontinuation niche governance. In Goulet, F. and Vinck, D. (eds) *Doing Without, Doing With Less: The New Horizons of Innovation*. Edward Elgar.

Stegmaier, P. (2017) The reflexive creation of exiting: The governance of discontinuation. Summer School "Perspectives on Innovation Society". TU Berlin, 17 June.

Stegmaier, P., Kuhlmann, S. and Visser, V. R. (2012a) Governance of the discontinuation of socio-technical systems. Jean Monnet Conference: The Governance of Innovation and Socio-Technical Systems in Europe: New Trends, New Challenges, International Workshop, Copenhagen Business School, Denmark.

Stegmaier, P., Visser, V. R. and Kuhlmann, S. (2012b) Governance of the discontinuation of socio-technical systems—An exploratory study of the incandescent light bulb phase-out. 4S/EASST Conference, Panel on the Governance of Innovation and Sociotechnical Systems: Design and Displacements – I, Copenhagen Business School, Denmark.

Stegmaier, P. and Kuhlmann, S. (2016) Capturing how obsolescent innovation must give way: The discontinuation governance heuristic and the phase-out of the incandescent light bulbs in the EU). STePS Annual Research Days, University of Twente, Enschede, 14 June.

Stegmaier, P., Kuhlmann, S. and Visser, V. R. (2014) The discontinuation of socio-technical systems as governance problem. In Borrás, S. and Edler, J. (eds) *Governance of Systems Change*. Edward Elgar.

Stegmaier, P., Visser, V. R. and Kuhlmann, S. (2021) The incandescent light bulb phase-out. Exploring patterns of framing the governance of discontinuing a socio-technical regime. *Energy, Sustainability and Society*, 11, 1–22. https://doi.org/https://doi.org/10.1186/s13705-021-00287-4.

Stegmaier, P., Joly, P.-B., Kuhlmann, S. and Stirling, A. (2023) *Technologies of Discontinuation – Towards Transformative Innovation Policies*. Edward Elgar.

Stephenson, J.R., *et al.* (2021) Energy cultures and national decarbonisation pathways. *Renewable and Sustainable Energy Reviews*, 137, 110592.

Stirling, A. (2014) Transforming power: Social science and the politics of energy choices. *Energy Research & Social Science*, 1, 83–95. https://doi.org/https://doi.org/10.1016/j.erss.2014.02.001.

Strauss, A. (1978) *Negotiations: Varieties, Contexts, Processes, and Social Order*. Jossey-Bass.

Streeck, W. and Thelen, K. (eds.). (2005) *Beyond Continuity: Institutional Change in Advanced Political Economies*. Oxford University Press.
Stuart, R.C. and Panayotopoulos, C.M. (1999) Decline and recovery in transition economies: The impact of initial conditions. *Post-Soviet Geography and Economics* 14(4), 267–280.
Sveiby, K. *et al.* (2012) *Challenging the Innovation Paradigm*. Routledge.
Tomlinson, J. (2000) *The Politics of Decline: Understanding Post-War Britain*. Routledge.
Turnheim, B. (2023) Destabilisation, decline and phase-out in transitions research. In Koretsky, Z. *et al.* (eds) *Technologies in Decline: Socio-Technical Approaches to Discontinuation and Destabilisation*. Routledge.
Turnheim, B. and Geels, F. (2012) Regime destabilisation as the flipside of energy transitions: Lessons from the history of the British coal industry (1913–1997). *Energy Policy*, 50, 35–49.
Turnhout, E. (2009) The rise and fall of a policy: Policy succession and the attempted termination of ecological corridors policy in the Netherlands. *Policy Sciences*, 42, 57–72.
van Asselt, H. (2014) Governing the transition away from fossil fuels: The role of international institutions. Stockholm Environment Institute Working Paper no. 2014-07. Stockholm Environment Institute – Oxford Centre.
van de Graaf, H. and Hoppe, R. (1996) *Beleid en Politiek: Een Inleiding tot de Beleidswetenschap en de Beleidskunde*. Coutinho.
van de Leemput, D. and van Lente, H. (2023) Caring for decline: The case of 16mm film artworks of Tacita Dean. In Koretsky, Z. *et al.* (eds) *Technologies in Decline: Socio-Technical Approaches to Discontinuation and Destabilisation*. Routledge.
van Leuven, A.J. and Hill, E.W. (2021) Legacy regions, not legacy cities: Growth and decline in city-centered regional economies. *Journal of Urban Affairs*. https://doi.org/10.1080/07352166.2021.1990775.
Villalonga, B. and McGahan, A.M. (2005) The choice among acquisitions, alliances, and divestitures. *Strategic Management Journal*, 26, 1183–1208.
Vogelsangs-Coombs, V. and Keller, L.F. (2013) *The government and governance of Ohio: Party politics and the dismantling of public administration*. *Administration & Society*, 47(4), 343–368.
Warmington, A. (1974) Obsolescence as an organizational phenomenon. *Journal of Management Studies*, 11(2), 96–114.
Weber, M. (1978). *Economy and Society: An Outline of Interpretive Sociology* (Vol. 2). University of California Press.
Weiner, K. and Will, C. (2016) Users, non-users and "resistance" to pharmaceuticals. In Hyysalo, S., Jensen, T.E. and Oudshoorn, N. (eds) *The New Production of Users: Changing Innovation Collectives and Involvement Strategies*. Routledge.
Weitzel, W. and Jonsson, E. (1989) Decline in organizations: A literature integration and extension. *Administrative Science Quarterly*, 34(1), 91–109. https://doi.org/www.jstor.org/stable/2392987.
Wyatt, S. (2003) Non-users also matter: The construction of users and non-users of the internet. In Oudshoorn, N. and Pinch, T. (eds) *How Users Matter: The Co-Construction of Users and Technology*. MIT Press.
Yesley, M.S. (2008) What's ELSI got to do with it? Bioethics and the Human Genome Project. *New Genetics and Society*, 27(1), 1–6.
Zwart, H. and Nelis, A. (2009) What is ELSA Genomics? Profile and challenges of an emerging research practice. *EMBO Reports*, 10, 540–544.

Streeck, W. and [illegible] (2005) [illegible] *Political Economies*. [illegible] University Press.
S[illegible] and [illegible] (1999) Declining [illegible] The impact of [illegible] *Post-Soviet Geography and Economics* [illegible]
S[illegible] (2018) *Challenging* [illegible]
Tomlinson, J. [illegible]
[illegible] H. (2013) [illegible] *Technological* [illegible] *Decarbonization* [illegible]
[illegible] L. and [illegible] (2012) [illegible] decline [illegible] energy [illegible] *Energy Policy* [illegible]
[illegible] (2000) [illegible] policy [illegible] the [illegible]
van [illegible] (2019) [illegible] the [illegible] and [illegible] of [illegible] *Coal* [illegible]
[illegible] (2010) [illegible]
[illegible] and [illegible] (2012) [illegible] *Industrial Relations* [illegible]
van [illegible] and [illegible] (2017) [illegible] regional [illegible] *Regional Studies* [illegible]
Vallentin, [illegible] (2005) The choice [illegible] and [illegible] *Journal* [illegible]
[illegible] (2013) [illegible] *Public Administration* [illegible]
W[illegible] (1974) [illegible] organizational [illegible] *Journal of* [illegible]
W[illegible] (1975) [illegible]
W[illegible] (2010) [illegible]
W[illegible] (2009) [illegible] and [illegible]
W[illegible] and [illegible] (2019) [illegible]
[illegible] (2008) [illegible]
[illegible] (2019) [illegible] and challenges [illegible]

PART II

Empirical explorations

5

DISCOURSES AROUND DECLINE

Comparing the debates on coal phase-out in the UK, Germany and Finland

Jochen Markard, Karoliina Isoaho and Linda Widdel

5.1 Introduction

Decline of unsustainable practices is a crucial process in sustainability transitions (Markard and Rosenbloom 2020; Rosenbloom and Rinscheid 2020). Only if problematic practices or technologies decline, and eventually vanish, is there a chance for ecosystems to recover and for socio-technical systems to become more sustainable. To tackle climate change, for example, massive reductions in the use of fossil fuels are required (IPCC 2022). If we want to remain within the 1.5°C target, power generation from coal and natural gas has to decline worldwide at a speed for which there are hardly any historic precedents (Vinichenko et al. 2021).

Decline can happen without policy intervention (e.g., DVDs being replaced by video streaming) but it may also be guided by public policy. An example for the latter is the policy driven phase-out and ban of incandescent light bulbs (Stegmaier et al. 2021). Especially when time is running out and negative consequences accumulate as in the case of climate change, it is essential for policymaking to act swiftly and to accelerate processes of decline.

A key approach to guide and accelerate decline is *phase-out policies*: 'governance interventions aimed at terminating specific technologies, substances, processes, or practices that are considered harmful' (Rinscheid et al. 2021: 27). Phase-out policies may specify a date by which the practice has to end, a path or steps toward that end, compensations for those negatively affected by the phase-out, and other details (ibid.). Phase-out policies have been implemented for toxic substances such as DDT (Maguire and Hardy 2009), products such as light bulbs (Stegmaier et al. 2021) or gasoline vehicles (Meckling and Nahm 2019) and technologies such as nuclear power (Markard et al. 2020).

Phase-out policies are typically very much contested, which is why the underlying politics, i.e., the political processes leading to phase-out, are very important

DOI: 10.4324/9781003213642-5

(Isoaho and Markard 2020; Rosenbloom 2018). In the case of coal, for example, environmental NGOs, social movements and scientists have been observed to argue in favor of phase-out, while utility companies, coal producers and unions seek to avert or delay phase-out (Brauers et al. 2020; Leipprand and Flachsland 2018). The struggle over phase-out policies is also very much a struggle over the legitimacy of the focal practice or technology (Markard et al. 2021). Only if the established technology loses its legitimacy can we expect widespread societal and political support to enact phase-out policies.

In this chapter,[1] we study political struggles over technology legitimacy and phase-out across three different countries. We focus on coal-fired power generation, which is one of the main global sources of CO_2 emissions and therefore key to tackling climate change (IEA 2021). We conceptualize coal as a *technological innovation system in decline*. To illuminate the unfolding political conflicts, we analyze the public discourses where arguments in favor of and against phase-out are expressed by a broad variety of actor groups and stakeholders (Hajer 1995; Hajer 2006). Tracing general discourse dynamics, major arguments (storylines) and the actor groups that mobilize them, provides key insights into the political processes leading to a phase-out decision.[2]

Our analysis includes three countries: the United Kingdom, Germany and Finland. We chose these countries because they all decided to phase out coal and have several 'macro-level' similarities which improve comparability: culture (Western European), societal values (sustainability and climate change are important), political system (parliamentary democracies), mature electricity/energy markets (low growth rates). At the same time, they vary in key dimensions including: relevance of coal for electricity supply and jobs, age of power plants, availability and progress of alternatives such as renewable energies, nuclear energy and natural gas. Two country case studies (UK, Germany) have already been published (Isoaho and Markard 2020; Markard et al. 2021). The Finnish case and the comparative perspective are novel.

Our article complements prior research on coal decline, which analyzed early stages of decline and regime destabilization (Turnheim and Geels 2012, 2013), debates and conflicts around phase-out (Leipprand and Flachsland 2018; Liersch 2022; Rosenbloom 2018) and strong resistance against phase-out (Stutzer et al. 2021; Trencher et al. 2020). We also add to an emerging line of comparative studies on coal decline (Diluiso et al. 2021), including cross-country comparisons (UK and Germany) focusing on justice concerns (Bang et al. 2022) or general hurdles and drivers (Brauers et al. 2020).

We make three contributions: the commonalities and differences we identify in the public discourse may help to inform policymaking in other places and cases; our comparative analysis and methodological learnings can inspire and inform related research; and our theoretical framing may help to widen the conceptual repertoire in transition studies.

Next, we introduce our theoretical background and review existing work. Section 5.3 provides a short overview of the three countries in terms of power generation, the role of coal and the phase-out decision. Section 5.4 introduces our approach, data sources and analysis. Section 5.5 presents the results. Section 5.6 discusses the findings and conclusions.

5.2 Theoretical background

Decline is an essential part of socio-technical transitions: as innovations and new system configurations emerge and diffuse more widely, established practices, structures and technologies decline. Eventually, they may even vanish. One example is the decline of large sailing ships, which were replaced by steamships in the course of the 19th century (Geels 2002). Examples in the realm of consumer products include video cassettes, DVDs or ('non-smart') mobile phones (Markard 2020).

In the literature on sustainability transitions, decline is receiving increasing attention (Köhler et al. 2019; Markard et al. 2020; Rosenbloom and Rinscheid 2020). One reason behind this is that ongoing transitions in energy or transport have entered a new phase of development, in which innovations such as renewable energies or electric vehicles diffuse rapidly and established technologies such as coal or conventional vehicles are in decline (Markard 2018). Another reason is that, in order to cope with urgent sustainability challenges such as climate change, public policies are needed to phase out problematic practices as quickly as possible to prevent further damage (IPCC 2021; Vinichenko et al. 2021).

In transition studies, decline processes have been analyzed from different perspectives.[3] With a focus on incumbent actors, Turnheim and Geels (2012) introduced the concept of industry or regime destabilization. As external pressures mount in an industry, or socio-technical regime, the provision of financial resources drops, legitimacy declines and the commitment of incumbent actors crumbles (ibid.). These dynamics may also include an increasing involvement of policy-making, which might eventually result in major policy changes, e.g. in the form of stricter regulations (Geels and Penna 2015), removal of policy support (Roberts 2017) or even phase-out (Brauers et al. 2020; van Oers et al. 2021).

Recently, scholars have also started to study decline from a technological innovation systems (TIS) perspective. The TIS approach highlights that different kinds of actors and institutions interact, thereby shaping the development of the focal technology (Bergek et al. 2008a; Markard 2020). While the framework was developed, and often used, to study the emergence of innovations (Bergek et al. 2008a; Markard and Truffer 2008), it can also be mobilized to capture processes of decline, which are a 'natural' part of technology or industry life cycles (Klepper 1997; Markard 2020).

A focal TIS is interacting with other systems in its context (Bergek et al. 2015; Markard and Hoffmann 2016). These include other technological innovation systems and sectors as well as broader 'societal' systems such as the policy system, the scientific system or civil society. All of these systems interact and can support or hinder the development of the focal system. For example, when a competing TIS grows (e.g., around electric vehicles) this typically has a negative effect on the focal

TIS (e.g., around conventional vehicles). The opposite holds for complementary TIS such as domestic coal mining. For example, Stutzer et al. (2021) have shown that coal mining companies, their shareholders and traditional mass media played an essential role in legitimizing the approval of a large new coal mine in Australia.

In order to study whether a TIS prospers or does not do well, scholars have suggested a set of TIS functions such as knowledge development, resource mobilization or market formation (Bergek et al. 2008a; Hekkert et al. 2007). While TIS functions have mostly been used to study emerging technologies, they can also be adapted and used to study the decline of a TIS (Bento et al., in review). One of these functions is about the *creation, or destruction, of legitimacy* (Bergek et al. 2008b; Markard et al. 2016). When an innovation emerges, actors that support it seek to create legitimacy, e.g. as they explain what it is about and mobilize arguments why it is needed (Aldrich and Fiol 1994; Binz et al. 2016). When technologies decline, the opposite happens, as actors argue why it should not be used any more. During decline, we see struggles of actors that seek to undermine legitimacy (e.g. highlighting the associated risks) and those that try to maintain or re-establish legitimacy (Isoaho and Markard 2020; Rosenbloom 2018).

In this study, we analyze struggles over legitimacy with the help of argumentative discourse analysis (Hajer and Versteeg 2005; see section 5.4.1 for more detail). This approach identifies statements of different actors about a focal issue (here: coal-fired power generation as an established technology). These statements, or *storylines*, highlight certain values, technology characteristics and relationships, while excluding others. With these storylines, actors frame a technology in a specific way, thereby shaping its legitimacy. Below, we compare how the discourse unfolds over time, which actors and actor groups are involved, what storylines are more frequent than others and who uses which storylines.

Figure 5.1 shows a simple conceptual framework that informed our analysis. The focal TIS (on coal-fired power generation) interacts with other TIS in its context. Here we depict those that are the most relevant later in the empirical analysis. The focal TIS is part of the larger electricity supply system for which different technologies are available (competing TIS). At the same time, there are complementary TIS such as those that are located upstream (coal mining) or downstream (carbon capture and storage technology) in the value chain (Andersen and Markard 2020). The TIS also interacts with energy consumption systems, policy and science systems, and civil society. In all of the aforementioned systems there are (different kinds of) actors that have an interest when it comes to coal phase-out. In the public discourse, with the storylines they use, actors are either seeking to destroy the legitimacy of the focal TIS or to uphold it. A decline of legitimacy eventually facilitates the policy decision to phase out the focal technology.

Our empirical analysis (and the results section) concentrates on the dashed arrows: the arguments actors use to influence the legitimacy of the focal technology. While most of these storylines are related to the focal technology, some also relate to competing and complementary TIS.

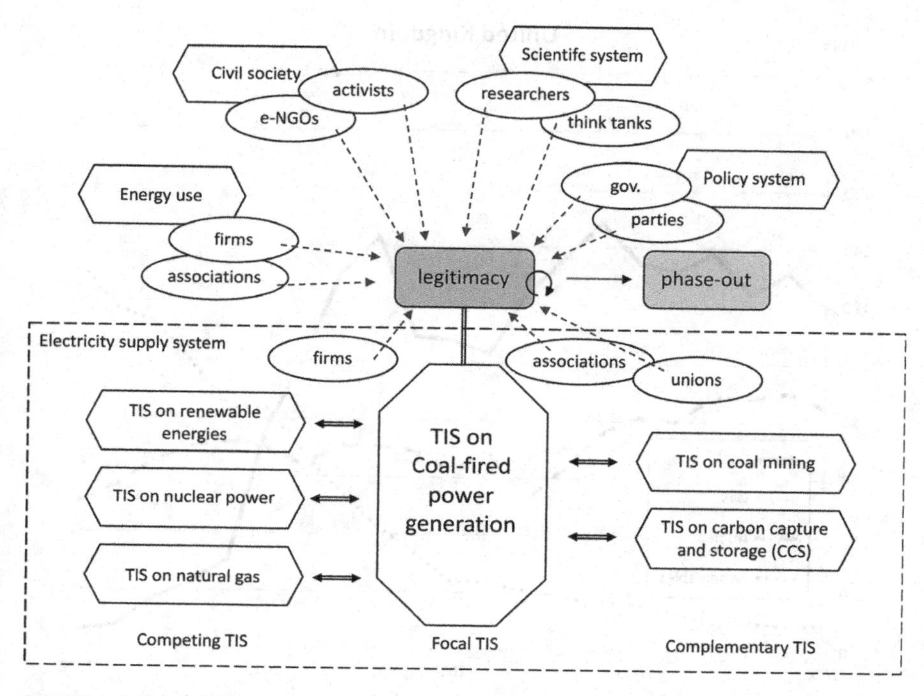

FIGURE 5.1 Focal TIS, context systems, actor groups and struggles over decline

5.3 Coal TIS and contexts: A brief overview across countries

This section provides a brief overview of the main characteristics and developments in the electricity supply system in the three countries and the developments around coal and phase-out.[4] Figure 5.2 displays four main sources of power generation: coal, nuclear, gas and renewables. Table 5.1 lists some key characteristics.

Coal decline is most advanced and almost completed in the UK, where it dropped from a 32% share of total power generation in 2000 to less than 2% in 2020. In this period, coal was primarily replaced by wind power and, to some extent, by natural gas. The UK is building new nuclear power plants. The country already saw an earlier wave of coal decline in the 1990s as a consequence of electricity market liberalization and the rapid expansion of gas-fired power plants (Turnheim and Geels 2012; Winskel 2002). In 2015, in the context of the Paris Climate Agreement, the British government pledged to phase out coal by 2025. In 2021, the goal was brought forward to 2024. At the time of the pledge, many coal power plants were already set for closure because they were not meeting EU emission regulations any more.

In Germany, coal decline was very moderate at first but it has accelerated recently. In 2013, power generation from coal had a share of 48%, which dropped to 23% in 2020.[5] The German coal phase-out was suggested by a commission in 2019 and passed through Parliament in 2020 (Markard et al. 2021). The phase-out

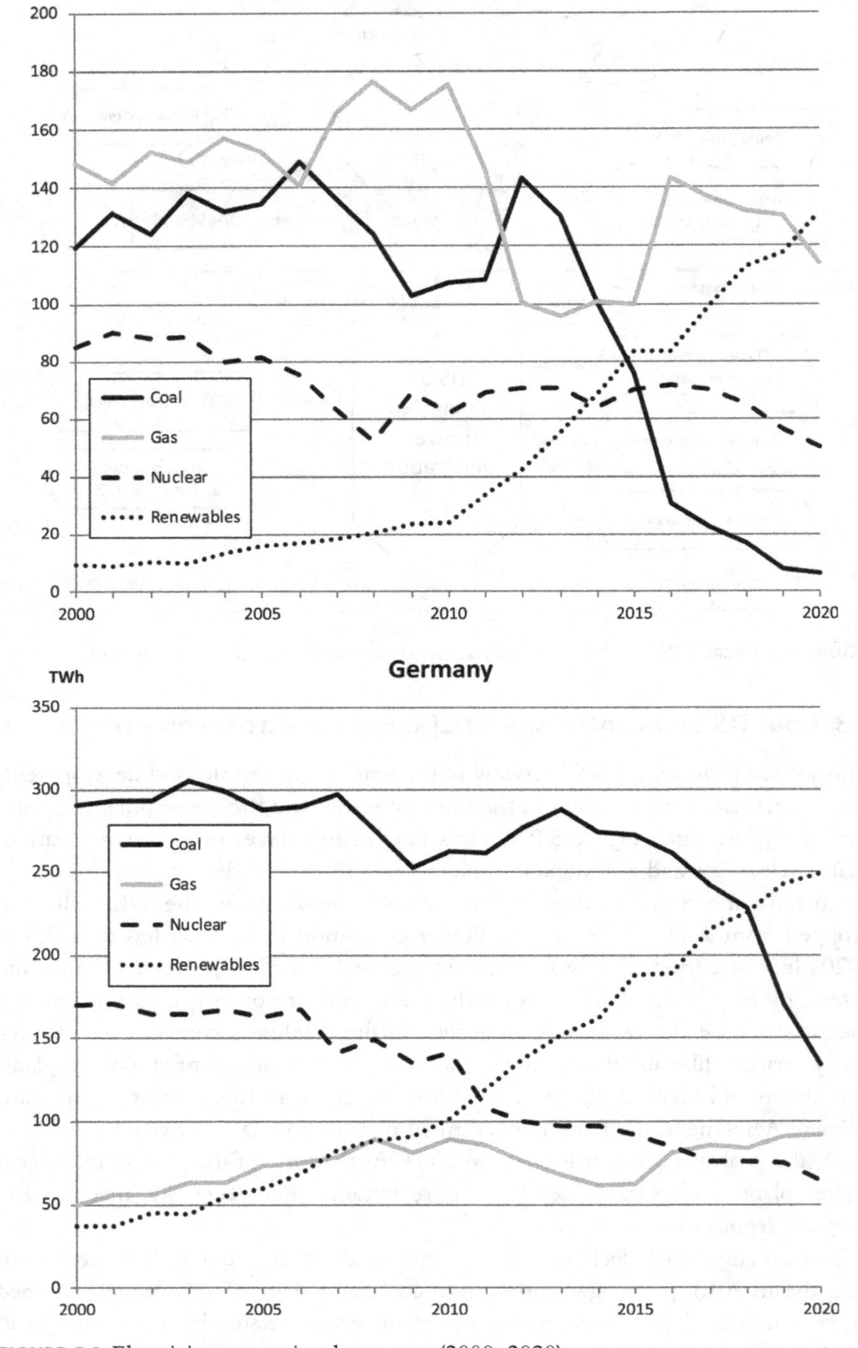

FIGURE 5.2 Electricity generation by source (2000–2020)

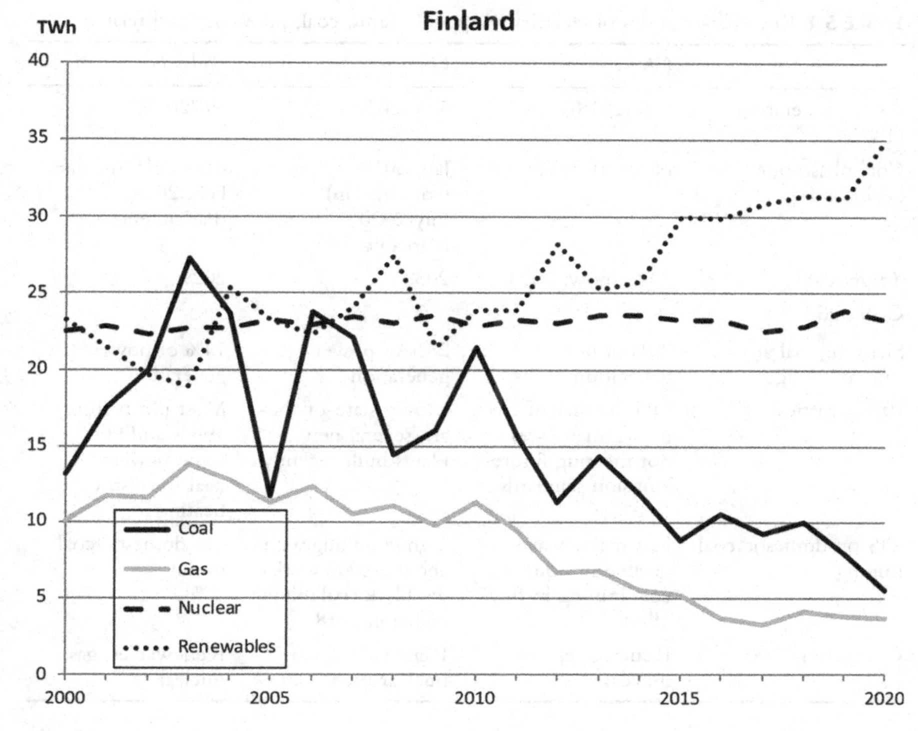

FIGURE 5.2 (Cont.)

deadline is 2038. Alongside coal, nuclear power is also in decline. The nuclear phase-out was decided in 2000 and accelerated in 2011 following the Fukushima accident. For decades, German energy policy has been characterized by very intense socio-political conflicts over the risks and costs of nuclear energy (Markard et al. 2020; Skea et al. 2013). Nuclear declined from 31% (2000) to 11% (2020) and will be phased out by the end of 2022. Both coal and nuclear have been primarily replaced by renewables (up from 7% to 40%) but also by natural gas (up from 9% to 15%, see Figure 5.2).

In Finland, coal-fired power generation has seen ups and downs in the early 2000s (due to fluctuations in energy/electricity prices and winter temperatures) and a moderate decline since around 2010. Plans to phase out coal were first announced by the government in 2016. In 2018, a phase-out proposal was submitted to Parliament and it passed in 2019. The phase-out deadline is 2029. In 2018, the share of coal for power generation was 14%. In Finland, district heating is very common and most district heating systems are connected to combined heat and power plants, in which coal is used, next to biomass or waste. Finland, unlike the other two countries, has important hydropower resources (around 20% of the power supply) but the country is also a net importer of electricity (15 TWh in 2020). Finland, similar to the UK, is expanding nuclear power.

TABLE 5.1 Key characteristics of electricity supply systems, coal, phase-out and related TIS

	UK	*Germany*	*Finland*
Power generation (TWh)	335 (2018)	574 (2020)	69 (2020)
Coal phase-out decision	Nov. 2015 (pledge)	Jan. 2019 (commission) July 2020 (Parliament)	Apr. 2018 (pledge) Feb. 2019 (Parliament)
Target date	2025 (now: 2024)	2038	2029
Coal TIS			
Share of coal at time of pledge	23% of power generation	28% of power generation	14% of power generation
Particularities	Old, inefficient coal plants, many were not meeting future emission standards	Mostly state-of-the-art, several new plants built recently	Most plants from 1980s and 90s, some newer; coal for district heating
TIS on domestic coal mining	Last major wave of decline in domestic coal mining in the 1980s	Lignite mining with about 20,000 workers; black coal mining ended in 2018	No domestic coal mining
Competing TIS	Renewables, gas, nuclear	Renewables, gas; nuclear phase-out	Renewables, gas, nuclear

5.4 Methods and data

Our analyses begin in 2000 for all three countries. At this time, concerns over climate change had been clearly formulated but coal phase-out was not an issue yet. The analyses end after the phase-out decisions had been made. For the UK, we collected data until the end of 2017, where we saw that the discourse cooled off after the decision. For Finland and Germany, we therefore set the cut-off date in the month or shortly after the phase-out passed parliament: April 2019 for Finland and July 2020 for Germany. Our results for the UK (Isoaho and Markard 2020) and Germany (Markard et al. 2021) have already been published. Here, we add the comparative perspective as well as unpublished data from Finland.

5.4.1 Methodological background

In our study, we analyze the public discourse as it is expressed in newspaper articles. Following earlier studies (e.g. Isoaho and Markard 2020; Leipprand et al. 2016; Rosenbloom et al. 2016), we build on Hajer's (1995, 2006) argumentative approach to discourse analysis. This approach grants arguments on a specific topic, brought forward by stakeholders, a key role in the political process. Through discourse, shared understandings of social and physical phenomena are created and conflicting views become apparent. A central concept in discourse analysis is the storyline: 'a condensed statement summarizing complex narratives, used by people as "short hand" in discussions' (Hajer 2006: 69). 'Through storylines, actors select certain aspects of the discourse while excluding others, thereby reducing the complexity of policy issues ... For example, the storyline "Coal is bad for the climate" condenses complex arguments on the release of CO_2 through the burning of coal, how CO_2 contributes to climate change and how climate change is bad for society and nature' (Markard et al. 2021: 317).

Discourse analysis has the capacity to reveal conflicts and argumentative reactions, in which some storylines are included and others omitted from the statements of actors (Isoaho and Karhunmaa 2019). The approach also acknowledges that actors' discursive positions are not constant but subject to change. Therefore, it allows the tracing of changes in the discourse over time.

Building on Hajer's concept of storyline, we take a quantitative approach to analyzing discourse, i.e., we count how often certain storylines were mentioned.[6] We do this to facilitate comparability across the three cases. The downside of this approach is that we provide less information on the (qualitative) content of storylines and how they were presented.

Our analysis uses archival data from nationwide newspapers. We focus on newspapers as data source because they are a central medium in which political arguments are exchanged; newspapers, and media more generally, can be viewed as a major environmental policymaking arena alongside more formal venues such as parliamentary debates (Boykoff and Boykoff 2007; Hansen 2010). Moreover, investigating news articles allows us to cover the discourses from both incumbent and niche actors. The media is often interested in highlighting conflicts and struggles to attract attention, and so newspaper articles are likely to contain more diverse actor interests than for example policy documents (Delshad and Raymond 2013).

5.4.2 Data sources and data collection

For the UK case study, we downloaded newspaper articles from the LexisNexis academic database. After a scoping phase with test-runs on several newspapers, *The Guardian* was chosen as the main source as it was the only nationwide quality newspaper available in the database that systematically covered energy and climate issues.[7] For Germany, we chose the daily editions of *Süddeutsche Zeitung* (*SZ*) and *Die Welt* as data sources. Both newspapers report about the German energy market on a regular basis and were accessible through LexisNexis and the Bavarian State Library's online archive.[8] Both newspapers have different ideological stances: *SZ* represents center-left and *Die Welt* conservative, market-liberal values. For the Finnish case study, we chose *Helsingin Sanomat*, the only nationwide newspaper in Finland, and news articles from *YLE*, Finland's nationwide public broadcasting company.[9] Both sources claim to be politically independent and liberally oriented (Teräväinen 2014). To collect relevant articles, we did both a general search for articles on coal used for power generation and a more specific search on coal phase-out.[10] After obtaining the data, we went through all articles and eliminated

TABLE 5.2 Overview of data sources

	UK	*Germany*	*Finland*
Source(s)	*The Guardian*	*Süddeutsche Zeitung* (*SZ*), *Die Welt*	*Helsingin Sanomat* (*HS*), *YLE*
Time period	2000 – Dec 2017	2000 – July 2020	2000 – April 2019
Number of articles	249	329 (*SZ*) 281 (*Die Welt*)	77 (*HS*) 91 (*YLE*)
Storylines coded	471	814	548

duplicates and false positives (e.g., articles about domestic politics, housing, manufacturing etc.). The final data set for the UK consisted of 249 articles (2000–2017), the German data set included 610 articles (2000–2020) and the Finnish 168 articles (2000–2019). See Table 5.2 for an overview.

5.4.3 Data analysis

We developed a common strategy for the analysis of storylines. First, two authors[11] examined a subset of articles (every second or third, depending on the sample size) and *inductively* derived storylines from the sample. Here we followed pre-defined steps: 1) identify key text passages where coal is discussed, 2) code value judgments or arguments related to coal, 3) identify and code actors that make these arguments.

After independent analysis, the results were then compared and discussed between the authors and finally consolidated into a list of storylines. Due to this bottom-up approach, storylines vary across countries. However, we also found many similarities (Table 5.3). Next, the coding rules for each storyline were discussed and aligned. Then the entire data set was coded. All coding was performed using the NVivo software package for qualitative analysis. In a final step, similar storylines were grouped together across countries and paraphrased. Table 5.3 lists all storylines sorted by their frequency.

Actors were first coded by their name and later assigned to actor groups. The groups were created inductively and updated as the analysis went on. Actor groups were aligned across the three case studies.

5.5 Results

We look into four different aspects of the discourses: general dynamic, the most prominent storylines, the most prominent actor groups and which group argues in favor of or against coal phase-out.

5.5.1 Discourse dynamics

We find several similarities but also differences in how the discourses unfolded over time in the three countries (Figure 5.3 and Table 5.4). In all countries, we see a clear

TABLE 5.3 Overview of storylines and number of codes

Storyline (and paraphrase)	*UK*	*Germany*	*Finland*
Bad for climate Coal is bad for the climate.	200	337	171
Coal not needed Coal can be replaced by other energy carriers, it is not needed to secure power supply.	56	56	161
Coal is reliable Coal is needed to keep the lights on.	38	101	43
CCS is a solution CCS technology is a solution to the climate problem, it can capture the emissions.	75	32	23
Bad for economy Coal decline and phase-out have a negative effect on jobs and / or regional economies.	-	68	26
Structural change needed Structural change is inevitable and there are new jobs in sustainable industries.	-	52	31
Coal is cheap	11	46	16
Not so bad Our emissions from coal do not matter at a global scale.	-	27	42
Health risk Burning coal pollutes the air and creates health risks.	45	14	-
Coal is expensive	-	38	16
CCS no solution CCS is too expensive and risky.	29	13	7
Coal is our identity We are a coal country, coal is part of who we are.	17	-	-
Coal ban problematic A ban leaves little time to develop alternatives.	-	-	12
No health risk Coal is not a health risk.	-	5	-

peak in media attention around the time when the coal phase-out was announced (or decided) and lively discourse activity in the years before.[12] While there were some sporadic articles on the topic in the early 2000s, more regular discourse activity started around 2007 (UK) or 2008 (Germany, Finland). In the UK we find two waves of attention, with a first major build-up around 2008–2009. These years are characterized by the implementation of the Climate Change Act (2008) and a specific debate around carbon capture and storage technology (CCS) as a means to decarbonize new coal power plants (Isoaho and Markard 2020). This debate largely disappeared in subsequent years as CCS turned out not to be viable. In Germany, we find a more or less

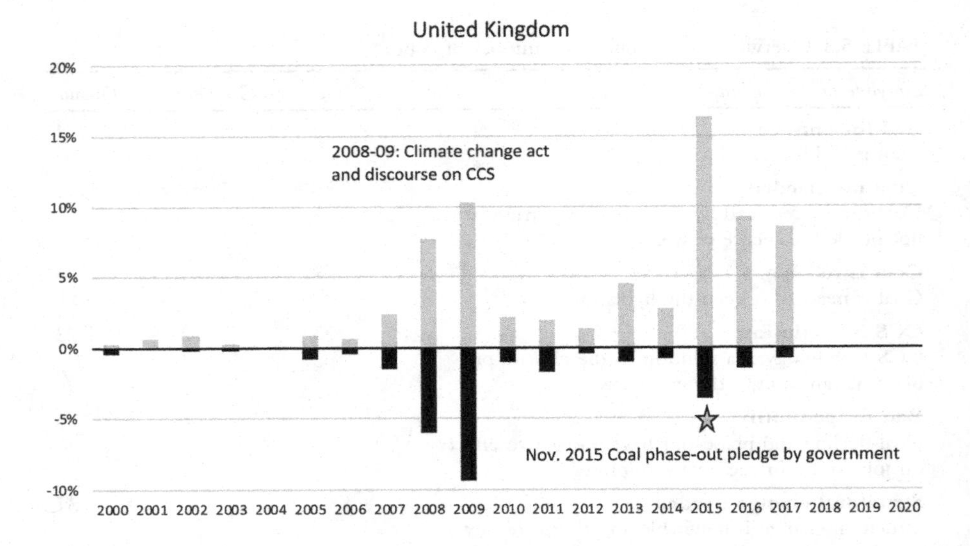

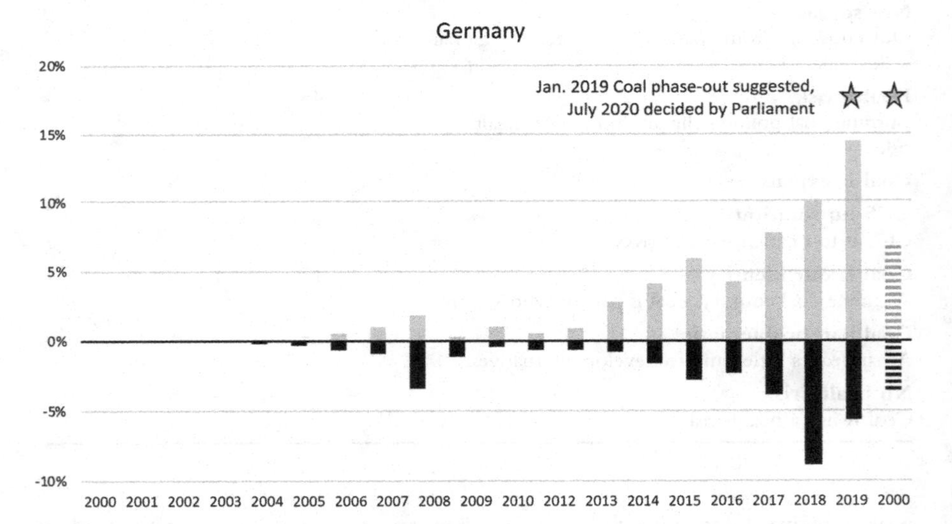

FIGURE 5.3 Discourse activity over time per country
Share of storylines in favor of phase-out (light grey, positive) and against (dark grey, negative). To calculate the share, we divided the number of storylines in a specific year by the overall number of storylines for the entire period for each country. The columns for the final year in Germany and Finland are shaded because data collection stopped in July and April, respectively.

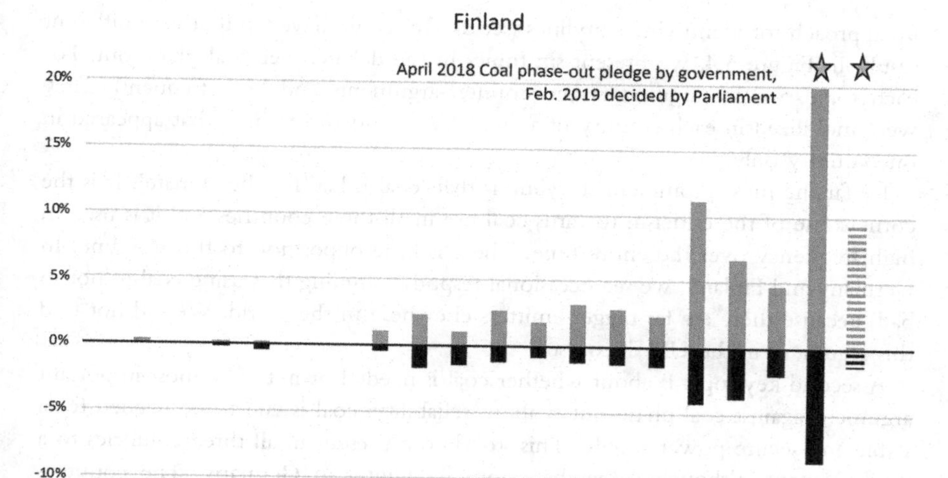

FIGURE 5.3 (Cont.)

steady build up with high discourse activity around the phase-out decision. In 2008, there was also a debate around CCS technology and announcements of pilot plants. In Finland, we see a steady but low discourse activity from 2008–2014, then a first peak in 2015 when there was an intense debate about a new coal power plant in Helsinki, then decreasing attention until a major peak in 2018, when the government proposed to phase-out coal.

The timespan from when discourses became more intense until the phase-out decision was about 9 years in the UK (2007–2015), 11 years in Finland (2007–2018) and more than 12 years in Germany (2008–2020).

Comparing storylines in favor (displayed as positive numbers) and against phase-out (negative numbers), we also find commonalities. In all three countries, pro phase-out storylines outnumber contra storylines. As a general development, we see that the early years are characterized by a more equal weight of pro/contra arguments,[13] while in later years, the discourses tilt in favor of pro phase-out arguments. This pattern is most clearly visible in the UK. In Germany, there are two years at the beginning of the debate in which contra arguments had a majority. Contra arguments also remain strong closer to the phase-out decision. The ratio of pro to contra storylines is highest in Finland (2.4), followed by the UK (2.3) and Germany (1.8). The high value in the UK might be an effect of using *The Guardian*, a left-leaning newspaper. In Germany, the center-left newspaper *SZ* has a ratio of 2.1 (pro vs contra), while the conservative *Die Welt* has a value of 1.6.

5.5.2 Which storylines are mobilized?

There are many similarities across the three countries with regard to the storylines mentioned (Table 5.3). This is an interesting finding in itself because our bottom-up approach for identifying storylines (see above) could have led to a very different result. In Figure 5.4, we present six topics in the debate over coal phase-out. For each topic, we list arguments and counter-arguments and how frequently they were mobilized in each country (if at all). We excluded storylines that appeared in one country only.

By far the most prominent storyline is that 'coal is bad for the climate'. It is the cornerstone of the criticism towards coal use in all three countries and it is used in high frequency over the entire time. There is little opposition to this storyline. In Germany and Finland, we see occasional responses arguing that using coal is 'not so bad' because there are far bigger emitters elsewhere in the world. We did not find this argument in the UK discourse.

A second key topic is about whether coal is needed or not. The most important argument against coal phase-out is about reliability: coal is said to be needed for a stable and secure power supply. This storyline is present in all three countries to a similar extent, although somewhat more frequently in Germany. The counter-argument is that 'coal is not needed'. It is argued, for example, that renewable energy sources can reliably replace it. This storyline was used very often in Finland, but much less so in Germany, where the share of coal for power generation was still rather high at the time of the discussion.

Another debate centers around the economic effects of phasing out coal. Some argue that a phase-out is 'bad for the economy', e.g., pointing to job losses in German coal mining regions. Others respond that structural changes are needed anyway in these regions and that there are also economic opportunities when developing technological alternatives to coal. These arguments appeared both in Germany and Finland but not in the UK, where jobs in coal mining had already been lost many years before.

A fourth topic is about carbon-capture and storage (CCS) technology. Some argue that CCS could be a viable approach to retain the emissions from coal-fired power plants ('clean coal'), while others were skeptical about the technological and financial viability of CCS and the associated risks. The CCS debate gained prominence in the UK around 2008–2009 and it was also associated with hopes to create a new industry around CCS and to become an international leader (and technology exporter) in this field. CCS was much less of an issue in the other two countries. However, CCS was also discussed in Germany in the same years. Vattenfall Europe built a CCS pilot plant in Eastern Germany[14] and also two other major German utilities, RWE and E.On, argued in favor of CCS technology. In Finland, there was little attention given to CCS at that time. The country saw a CCS debate 10 years later in 2018, when the Finnish utility Fortum announced a pilot project in Norway.

Another discussion addressed the costs of coal-fired power generation. Opponents of phase-out argued that coal should be kept because it is cheap, while those in favor of phase-out pointed to the costs that are not accounted for.

Despite many similarities, there were also differences between the countries. For example, the argument that there are health risks associated with burning coal was

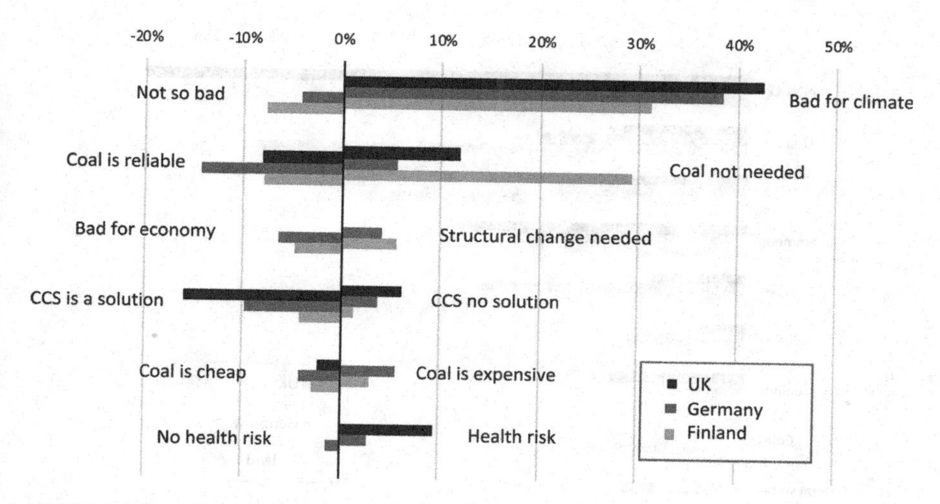

FIGURE 5.4 Storylines and responses for each country
Share of specific storylines in relation to all storylines for the respective country; pro (contra) phase-out: positive (negative) values; no data: storyline was not mentioned.

regularly mentioned in the UK but it was hardly mentioned in Germany and not at all in Finland. In addition, we found a pro-coal storyline around 'coal is part of our national identity' that was unique for the UK. In Finland, we found a couple of arguments that said a coal ban would be problematic, e.g., as there is no flexibility and it does not leave enough time to develop alternatives.

5.5.3 Which actor groups are present in the discourse?

The discourse on coal phase-out is shaped by a broad variety of actors and actor groups. There are six major groups of actors that occur in all three countries, although to varying degrees (Figure 5.5).

Overall, environmental NGOs and climate activists present the most dominant actor group. They are by far the most prominent actor group in the UK[15] and also very prominent in Germany. In Finland, however, they were a much less important voice in the debate. A second important group is electric utilities. They were the most active voice in Finland and had an average discourse activity in Germany. In the UK, they were the least prominent group. Scientific experts and think tanks feature prominently and at a similar level in all three countries and so do government actors. Parties played a prominent role in Germany, while they were less important in the two other countries. Industry actors were most prominent in the UK, of some relevance in the German debate and of hardly any importance in Finland.

Some groups of actors only appeared in one country. In Germany, we found some federal states (Länder) with strong voices and also labor unions. Both are related to lignite mining, which is concentrated in a few states with jobs at stake. In

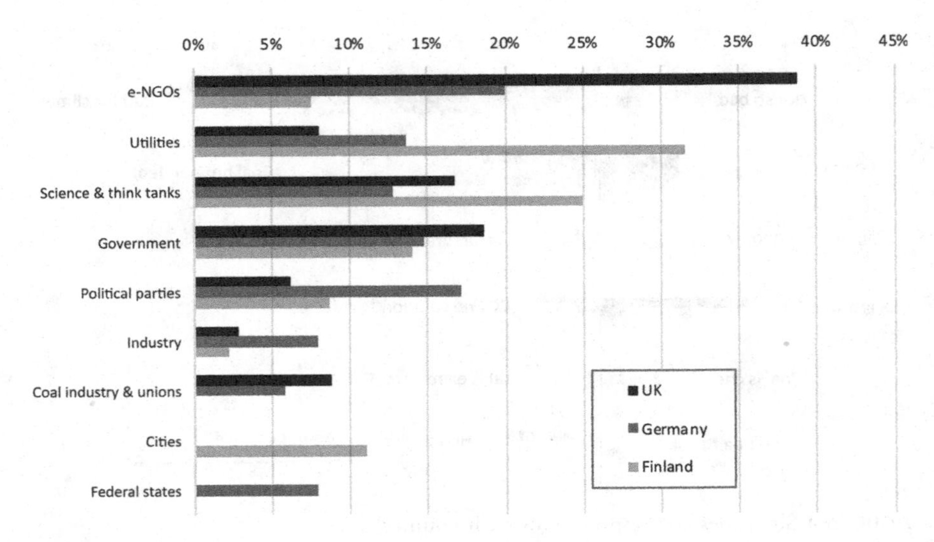

FIGURE 5.5 Main groups of actors for each country
Share of storylines mentioned by members of a specific actor group in relation to all storylines in the respective country.

Finland, cities also had a voice in the debate. Here, most coal power plants are owned by municipal utilities. The plants supply cities with electricity and heat (district heating).

In summary, there are fewer similarities in terms of actor groups across countries than in terms of storylines. In other words, similar arguments and counter-arguments are made—in part by the same groups of actors, in part by other groups.

5.5.4 Which actor groups mobilize which storylines?

Not surprisingly, environmental NGOs and climate activists were the key groups of actors that argued in favor of coal phase-out (Figure 5.6). This pattern holds across all countries, even though NGOs had a much stronger voice in the UK than in Germany and Finland (see above). The main argument of these actors was about climate change. They also argued that coal is not needed. In the UK and Germany, some NGOs were in favor of CCS at the beginning.

Science experts and think tanks also spoke mostly in favor of phase-out. Their main arguments were climate change, alternative energy supply options (coal not needed) and, to a lesser extent, that structural change is needed anyway and that there are also economic opportunities in coal phase-out. In addition, there were several voices that saw a merit in CCS technology, especially in the UK and Finland.

Both of these groups were quite homogeneous in their positions and did not change their arguments over time (except for the pro CCS arguments). Government actors, in contrast, were less homogeneous. We find, for example, environment ministers often arguing in favor of coal phase-out, while other government officials argued in favor of CCS (especially in the UK) or raised issues around job losses and potential negative economic impacts of phase-out (primarily Germany and Finland). Also, storylines from government actors shifted over time, from pro-coal to pro-phase-out.

Utility companies, coal industry and unions, as well as industry actors, show somewhat contrasting positions to the aforementioned groups. They mobilized more storylines against coal phase-out than in favor and their two most central arguments were security of supply (utilities and coal industry), job losses (coal industry) and costs of supply, i.e. coal being cheap (primarily industry). This pattern is more pronounced in Germany and the UK than in Finland, where coal industry and coal-related labor unions did not play a role in the debate. In Finland, we also found many utilities in favor of phase-out, especially in the year when it was finally decided.

A more detailed analysis of the main opposing storylines (Figure 5.7) shows several similarities between the UK and Germany: climate change as the main argument pro phase-out, primarily mobilized by e-NGOs, science actors and parts of government, plus 'coal not needed' as an additional but much less prominent argument. In Finland, both climate and 'not needed' are of equal importance with all four actor groups mobilizing them (but e-NGOs comparably less). On the 'pro-coal' side we find reliability as the key storyline in all three countries, mostly used by utilities. The UK is an exception here with CCS playing a major role (see above) and the government having quite a prominent voice. In fact, the UK government was the main actor defending a continued use of coal in the early years of the UK discourse.

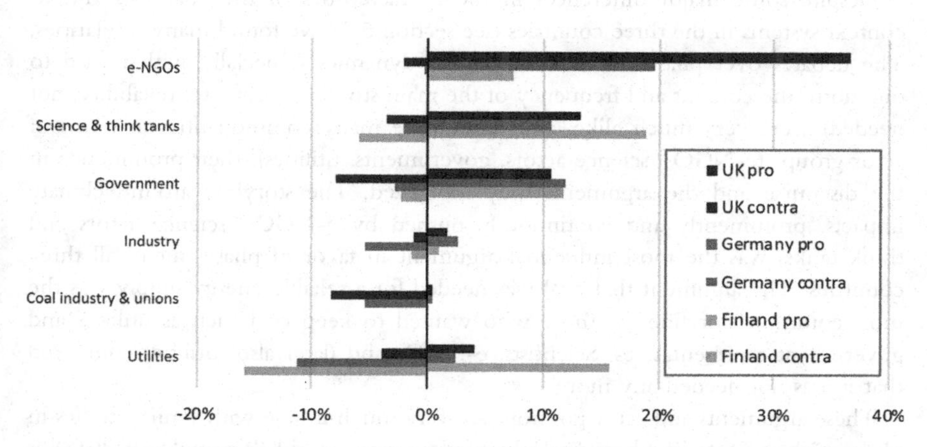

FIGURE 5.6 Main actor groups and whether they were in favor of or against coal phase-out

Share of storylines mentioned by members of a specific actor group (ranked from pro to contra phase-out) in relation to all storylines in the respective country. Pro (contra) phase-out: positive (negative) values.

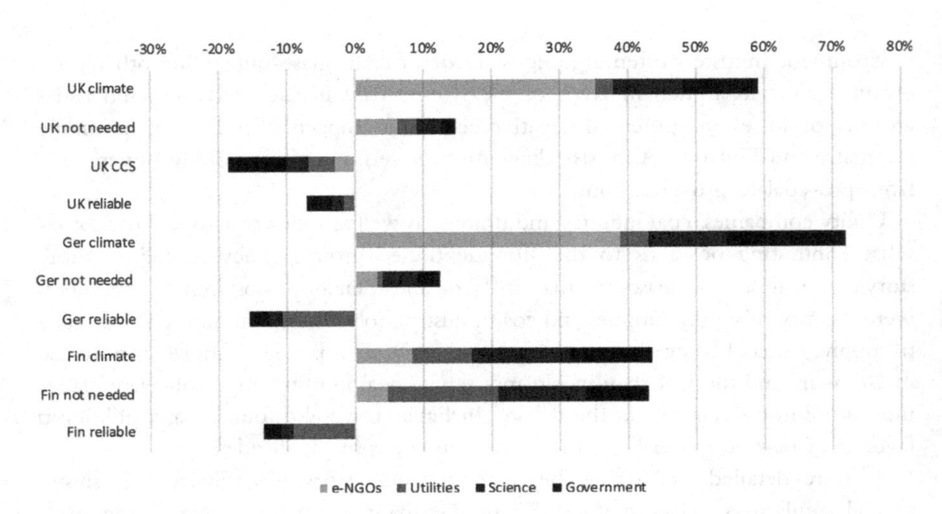

FIGURE 5.7 Most frequent storylines and actor groups for each country
Share of storylines mentioned by a specific actor group in relation to all codes for those storylines and actor groups.

5.6 Discussion and conclusion

This study set out to analyze the public discourses around coal phase-out across three different countries. Inspired by Rosenbloom (2018), we wanted to explore how discourse dynamics, storylines, actors and actor positions in favor or against phase-out compare.

Despite some major differences in the characteristics of the coal TIS and its context systems in the three countries (see section 5.3), we found many similarities. The debates over phase-out showed similar dynamics (especially with regard to duration), the content and frequency of the main storylines (climate, reliability, not needed) were very much alike, and there were many commonalities in terms of actor groups (e-NGOs, science actors, governments, utilities), their prominence in the discourse and the arguments they mobilized. The storyline around climate impacts, prominently and continuously pushed by e-NGOs, science actors and think tanks, was the most influential argument in favor of phase-out in all three countries. The argument that coal was needed for a reliable energy supply was the most common storyline for those who wanted to keep coal such as utilities and government representatives. Scientists, e-NGOs and (later also) utilities countered that it was not needed any more.

These arguments and actor positions are very much in line with similar studies in places such as Ontario where coal phase-out was successful (Rosenbloom 2018) as well as places such as Japan or Australia where it was not (Trencher et al. 2020; Stutzer et al. 2021). Vested interests related to coal mining and coal-based power generation, especially if they are politically well connected, are clearly a major obstacle to phase-out (ibid.; Brauers et al. 2020). A topic that seems to differ across places is the debate about health issues caused by emissions from coal-fired power plants. In Ontario, health was mentioned as a critique as frequently as climate change, while in our cases it only played some role in the UK and there it was much less frequent than the climate argument.

In our sample, phase-out debates lasted from 9 years (UK) to 12 years (Germany). This is similar to the findings from Ontario where it took about 8 years (Rosenbloom 2018). However, there were some differences in dynamic: we found two waves in the UK, a steady rise of attention in Germany and two phases in Finland. In Ontario, the dynamic was yet different with a build-up until an original phase-out date, which was then postponed for 4 years. These dynamics seem to be very much influenced by case-specific developments (e.g., anti-coal protests, CCS pilot plants, policy announcements), but we also found some imprint of international events such as the climate conferences.

We also saw differences with regard to actor groups. In the UK, utilities were rather silent and e-NGOs had a comparably strong voice. This may also be due to the data source. As a left-leaning newspaper, *The Guardian* can be expected to report more frequently on anti-coal protests or views of e-NGOs than other news outlets. Finland almost shows the opposite picture with e-NGOs being rather silent, while (municipal) utilities were the most prominent. The latter can be explained by the importance of coal for municipal district heating (Karhunmaa 2019).

Another difference lies in the economic repercussions of coal phase-out. This was most often mentioned in Germany, especially with regard to jobs in lignite mining. In Finland, this argument came up as well and it was also reported by Rosenbloom (2018). The counter-argument that structural changes are needed anyway only occurred in Germany and Finland.

Many of these differences relate to the characteristics of national contexts (Table 5.4). For example, whether coal phase-out is successful and how long it will take clearly depends on the stability of the existing TIS around coal—an idea that is also reflected in the concept of regime destabilization (Turnheim and Geels 2012). This stability depends on a broad range of factors, e.g., whether coal covers a major share of a nation's energy supply (Table 5.1), whether plants are old or new, whether it is relevant for local jobs or for energy security reasons. Future studies might want to explore the issue of TIS stability, and regime strength, in a more systematic way. For example, actors in the coal TIS only showed little innovation effort (e.g., in the form of CCS technology) to avert looming decline. This might be different in other cases, e.g., car manufacturers seeking to build a lifeline for combustion technology with plug-in hybrids.

Finally, a related dimension centers around the availability of power supply alternatives, or competing TIS. If these are mature and can be expanded swiftly (like natural gas and renewables in the UK), coal phase-out can happen quickly compared to cases such as Germany, where nuclear phase-out already puts a strain on the electricity system and renewable energy expansion (especially wind energy) has been confronted with resistance by local initiatives and policymakers (Bues

TABLE 5.4 Summary of phase-out decisions, discourse characteristics and TIS features

	UK	*Germany*	*Finland*
Status of coal phase-out			
Status	Phase-out almost complete	Phase-out has begun	Phase-out has begun
Expected duration[i1]	2015–2024 (moved up from 2025); 9 years	2020–2038; 18 years	2019–2029; 10 years
Discourse characteristics			
Discourse dynamics (5.1)	Duration[ii2]: 9 years; two waves; first wave with debate around CCS	12 years; slow build-up to intense debate over 4 years	11 years; 7 years with moderate discourse plus 4 years of intense debate
Most frequent storylines (5.2)	Coal bad for climate; CCS is a solution; coal not indispensable; coal reliable	Coal bad for climate; coal reliable; phase-out bad for economy; coal not indispensable	Coal bad for climate; coal not indispensable; coal not so bad; coal reliable
Most prominent actor groups (5.3)	NGOs (162), government (80), science (77), industry associations (58)	NGOs (157), parties (126), government (107), utilities (96)	Utilities (140), science (90), government (64), cities (46)
Actors in favor of / against phase-out (5.4)	In favor: e-NGOs, science; Against: Industry, utilities; Changing: government	In favor: e-NGOs, science, government; Against: Industry, utilities; Changing: –	In favor: e-NGOs, science, government; Against: - Changing: Utilities
Coal TIS			
Relevance for energy system (3)	Moderate (at time of phase-out decision), Low (today)	High (also due to nuclear phase-out and slow growth of renewables)	Moderate (also due to district heating)
Techno-economic (3)	Existing plants were old	Recent investments into new coal power plants	Most plants from 1980s and 90s; district heating
Context			
Electricity supply system (3)	Utilities weakened by liberalization	Large influential utility companies	Influential municipal utility companies
Complementary TIS (3)	Domestic mining declined in 1990s and 2000s; CCS was not successful	Domestic lignite mining still ongoing	No mining, no CCS
Competing TIS (3)	Strong political support for nuclear; support for natural gas; moderate support for wind energy	Nuclear phase-out under way (completed in 2022); waning support for renewables, local resistance	Strong political support for nuclear; slowly increasing support for renewables (e.g. biomass for district heating)

2020). Despite these challenges, the new German government, a coalition of Social Democrats, the Green Party and Liberals, wants to move coal phase-out ahead to 2030.[16] Also in Finland, the city of Helsinki has announced the closure of its coal power plant 2 years ahead of schedule.[17]

Our study has shown that discourse analysis can generate important insights into the political struggles that surround technology decline and phase-out decisions. Our analysis of public discourses can nicely complement the study of more formal spheres of interaction such as parliamentary debates (Leipprand et al. 2016; Müller-Hansen et al. 2021). Discourse analysis has clear strengths when it comes to identifying (conflicting) values and ideas, which is why it lends itself to studying politics. At the same time, it cannot capture all dimensions of transition processes. For example, techno-economic and material aspects such as the age of power plants, existing infrastructures or the performance of competing technologies also play into phase-out decisions.

We have demonstrated how discourse analysis can be mobilized to systematically compare different cases. There are some caveats though. To fit all in one article, we focused on a quantitative comparison (here: counting storylines). This helped us to condense information and to facilitate comparability but it came at the expense of more detailed case insights. Future studies may find ways of combining quantitative and qualitative elements of discourse analysis in comparative research designs. Discourse analysis can also be used to delve deeper into the politics of transitions, e.g. identifying coalitions of actors who mobilize the same storylines (Lowes et al. 2020; Markard et al. 2021).

Another analytical challenge is about developments unfolding at different times and largely asynchronously in different places. Originally, we wanted to address this by distinguishing (and then comparing) different phases of development as in the single case studies (Isoaho and Markard 2020; Markard et al. 2021). This would have required a general 'theory' on discourse dynamics, similar to the issue life cycle approach used by Penna and Geels (2012). As it turned out to be more complex than anticipated, we decided to drop the phases and leave some general theory as a topic for future research.

A third issue for improvement is about data sources. In general, newspaper articles are clearly suitable for discourse analytical purposes. However, it is important to have a balanced selection of sources. The fact that, for the UK, we only drew from the left-leaning *Guardian* is clearly a limitation of our study. Learning from this, we widened our selected newspapers for the other two cases. Future studies might also want to consider additional sources such as social media. Especially for larger amounts of data, automated processes for text analysis such as natural language processing create new opportunities.

Let us conclude with a brief conceptual reflection. Applying the TIS framework to technologies in decline is new, and for some it might still be counter-intuitive because there is not necessarily much innovation in decline. At the same time though, the TIS approach provides a generic, systems-based framework to capture technology dynamics (including expansion, stagnation or decline), which has the potential to be applied widely in transitions research (Markard 2020). With the increasing complexity of the low-carbon energy transition (e.g., as more and more actors, technologies and sectors become involved), the TIS framework might provide crucial building blocks to analyze the dynamics and interactions of a multitude of different technologies in different stages of development (Andersen and Markard 2020; Markard 2020; Rosenbloom 2020). Analyzing TIS dynamics and structures across places[18] (Binz and Truffer 2017) or value chains (Andersen and Markard 2020; Ulmanen and Bergek 2021), understanding various kinds of context systems (Bergek et al. 2015; Ulmanen and Bergek 2021) and identifying key processes of TIS decline (Bento et al., in review) will be important contributions to this larger research agenda.

Notes

1 We thank Julia Bachmann, Sakari Höysniemi, Kamilla Karhunmaa, Christof Knoeri, Zahar Koretsky, Adrian Rinscheid, Peter Stegmaier, Bruno Turnheim and Amanda Williams for their support and comments on earlier drafts. If it were not for the inspiring discussions with Bruno and his patience with the many deadlines Jochen failed to meet, this work might not have seen the light of day. We also got valuable feedback when presenting an earlier version at the International Conference for Sustainability Transitions (IST) in Ottawa, June 23–26, 2019. Jochen Markard acknowledges funding from the Norwegian Research Council (Conflicting Transition Pathways for Deep Decarbonization, Grant number 295062/E20) and from the Swiss Federal Office of Energy (SWEET programme, PATHFNDR consortium).

2 Discourse analysis primarily captures ideational and value-related dimensions of transitions and there is a risk of overlooking material and economic aspects. In the final section, we therefore also discuss broader 'TIS features' to better understand the different approaches to phase-out.

3 See also Koretsky (2023) for a conceptual discussion.

4 Note that the recent developments around the war in Ukraine and the shifting geopolitics of energy supply will certainly affect the future use of, and political decisions related to, natural gas and coal. For more detailed insights into the coal phase-out in the UK and Germany see Brauers et al. (2020), Isoaho and Markard (2020) or Markard et al. (2021).

5 Note that power generation and consumption in 2020 was lower than usual due to the Covid-19 pandemic.

6 This analysis is based on reading, interpretation and manual coding of the articles.

7 Note that *The Guardian* is a left-leaning newspaper, so it is likely that the voices of for example environmental NGOs are reported more frequently than in other outlets.

8 We covered articles until January 2019 to include the phase-out decision of the German coal commission.

9 The Finnish sources could only be accessed through source specific databases, which did not support Boolean operators. We used the same keywords separately and tried to mimic the search string as well as possible with different combinations.

10 These are the exemplary search strings for the UK. General: (GEOGRAPHIC(UK) AND (decline w/p coal) OR (phase-out w/p coal) AND (electricity OR power); Specific: GEOGRAPHIC(UK) AND HLEAD(coal) AND LENGTH>500 AND (electricity OR power OR carbon OR decarbon! OR decline OR phase-out)

11 Only one author for the Finnish case.

12 Note that data for Germany and Finland is truncated in the last reported year. We stopped data extraction for Finland in April 2019, and for Germany in July 2020, when the coal phase-outs were decided in parliament.

13 In Germany in 2008, we even see contra storylines outnumbering pro arguments.

14 The pilot plant was closed in 2014.
15 Note that NGOs might get more of a voice in the left-leaning *Guardian* than in more conservative newspapers.
16 German parties agree on 2030 coal phase-out in coalition talks. *Reuters*, Nov. 2021, www.reuters.com/business/cop/exclusive-germanys-government-in-waiting-agrees-phase-out-coal-by-2030-sources-2021-11-23/, accessed April 30, 2022.
17 Helsinki to shut down coal-fired power plant 2 years ahead of schedule. *YLE News*, June 2021, https://yle.fi/news/3-11993952, accessed April 30, 2022.
18 While we have analyzed country-level developments as largely independent, future research should also address how, e.g., phase-out decisions in one place affect TIS dynamics elsewhere.

References

Aldrich, H.E. and Fiol, C. M (1994) Fools rush in? The institutional context of industry creation. *Academy of Management Review*, 19, 645–670.

Andersen, A.D. and Markard, J. (2020) Multi-technology interaction in socio-technical transitions: How recent dynamics in HVDC technology can inform transition theories. *Technological Forecasting and Social Change*, 151, 119802.

Bang, G., Rosendahl, K.E. and Böhringer, C. (2022) Balancing cost and justice concerns in the energy transition: comparing coal phase-out policies in Germany and the UK. *Climate Policy*. https://doi.org/10.1080/14693062.2022.2052788.

Bento, N., Nunez-Jimenez, A. and Kittner, N. (in review) Decline processes in technological innovation systems: lessons from energy technologies. *Research Policy*.

Bergek, A., Jacobsson, S., Carlsson, B., Lindmark, S. and Rickne, A. (2008a) Analyzing the functional dynamics of technological innovation systems: A scheme of analysis. *Research Policy*, 37, 407–429.

Bergek, A., Jacobsson, S. and Sanden, B.A. (2008b) 'Legitimation' and 'Development of external economies': Two key processes in the formation phase of technological innovation systems. *Technology Analysis & Strategic Management*, 20, 575–592.

Bergek, A., Hekkert, M.P., Jacobsson, S., Markard, J., Sanden, B.A. and Truffer, B. (2015) Technological innovation systems in contexts: Conceptualizing contextual structures and interaction dynamics. *Environmental Innovation and Societal Transitions*, 16, 51–64.

Binz, C. and Truffer, B. (2017) Global innovation systems: A conceptual framework for innovation dynamics in transnational contexts. *Research Policy*, 46, 1284–1298.

Binz, C., Harris-Lovett, S., Kiparskyd, M., Sedlak, D.L. and Truffer, B. (2016) The thorny road to technology legitimation – Institutional work for potable water reuse in California. *Technological Forecasting and Social Change*, 103, 249–263.

Boykoff, M.T. and Boykoff, J.M. (2007) Climate change and journalistic norms: A case-study of US mass-media coverage. *Geoforum*, 38, 1190–1204.

Brauers, H., Oei, P.-Y. and Walk, P. (2020) Comparing coal phase-out pathways: The United Kingdom's and Germany's diverging transitions. *Environmental Innovation and Societal Transitions*, 37, 238–253.

Bues, A. (2020) *Social Movements against Wind Power in Canada and Germany: Energy Policy and Contention*. Routledge.

Delshad, A. and Raymond, L. (2013) Media framing and public attitudes toward biofuels. *Review of Policy Research*, 30, 190–210.

Diluiso, F. *et al.* (2021) Coal transitions – Part 1: A systematic map and review of case study learnings from regional, national, and local coal phase-out experiences. *Environmental Research Letters*, 16, 113003.

Geels, F.W. (2002) Technological transitions as evolutionary reconfiguration processes: A multi-level perspective and a case-study. *Research Policy*, 31, 1257–1274.

Geels, F.W. and Penna, C.C.R. (2015) Societal problems and industry reorientation: Elaborating the Dialectic Issue LifeCycle (DILC) model and a case study of car safety in the USA (1900–1995). *Research Policy*, 44, 67–82.

Hajer, M. (1995) *The Politics of Environmental Discourse*. Oxford University Press.

Hajer, M. (2006) Doing discourse analysis: Coalitions, practices, meaning. In Van Den Brink, M. and Metze, T. (eds) *Words Matter in Policy and Planning*. Netherlands Graduate School of Urban and Regional Research.

Hajer, M. and Versteeg, W. (2005) A decade of discourse analysis of environmental politics: Achievements, challenges, perspectives. *Journal of Environmental Policy and Planning*, 7, 175–184.

Hansen, A. (2010) *Environment, Media and Communication*. Routledge.

Hekkert, M., Suurs, R.A.A., Negro, S., Kuhlmann, S. and Smits, R. (2007) Functions of Innovation Systems: A new approach for analysing technological change. *Technological Forecasting and Social Change*, 74, 413–432.

IEA (2021) *Net Zero by 2050: A Roadmap for the Global Energy Sector*. International Energy Agency.

IPCC (2022) Summary for Policymakers. In Shukla, P. R. et al. (eds) *Climate Change 2022: Mitigation of Climate Change. Contribution of Working Group III to the Sixth Assessment Report of the Intergovernmental Panel on Climate Change*. Cambridge University Press, Cambridge, New York.

Isoaho, K. and Karhunmaa, K. (2019) A critical review of discursive approaches in energy transitions. *Energy Policy*, 128, 930–942.

Isoaho, K. and Markard, J. (2020) The politics of technology decline: Discursive struggles over coal phase-out in the UK. *Review of Policy Research*, 37, 342–368.

Karhunmaa, K. (2019) Attaining carbon neutrality in Finnish parliamentary and city council debates. *Futures*, 109, 170–180.

Klepper, S. (1997) Industry life cycles. *Industrial and Corporate Change*, 6, 145–182.

Köhler, J. *et al.* (2019) An agenda for sustainability transitions research: State of the art and future directions. *Environmental Innovation and Societal Transitions*, 31, 1–32.

Koretsky, Z. (2023) Dynamics of technological decline as socio-material unravelling. In Koretsky, Z. *et al.* (eds) *Technologies in Decline: Socio-Technical Approaches to Discontinuation and Destabilisation*. Routledge.

Leipprand, A. and Flachsland, C. (2018) Regime destabilization in energy transitions: The German debate on the future of coal. *Energy Research and Social Science*, 40, 190–204.

Leipprand, A., Flachsland, C. and Pahle, M. (2016) Energy transition on the rise: discourses on energy future in the German parliament. *Innovation: The European Journal of Social Science Research*, 30, 283–305.

Liersch, C. and Stegmaier, P. (2022) Keeping the forest above to phase out the coal below: The discursive politics and contested meaning of the Hambach Forest. *Energy Research & Social Science*, 89, 102537.

Lowes, R., Woodman, B. and Speirs, J. (2020) Heating in Great Britain: An incumbent discourse coalition resists an electrifying future. *Environmental Innovation and Societal Transitions*, 37, 1–17.

Maguire, S. and Hardy, C. (2009) Discourse and deinstitutionalization: The decline of DDT. *Academy of Management Journal*, 52, 148–178.

Markard, J. (2018) The next phase of the energy transition and its implications for research and policy. *Nature Energy*, 3, 628–633.

Markard, J. (2020) The life cycle of technological innovation systems. *Technological Forecasting and Social Change*, 153, 119407.

Markard, J. and Hoffmann, V.H. (2016) Analysis of complementarities: Framework and examples from the energy transition. *Technological Forecasting and Social Change*, 111, 63–75.

Markard, J. and Rosenbloom, D. (2020) A tale of two crises: COVID-19 and climate. *Sustainability: Science, Practice and Policy*, 16, 53–60.

Markard, J. and Truffer, B. (2008) Technological innovation systems and the multi-level perspective: Towards an integrated framework. *Research Policy*, 37, 596–615.

Markard, J., Wirth, S. and Truffer, B. (2016) Institutional dynamics and technology legitimacy: A framework and a case study on biogas technology. *Research Policy*, 45, 330–344.

Markard, J., Bento, N., Kittner, N. and Nunez-Jimenez, A. (2020) Destined for decline? Critically examining nuclear energy with a technological innovation systems perspective. *Energy Research & Social Science*, 67, 101512.

Markard, J., Rinscheid, A. and Widdel, L. (2021) Analyzing transitions through the lens of discourse networks: Coal phase-out in Germany. *Environmental Innovation and Societal Transitions*, 40, 315–331.

Meckling, J. and Nahm, J. (2019) The politics of technology bans: Industrial policy competition and green goals for the auto industry. *Energy Policy*, 126, 470–479.

Müller-Hansen, F., Callaghan, M.W., Lee, Y.T., Leipprand, A., Flachsland, C. and Minx, J. C. (2021) Who cares about coal? Analyzing 70 years of German parliamentary debates on coal with dynamic topic modeling. *Energy Research & Social Science*, 72, 101869.

Penna, C.C.R. and Geels, F.W. (2012) Multi-dimensional struggles in the greening of industry: A dialectic issue lifecycle model and case study. *Technological Forecasting and Social Change*, 79, 999–1020.

Rinscheid, A., Rosenbloom, D., Markard, J. and Turnheim, B. (2021) From terminating to transforming: The role of phase-out in sustainability transitions. *Environmental Innovation and Societal Transitions*, 41, 27–31.

Roberts, J. (2017) Discursive destabilisation of socio-technical regimes: Negative storylines and the discursive vulnerability of historical American railroads. *Energy Research & Social Science*, 31, 86–99.

Rosenbloom, D. (2018) Framing low-carbon pathways: A discursive analysis of contending storylines surrounding the phase-out of coal-fired power in Ontario. *Environmental Innovation and Societal Transitions*, 27, 129–145.

Rosenbloom, D. (2020) Engaging with multi-system interactions in sustainability transitions: A comment on the transitions research agenda. *Environmental Innovation and Societal Transitions*, 34, 336–340.

Rosenbloom, D. and Rinscheid, A. (2020) Deliberate decline: An emerging frontier for the study and practice of decarbonization. *Wiley Interdisciplinary Reviews: Climate Change*, 11, e669.

Rosenbloom, D., Berton, H. and Meadowcroft, J. (2016) Framing the sun: A discursive approach to understanding multi-dimensional interactions within socio-technical transitions through the case of solar electricity in Ontario, Canada. *Research Policy*, 45, 1275–1290.

Skea, J., Lechtenböhmer, S. and Asuka, J. (2013) Climate policies after Fukushima: Three views. *Climate Policy*, 13, 36–54.

Stegmaier, P., Visser, V.R. and Kuhlmann, S. (2021) The incandescent light bulb phase-out: exploring patterns of framing the governance of discontinuing a socio-technical regime. *Energy, Sustainability and Society*, 11, 14.

Stutzer, R., Rinscheid, A., Oliveira, T.D., Mendes Loureiro, P., Kachi, A. and Duygan, M. (2021) Black coal, thin ice: The discursive legitimisation of Australian coal in the age of climate change. *Humanities & Social Sciences Communications*, 8, 178.

Teräväinen, T. (2014) *Representations of Energy Policy and Technology in British and Finnish Newspaper Media: A Comparative Perspective*. Public Understanding of Science.

Trencher, G., Rinscheid, A., Duygan, M., Truong, N. and Asuka, J. (2020) Revisiting carbon lock-in in energy systems: Explaining the perpetuation of coal power in Japan. *Energy Research & Social Science*, 69, 101770.

Turnheim, B. and Geels, F.W. (2012) Regime destabilisation as the flipside of energy transitions: Lessons from the history of the British coal industry (1913–1997). *Energy Policy*, 50, 35–49.

Turnheim, B., and Geels, F.W. (2013) The destabilisation of existing regimes: Confronting a multi-dimensional framework with a case study of the British coal industry (1913–1967). *Research Policy*, 42, 1749–1767.

Ulmanen, J. and Bergek, A. (2021) Influences of technological and sectoral contexts on technological innovation systems. *Environmental Innovation and Societal Transitions*, 40, 20–39.

Van Oers, L., Feola, G., Moors, E. and Runhaar, H. (2021) The politics of deliberate destabilisation for sustainability transitions. *Environmental Innovation and Societal Transitions*, 40, 159–171.

Vinichenko, V., Cherp, A. and Jewell, J. (2021) Historical precedents and feasibility of rapid coal and gas decline required for the 1.5°C target. *One Earth*, 4, 1477–1490.

Winskel, M. (2002) When systems are overthrown: The 'dash for gas' in the British electricity supply industry. *Social Studies of Science*, 32, 563–598.

6

MAPPING THE TERRITORIAL ADAPTATION OF TECHNOLOGICAL TRAJECTORIES

The phase-out of the internal combustion engine

Daniel Weiss and Philipp Scherer

6.1 Introduction

In the sustainability transition literature, many scholars use the technological innovation system approach (TIS) to study technology dynamics from a systemic perspective. Recently, the framework was expanded to study the decline of technologies in the TIS life cycle framework (Markard 2020, Markard et al. 2023). The latter is especially important in sectors where successful transitions depend not only on the emergence of sustainable technologies but also the phase-out of unsustainable ones. Prime examples hereof are the energy and mobility sector. Although a few empirical studies already investigated the different dimensions of the TIS life cycle (Isoaho and Markard 2020, Markard et al. 2020; Markard et al. 2023), we still lack a complete understanding of TIS decline, especially considering the spatial development perspective (Binz et al. 2014). Acknowledging this gap, our study wants to add to this emerging literature by focusing on the technological dimension of the TIS life cycle and investigating how the dominant technological trajectory in the TIS changes as an adaptation of the incumbents to resist a potential decline. The latter is initiated by changes in the context and structure of the TIS, such as changing consumer preferences, a rise in the price of critical commodities, technological competition or loss of policy support (Markard 2020).

We rest empirical analysis on the potential phase-out of the internal combustion engine (ICE) as a vital part of a successful sustainability transition in the mobility sector. In particular, the ICE has been facing increasing decline pressures due to environmental concerns, a rise in fuel prices and strengthening competition from electric powertrains. This has compelled car manufacturers to reconsider their business models and technology choices to improve the performance and efficiency of the ICE using gasoline, diesel or hybrid powertrains (Oltra and Saint Jean 2009, Sushandoyo et al. 2012, Dijk et al. 2013). Interestingly, we observe historically

DOI: 10.4324/9781003213642-6

differentiated propensities towards different powertrain technologies across major car markets: a high share of diesel cars in the EU compared to the uprise of the hybrid powertrain in the USA and Japan (Oltra and Saint Jean 2009, Cames and Helmers 2013, Bohnsack et al. 2015). This observation follows the spatial development perspective, which argues against considering TIS as globally homogenous systems but rather as a multi-scalar setup of heterogenous territories. The latter represent interconnected subsystems on different spatial scales, characterized by regional or national circumstances that influence the local development of a TIS (Binz et al. 2014, Bergek et al. 2015, Binz and Truffer 2017). Accordingly, our study investigates how context changes over time and space affect the technological adaptation process of the TIS incumbents against decline, represented by the dominant trajectories in different territories and time periods. Thus, rather than investigating the technological choices of individual companies, we mainly focus on a meso-level analysis to reveal differences in the technology dynamics induced by changes in the surrounding environment of the TIS.

In order to examine changes in the dominant technological trajectories and their underlying knowledge search processes in the ICE-TIS, we apply the method of main path analysis to patent citation networks (Hummon and Doreian 1989, Verspagen 2007). Contrary to similar TIS studies (e.g., Negro et al. 2008, Islam and Miyazaki 2009, Dewald and Fromhold-Eisebith 2015), this method allows us to track the changes in the direction of technological development quantitatively. We begin our analysis by considering the changes in the USA ICE-TIS over time. Our results indicate that until 2003 the prevalent knowledge search processes were concerned with improvements in fuel injection and fuel filter systems to cope with the pressures in the US territory. In more recent years, research activities shifted towards the hybrid powertrain, which can be depicted as a "median strategy" between fully electric and diesel/gasoline cars that allows car manufacturers to exploit complementarities between the environmental constraints, demands of customers and their own technological competencies (Oltra and Saint Jean 2009). Taking account of the spatial development perspective (see below), our results reveal a differentiated adaptation process across TIS territories. In particular, while we observe a diesel-focused dominant trajectory in the EU, we observe search processes dedicated to the hybrid powertrain in the USA and Japan. Considering that, we highlight policy regulation as one of the key differences in the territorial TIS structures causing this disparity. The interpretations of our results remain the same when using alternative main path algorithms (Liu et al. 2019).

Overall, by recognizing the path-dependent and territory-specific structure of knowledge search processes in the TIS, our results capture an ongoing commitment of European firms towards diesel technology research. These insights are based on patent data until the beginning of 2019. To our knowledge, this insight has not yet been discussed by the related empirical studies which do not consider the evaluation of the patent citation network structure (e.g., Frenken et al. 2004, Oltra and Saint Jean 2009, Sushandoyo et al. 2012, Köhler et al. 2013, Liesenkötter and Schewe 2014, Borgstedt et al. 2017).

Our results are helpful to understand how the technological dimension of a mature TIS adapts to emerging decline pressures. We show that it is necessary to examine developments not only over time but also over spatially differentiated territories of the focal TIS. Furthermore, our study highlights the importance of transnational linkages and spillover effects that can lead to industry-wide changes conditioned by the match between the circumstances in key TIS territories and firm capabilities.

Finally, we discuss the implications of spatially heterogeneous TIS life cycles for our understanding of technology decline and policymaking. In doing so, we emphasize the role of international policy diffusion, competitive pressure among multinational firms and positive externalities between technological trajectories to accelerate the transition to sustainable mobility.

This chapter is structured as follows. In the following section, we introduce our theoretical framework, including the spatial development perspective on TIS. Then we describe the ICE as our empirical case. Next, we explain the main path analysis as our empirical method and outline the associated results. Lastly, we end the chapter by discussing the implications and limitations of our study.

6.2 The concept of TIS, technological trajectories, territories and context

The innovation system approach emerged as both the science and policy community began to consider technological change and the underlying innovation process as the outcome of interactive learning processes involving various actors, networks and institutions. As a heuristic concept based on evolutionary economics, the innovation system approach considers all actors and subsystems that are directly and indirectly involved in innovation processes (Hekkert et al. 2007). Thus, it emphasizes not only the competence of individual firms to produce innovations but also the importance of the institutional environment as well as its influence on the interaction between actors and networks. While the national or sectoral innovation systems are focused on the innovation performance of a particular country or sector, our study is based on technological innovation systems (TIS), which are centered around a focal technology. A TIS consists of an interacting set of actors, networks, infrastructures and institutions. In comparison with national or regional innovation systems, TIS have no definitive territorial boundary and can enclose various heterogeneous regions, industries, sectors and countries (Carlsson and Stankiewicz 1991, Bergek et al. 2008). The TIS itself is situated in a larger context that entails all structural elements outside of its boundaries but can indirectly influence its further development (Bergek et al. 2015).

More recently, scholars have argued that TIS should not be considered as globally spread out homogenous systems (Hekkert et al. 2007), but rather in favor of a spatial development perspective that regards TIS as a multi-scalar setup of heterogenous territories that are interlinked with each other (Binz et al. 2014). In particular, TIS territories describe subsystems on different spatial scales that are characterized by distinguished circumstances influencing the local technological development opportunities. These can comprise specific national or regional context circumstances such as formal regulations, social preferences, competitive pressures and local infrastructures, which influence the development of local TIS structures. Amongst each other, TIS territories are connected through meaningful linkages and couplings such as the multinational value-chain of the focal technology, multiregional actor-networks or supranational technology standards (Binz et al. 2014, Bergek et al. 2015, Binz and Truffer 2017). This provides scholars with a more spatially sensitive perspective that highlights not only the heterogeneous territorial dynamics of TIS but also the interdependencies between different spatial scales. Notably, this depicts a more comprehensive view than the national or regional innovation system frameworks, which merely focus on dynamics on one spatial scale while treating the influence of other scales as exogenous factors (Binz et al. 2014, Binz and Truffer 2017).

The focal technology of a TIS develops along a dominant path-dependent technological trajectory (Dosi 1982), determining the direction of knowledge search over time in the TIS (Carlsson and Stankiewicz 1991, Hekkert and Negro 2009, Hekkert et al. 2011, Bergek et al. 2015). Technological trajectories are directly shaped by historical choices, which are in turn determined by the specific circumstances surrounding the development of a technology. In accordance, the trajectory of the focal technology is dependent on factors in the focal TIS and its context, including territorial circumstances (Jacobsson and Johnson 2000, Markard and Truffer 2008, Hekkert and Negro 2009, Hellsmark et al. 2016).

Besides the prevailing focus on the emergence of sustainable technologies, Markard (2020) proposed the TIS life cycle framework to emphasize the decline of mature TIS as an integral part of successful sustainability transitions. In particular, the decline phase is often induced by changes in the context, e.g., shifts in societal expectations about the future, major economic crises or emerging competing technologies, which can ultimately lead to the death of the focal technology. However, incumbent actors can resist these decline pressures and delay or even interrupt a potential technology phase-out. Markard proposed several dimensions of decline, which are somewhat different from the traditional TIS functions of Hekkert et al. (2007) or Bergek et al. (2008), to track such an adaptation process of a mature TIS. These include changes in the size and actor base, the institutional structure and the technological performance and variation in the TIS over time. Although recent empirical studies started to gain insights into the decline process using the TIS framework (Isoaho and Markard 2020, Markard et al. 2020), we still lack a deep understanding of the individual dimensions, especially considering the spatial development of TIS. Thus, we want to add to this uprising literature by focusing on the technological dimension to investigate how the direction of technological development adapts to increasing decline pressures. In particular, this is reflected by changes in the dominant technological trajectory in the mature TIS, as the incumbent's expectations about novel development paths trigger a knowledge search process to improve the performance of the focal technology (Hekkert and Negro 2009). Besides changes over time, we expect territorial differences in the adaptation process of the dominant technological trajectory considering the spatial development perspective. Accordingly, the territorially dominant trajectories reflect which technological options the incumbents are focusing on to cope with the local decline pressures.

Next, we will describe the internal combustion engine as our empirical case for a mature TIS facing decline pressures.[1] In particular, we focus on the car industry as one of the most relevant industries for ICE development (Breuer and Zima 2017).

6.3 A short history of the internal combustion engine

Since the 1910s, the internal combustion engine has been the dominant technology in serving our mobility needs, especially regarding private and commercial vehicles (Breuer and Zima 2017). However, from 1960 onwards, several governments in the USA, Japan and Europe began to implement the first emission regulations triggered by changes in the context of the ICE-TIS, like public concerns about the adverse effects of emissions on human health and commencing air pollution problems (Pucher et al. 2017). In addition, oil prices increased sharply during the 1970s, which in turn raised the price of fuel as a critical commodity for the ICE-TIS (Furr and Snow 2012, Sushandoyo et al. 2012, Dijk et al. 2013).

In the 1990s, even more far-reaching regulations like the Zero Emission Vehicle (ZEV) program in California, the European Emission Standards in the European Union and the Japan Clean Air Program were implemented because of heightened concerns about climate change and pollution. Besides the continuing increase in fuel prices, governments began to implement supporting policies for electric vehicles and established low-emission zones (Sushandoyo et al. 2012, Dijk et al. 2013, Seiffert et al. 2017). Most recently, we observe a stark increase in the popularity of electric vehicles and a culminating aversion to combustion vehicles as more and more governments are aiming to abandon ICE vehicles completely (Wappelhorst 2020).

The literature suggests three possible technological pathways for the ICE-TIS to cope with these pressures. In particular, besides following the current dominant design of the ICE with (incremental) improvements of diesel and gasoline engines as two possible pathways, many also regard hybrid powertrains as a promising way forward (Pucher et al. 2017, Schäfer et al. 2017). In fact, by using an additional electrical engine, hybridization can reduce fuel consumption by up to 30%, depending on the specific configuration (series, parallel or power split/series-parallel hybrid), the type of ICE used (gasoline or diesel engine) and the degree of hybridization (micro, mild or full hybrid).

Considering these three possible development paths for the ICE, there have been significant differences in the propensity towards a specific powertrain technology across the major car markets (Oltra and Saint Jean 2009, Cames and Helmers 2013, Bohnsack et al. 2015). For example, while diesel vehicles take up an average market share of roughly 40%, their share has been negligible in the USA and Japan with a share of around 1–3% (Oltra and Saint Jean 2009, Cames and

Helmers 2013, Bohnsack et al. 2015). Similarly, hybrid cars experienced relatively strong sales in the USA and Japan compared to the EU market (Sushandoyo et al. 2012, ACEA 2018).

Against this backdrop, it still seems unclear which powertrain technology is chosen by the incumbent manufacturers to facilitate the ICE-TIS to overcome the decline pressures in terms of aggravating regulations and technological competition from alternative powertrains. Moreover, as illustrated by the differences in the propensity towards specific powertrain technologies, there seem to be important territory-specific context circumstances on the national level that influence the adaptation process to the decline pressures. Accordingly, the mature ICE-TIS seems to have developed structural couplings with national circumstances, which influence the spatial development of the technology, e.g., through national regulations, consumer preferences and post-tax fuel prices (Berggren and Magnusson 2012, Dijk et al. 2013, Bohnsack et al. 2015). Thus, although the spatial development perspective is not necessarily confined to a specific spatial scale (Binz et al. 2014), our empirical analysis will specifically focus on national territories. In particular, we consider the USA, EU and Japan as three major national territories in the ICE-TIS not only because of the number of vehicles sold in these markets but also their historical importance for the technological development of the ICE in terms of R&D expenditures and patents granted (Cames and Helmers 2013, Dijk et al. 2013, ACEA 2019).

Our analytical framework is summarized in Figure 6.1. Specifically, the developments in the context of the ICE-TIS entail important environmental and societal trends that affect the USA, the EU and Japan similarly. However, we expect

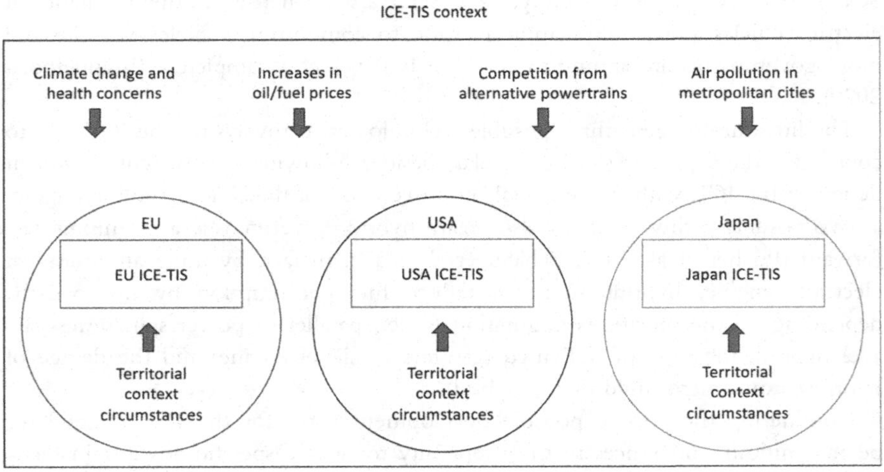

FIGURE 6.1 Analytical framework with three ICE-TIS territories and context circumstances

Source: Taken from Weiss and Scherer (2022)

different changes in the territorial TIS structures as a response to the context changes due to the territorial context circumstances, which, in turn, shape the dominant technological trajectory in the respective territory. Given this stylized framework, we will specifically focus on the differences on the national/supranational level in terms of broader political and historical changes to make our empirical analysis more manageable.

We will use the ICE as our empirical case to gather evidence for our two research questions: First, is there an adaptation of the technological trajectory of the ICE over time due to transformational pressures? Second, do we observe differences in the adaptation of the trajectories across different territories in the ICE-TIS?

6.4 Methodology

In this study, we apply a main path analysis to patent citation networks to reveal the dominating technological trajectories (Verspagen 2007). Patents are used as an innovation throughput indicator to capture technological knowledge, and their associated citations constitute an important source of knowledge for future patents, where the relevant knowledge flows from cited to citing patents (Grupp 1997). However, patent statistics are prone to several drawbacks. In particular, the number of patents and citations are significantly influenced by country, sector or firm specificities, which can bias analysis results and make them incomparable across different domains (Oltra and Saint Jean 2009, Martinelli 2012, Jaffe and Rassenfosse 2017, Noailly and Shestalova 2017). Compared to the usual methods to analyze patents and their citations networks, the main path analysis overcomes these drawbacks, as the procedure is much more focused on the flow of knowledge through the network, which also depends on the number of indirect citations and how the specific patent is embedded in the network structure (Mina et al. 2007, Fontana et al. 2009, Liu et al. 2019).

In general, a patent citation network consists of a set of vertices representing the patents and directed arcs denoting the citations between two patents. They are a representation of directed acyclical graphs in which the network is directed forward in time. Consequently, when following the direction of the citations, no patent can be visited twice (Hummon and Doreian 1989, Verspagen 2007, Batagelj 2003, Liu et al. 2019). In our network, we define patents that have forward citations but no backward citations as start points and the ones that only have backward citations as endpoints (Martinelli 2012).

For main path analysis, a traversal weight is assigned to each citation link, which considers not only the number of citations of a patent but also its position in the over network (Hummon and Doreian 1989, Verspagen 2007, Fontana et al. 2009, Liu et al. 2019). Following Batagelj (2003), we apply the search path count (SPC) for weighting our patent citation network. To identify the dominant trajectories based on the SPC, we rely on the global main path algorithm. The latter selects the citation chain from a start to an endpoint with the highest overall sum of the traversal weight indicator (De Nooy et al. 2018). Most significantly, this path is seen as the critical path of knowledge flow in the network, constituting the dominant technological trajectory and knowledge search process (Verspagen 2007).

To answer our research questions empirically, we employ two methodological procedures based on main path analysis. Firstly, we use the time-based approach that calculates the global main path in different time periods. The procedure used in the literature is to fix the starting year while alternating the ending year of each period (Verspagen 2007, Fontana et al. 2009, Barberá-Tomás et al. 2011, Martinelli 2012, Liu et al. 2019). Secondly, we calculate the dominating trajectories for each territory by using the patents from the national patent offices of the USA and Japan. For the EU we included patents from the EPO and the national patent offices of the member countries. The underlying assumption is that firms only consider filing for protection in countries in which they expect potential market value and development opportunities for the invention (Putnam 1996, Dechezleprêtre et al. 2015).

We extracted granted patents for the ICE-TIS from the DOCDB database provided by the European Patent Office (EPO)[2] by using the IPC codes depicted in Table 6.1. As the IPC codes are well-defined for the ICE, we prefer this method over extracting patents by means of relevant keywords since the latter involves several drawbacks that make an exact delineation more difficult (Oltra and Saint Jean 2009).

To make our dataset more manageable, we only include granted patents to eliminate multiple counting and one patent per patent family, as the latter depicts a group of closely related patents that share basically the same core technology (Grupp 1997, Noailly and Shestalova 2017). After correcting for cyclical components and deleting isolated patents, we end up with 221,700 patents, covering the period from 10 January 1901 until 31 January 2019, with 323,374 forward citations. Next, we extract 123,515 patents with 209,615 forward citations for the USA, 44,977 patents with 40,302 forward citations for the EU,[3] and 51,714 patents with 51,330 forward citations for Japan. Notably, this disparity in the number of patents and associated citations illustrates the differences in the incentives to disclose prior knowledge due to specific national patent laws (Schmoch and Grupp 1990).

TABLE 6.1 IPC codes for internal combustion engines

IPC patent class	*Definition*
F02B	Internal-combustion piston engines, combustion engines in general
F02D	Controlling combustion engines
F02F	Cylinders, pistons or casings for combustion engines, arrangements of sealings in combustion engines
F02M	Supplying combustion engines in general with combustible mixtures or constituents thereof
F02P	Ignition, other than compression ignition, for internal-combustion engines, testing of ignition timing in compression-ignition engines

Taken from Aghion et al. (2016)

For a subsequent interpretation and discussion of our main path analysis results, we will draw on secondary literature on national/supranational political and historical developments specifically related to the ICE in the three TIS territories studied.

6.5 Results

Figure 6.2 depicts the time evolution of the global main path for the USA ICE-TIS, focusing on the three most recent trajectories from 1978 to 2019. The development paths are ordered horizontally according to the periods in which the trajectories were calculated, with the left trajectory being the earliest one. Moreover, patents are ordered according to their granted date on the vertical axis, with the oldest patents at the bottom. Furthermore, to give a more concise picture, we excluded the endpoints of each trajectory since they do not contribute to the SPC weighting (Batagelj 2003, Verspagen 2007). We calculated the first trajectory on the left for the time periods 1978 to 1995, which entails patents about ICE fuel injection and fuel metering systems, focusing on gasoline-manifold fuel injection. In fact, it represents the knowledge search process during the period of changeover from carburetors to more efficient fuel injection systems (Dijk and Yarime 2010, Furr and Snow 2012).

The companies with the most patents in the trajectory, indicating a major partaking in the knowledge search process, are Bosch from Germany, General Motors from the USA and Weber SRL from Italy.

The second trajectory, covering the period from 1996 to 2002, comprises patents about fuel tank filters for gasoline engines, whose improvement became necessary to allow for further emissions reductions through higher fuel injection pressures, shorter injection duration and ethanol fuel (Trautmann 2017). The companies with the most patents in this trajectory are Nifco from Japan as well as General Motors and Kuss Corp from the USA. Notably, these two trajectories represent the historical observation that in the last decades, car emissions were mostly reduced by improvements in fuel injection systems (Pucher et al. 2017).

In contrast, the third trajectory, which prevails from 2003 to the beginning of 2019, is focused on different hybrid powertrain solutions. By and large, this knowledge search process constitutes a notable shift from the previous ones, as they still rely on the dominant ICE design. In fact, the hybrid powertrain trajectory can be regarded as a "median strategy" that constitutes a compromise between fully electric and diesel/gasoline cars (Oltra and Saint Jean 2009). Reflecting the lead of Japanese manufacturers, the hybrid trajectory is dominated by firms like Toyota and Aisin while expelling domestic manufacturers who were present in the prior trajectories (Sushandoyo et al. 2012).

Next, we depict the global main paths for the USA, Japan and EU in Figure 6.3 to examine territorial differences in the trajectories of the ICE-TIS. For the USA, we observe the same hybrid vehicle trajectory as in the previous figure. As expected, we also observe a hybrid-focused knowledge search for Japan in the middle of the figure. Again, the associated patents are dominated by Japanese firms. Although

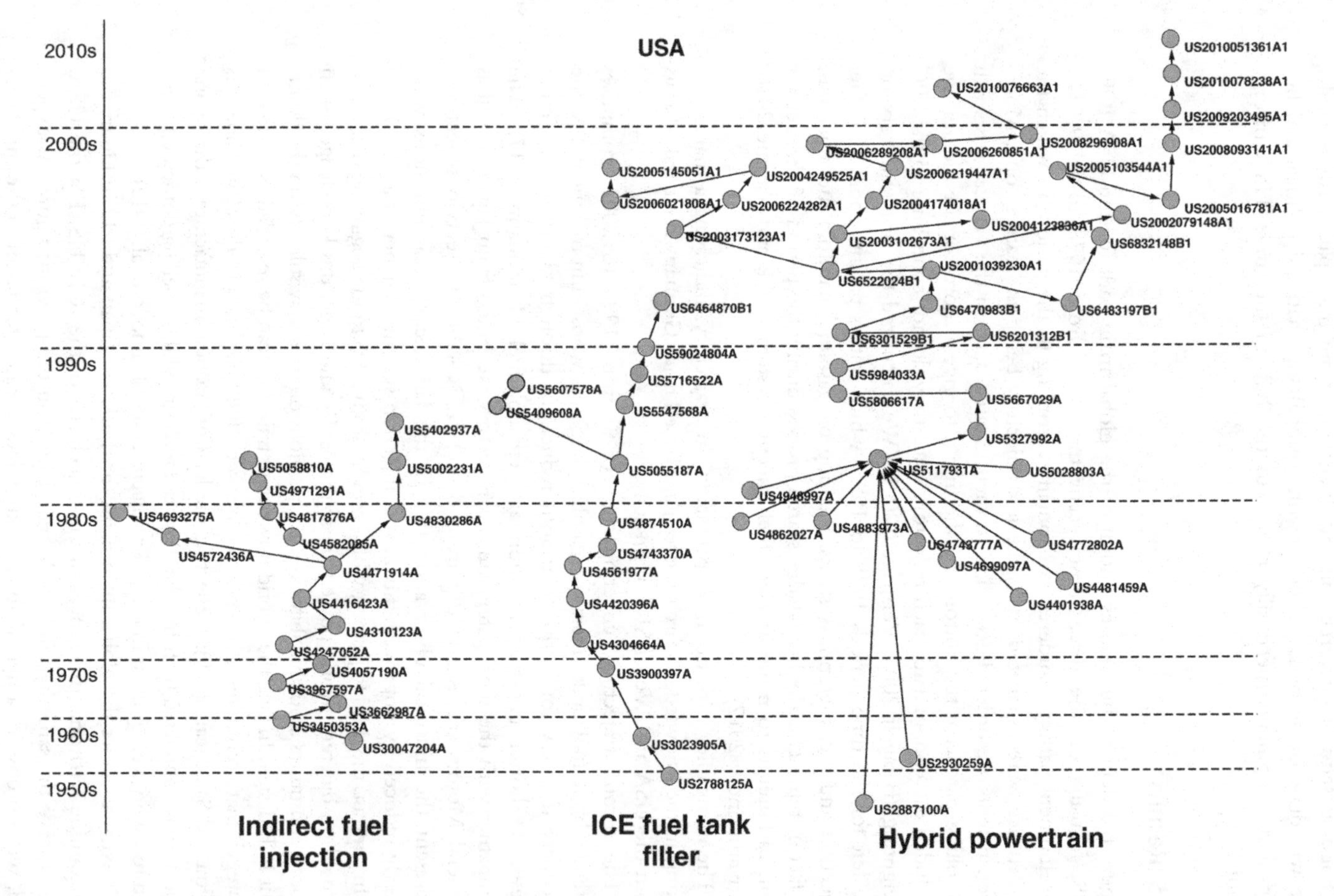

FIGURE 6.2 Dominant technological trajectories of the USA ICE-TIS. The circular nodes represent the patents, and the arrows depict forward citations from the cited to the citing node

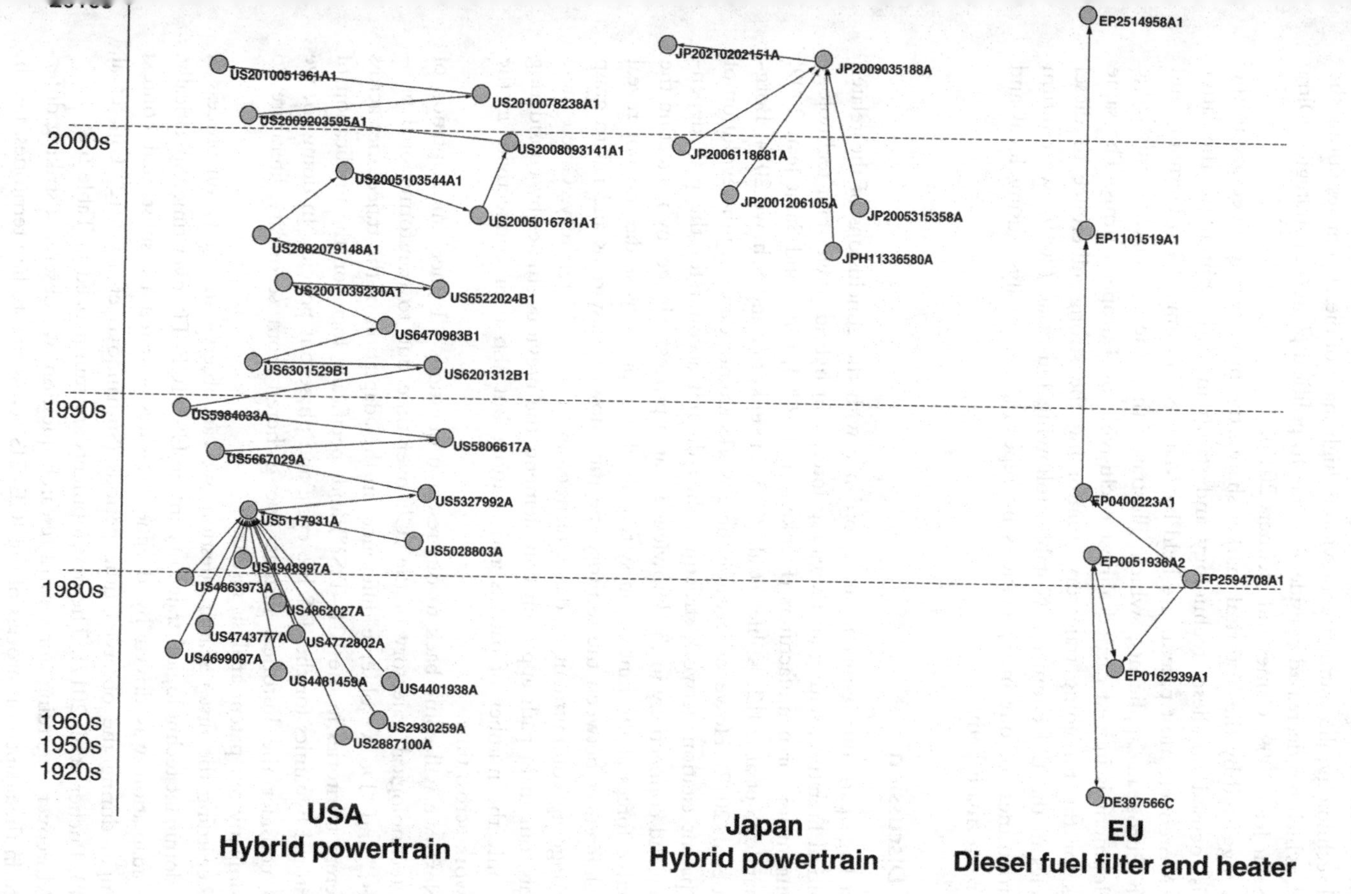

FIGURE 6.3 Dominant technological trajectories across three territories of the ICE-TIS. The circular nodes represent the patents, and the arrows depict forward citations from the cited to the citing node

Taken from Weiss and Scherer (2022)

not shown in the figure, the foregoing trajectory in Japan was concerned with diesel technology, illustrating the previously high share of diesel vehicles during the 1990s, which has decreased considerably due to political phase-out strategies (Oltra and Saint Jean 2009, Cames and Helmers 2013).

As suggested by the historical market shares, the most recent trajectory in the EU is focused on diesel technology and stands in stark contrast to the latter two. Moreover, most patents are held by Mann & Hummel from Germany and other European applicants, which illustrates that the associated knowledge development is indeed carried almost exclusively by European firms. The same holds for the foregoing trajectory, reinforcing the strong adherence to diesel vehicles in the EU countries. Notably, following Liu et al. (2019), we confirm the robustness of our interpretations by repeating our analysis using local and key route main paths.

6.6 Discussion

To a large part, our results are in accordance with the conclusions of the related empirical literature, stating an increasing interest in hybrid powertrain technology among major car manufacturers (Frenken et al. 2004, Oltra and Saint Jean 2009, Sushandoyo et al. 2012, Köhler et al. 2013, Liesenkötter and Schewe 2014, Borgstedt et al. 2017). However, because these works are not capturing the evolution of the patent citation network structure, they do not account for the path dependency and connectivity in the knowledge search process. Hence, by neglecting the selective forces at the innovation system level, the previous studies do not reveal the differences between the territory-specific knowledge systems and fail to grasp the ongoing commitment to diesel technology in the EU territory. Given these points, our main path approach complements the previous studies by considering not only the number of patents and citations but also their connectivity in the network across time.

Next, we will come back to our research questions ("Is there an adaptation of the technological trajectory of the ICE over time due to transformational pressures?" and "Do we observe differences in the adaptation of the trajectories across different territories in the ICE-TIS?") and specifically focus on the implication of territorial dynamics for the TIS life cycle and phase-out policies. In doing so, we also consider the historical insights we gathered from secondary literature to accompany our patent analysis results.

Regarding the first research question, we can observe an adaptation process of the dominant technological trajectory in the USA ICE-TIS over time. Specifically, the adaptation was driven by decline pressures initiated by substantial context changes during the observed time periods (Sushandoyo et al. 2012, Dijk et al. 2013, Pucher et al. 2017). The developments are summarized in Table 6.2.

Moreover, regarding our second research question, we observe a striking difference in the adaptation process of the ICE-TIS across the studied territories, i.e. the EU, Japan and the USA. In line with the TIS spatial development literature, these

TABLE 6.2 Evolution of the dominant technological trajectory of the USA ICE-TIS

Time period	*Trajectory*	*Main objective*	*Changes in TIS*	*Changes in context*
1980–1995	Indirect fuel injection	Changeover from carburetors to fuel injection systems.	First ICE emission regulations (e.g. Clean Air Act).	Beginning concerns about adverse effects of combustion engine emissions on human health, worsening air pollution in metropolitan cities, increase in oil and fuel prices.
1996–2002	ICE fuel tank filter	Increasing requirements for fuel filters due to advancing fuel injection systems.	Extension of emission regulations (e.g. Low-Emission Vehicle Program).	Increasing concerns about adverse effects of combustion engine emissions on the climate, continuing increase in fuel prices, beginning interest in alternative powertrains.
2003–2019	Hybrid powertrain	Reduction of fuel consumption and emissions.	Aggravating emission regulations and implementation of low/zero-emission zones.	Continuing climate change concerns and high fuel prices, uprising competition from fully electric powertrains.

Source: Taken from Weiss and Scherer (2022)

findings illustrate how different territorial context circumstances can have a major impact on the development of territorial TIS structures.

In our case, the decline pressures induced by the context changes provoked distinguished changes in the ICE regulations, which are seen as the major cause of this disparity in technological development. As summarized in Table 6.3, the EU legislation favored the progression of diesel engines and stood in contrast to the emission regulations in the USA and Japan that have been more advantageous to hybrid vehicles. In particular, the USA followed a fuel-neutral approach that offered no incentives for customers to buy diesel cars. Japan even aimed to phase-out diesel cars and began to impose stricter emission standards than for gasoline cars in the early 2000s (Hascic et al. 2009, Díaz et al. 2017). In contrast, diesel cars have been facing less stringent emissions regulations in the EU than gasoline ones, e.g., the latter faced threshold values for nitrogen oxide and hydrocarbons that were 40% higher (Cames and Helmers 2013).

While this constituted a considerable barrier for European manufacturers to export their diesel technology outside of the EU, Japanese manufacturers were able to capitalize on their technological leadership and initiate the diffusion of the

TABLE 6.3 Overview of the contributions

Territory	*Trajectory*	*Territory specific adaptation of TIS structures to context changes*
EU	Diesel fuel filter	Emission regulations put diesel at an advantage over gasoline and hybrid powertrains. National support programs for diesel engines. Lower fuel prices for diesel vehicles.
Japan	Hybrid powertrain	National policy programs favoring hybrid cars and aiming to phase-out diesel cars.
USA	Hybrid powertrain	National policy programs favoring hybrid cars. Fuel-neutral ICE regulation approach that inhibits cost advantage of diesel cars.

Source: Taken from Weiss and Scherer (2022)

hybrid powertrain outside of Japan (Åhman 2006, Pohl and Yarime 2012, Bohnsack et al. 2015). In particular, firms like Toyota and Honda were the first to develop and sell hybrid vehicles due to domestic support programs, which allowed them to exploit first-mover advantages as they entered the US market. This expansion strategy was triggered by the Californian Zero Emission Vehicle regulation, aiming to increase the share of vehicles with zero emissions during operation,[4] including hybrids, which equally affected Japanese manufacturers due to their large market shares in the US market. Subsequently, the massive success of Japanese hybrids like the Toyota Prius I incentivized other manufacturers to develop their own hybrid vehicles in the late 2000s (Pohl and Yarime 2012, Bergek et al. 2013, Dijk et al. 2013). Moreover, the ZEV program triggered an international policy diffusion, as other governments followed suit and implemented more ambitious policies towards sustainable mobility (Bohnsack et al. 2015).

Continuing this trend, we observe more and more national and local governments issuing phase-out targets for ICE vehicles, with the most ambitious being Norway aiming for the year 2025. Simultaneously, we observe increasing support measures for electric vehicles, including hybrid ones. Following this, large carmakers like Volkswagen and Audi began to abandon ICE development to focus solely on the production of electric vehicles (Berggren et al. 2015, Wappelhorst 2020). Corresponding with the trajectories we revealed in the USA and Japan, this is reflected in the decreasing ICE sales, which are mirrored by increasing sales of hybrid and electric cars. In the EU, this is particularly visible in the strong decline in diesel sales from the record high of nearly 6 million cars in 2016 to about 4 million cars in 2019 (ACEA 2021). Notably, this conflicts with the revealed dominant diesel trajectory and suggests that a hybrid trajectory could take over in future patent data. Taking a broader outlook, we could also see a switch to trajectories that are concerned with other ICE applications, e.g., in heavy-duty transportation, ships or trains, where the competition by alternative powertrains is still less fierce (Reitz et al. 2020). In this case, a decline of the ICE could be averted or delayed not only by focusing on a new powertrain solution like the hybrid powertrain, but also by switching to applications or sectors with an ongoing competitive advantage of the ICE, including niche markets like motorsport (Markard 2020).

Although this brief sketch of historical developments drawn from secondary literature is certainly not complete (see Bohnsack et al. 2015 for an extensive review), it allows us to describe several important considerations from our patent main path analysis for the TIS life cycle framework and phase-out policies. Accordingly, our analysis suggests that within the TIS life cycle the adaptation of the technological dimension is influenced by the developments in the TIS and its context during a certain time period and, maybe more importantly, by the spatial dimension. The latter includes not only the formation of territory-specific trajectories due to local technological opportunities but also transnational linkages or spillover effects that lead to industry-wide changes like the market entrance of Japanese hybrids in the USA (Quitzow 2015). This exemplifies that foreign firms can overtake the knowledge search process if there is a particular match between newly implemented policies and firms' technological capabilities. In turn, the entrance into lead markets like California can then initiate a runner-up diffusion of more sustainable technology solutions due to competitive pressure (Bohnsack et al. 2015). This resonates with the notion of automobiles as a production-led global innovation system (Binz and Truffer 2017), where strong regional couplings can create a local hype that leads to international knowledge exchange and exports (Bathelt et al. 2004).

6.7 Conclusion

In conclusion, the presented results suggest that spatially heterogenous TIS life cycles must be considered in future empirical and theoretical TIS research. We illustrated their importance by revealing different dominant technological trajectories across the investigated TIS territories, shaping the local innovation processes and the resulting product offering to consumers. Therefore, TIS territories will likely have different solutions based on the focal technology to resist the decline pressures, also implying different chances of survival. Based on this insight, a technology could be completely abandoned in one territory while still thriving in another, as we observed with diesel technology in Japan and the EU. This insight has to be confirmed and complemented using other indicators and methodologies to investigate technological development across territories like media data (Weiss and Nemeczek 2021), publication data (Binz et al. 2014), regression analysis (Berg et al. 2019) or system modeling (Walrave and Raven 2016). This study also exemplified spillover and feedback loops that suggest that the importance of specific territories for the overall development dynamics depends not only on the market size but also on the interdependence between the territorial context circumstances and firms' technological capabilities. Notably, this is especially important for the decline process of mature TIS, which are characterized by large multinational firms with institutionalized relation to the (territorial) context (Markard 2020). Accordingly, the decline process can be initiated or accelerated by developments in key territories as subsystems of the TIS that pressure important actors into a certain technological trajectory due to local technological opportunities.

From a policy standpoint, this emphasizes that policymakers have to consider different territories as subsystems of a TIS to take account of their local specificities as well as potential spillovers and feedback loops between them. The latter in particular can initiate a change through the whole TIS by means of international policy diffusion or competitive pressure among firms through common value-chains or multiregional actor networks (Bergek et al. 2015, Bohnsack et al. 2015, Binz and Truffer 2017). As the identified trajectories reflect the adaptation efforts in the respective territories, they reveal which of the possible powertrain options the incumbents are focusing on to cope with the local decline pressures. If national and supranational policymakers want to enforce a faster phase-out, they should pursue a place-sensitive approach with specific ICE regulations in line with the territorially dominant trajectories (Rohe and Chlebna 2021). Such an approach should also consider technological leaders and their involvement in certain markets to incentivize change or diffusion of their technological innovation strategy. In turn, progressive policies in important territories of the TIS could possibly actuate the establishment of an international sustainable powertrain trajectory if international policy coordination is not feasible. In the case of the mature ICE-TIS, the phase-out process is characterized by a range of different trajectories incumbents pursue to adapt further and prolong the technology's market dominance. Considering studies on spillover effects between these different powertrain TIS, i.e., the knowledge development in one powertrain is beneficial or detrimental to another, we can assume that the hybrid trajectory has the largest amount of positive spillovers to sustainable powertrains like battery-electric and fuel cell vehicles (Phirouzabadi et al. 2020). Hence, policymakers could abandon diesel and gasoline cars and focus their ICE phase-out strategies exclusively on hybrid as the dominant ICE design, which would not only offer incumbents further market opportunities but also facilitate the development of sustainable powertrains through maximizing positive knowledge spillovers. This could lead to fewer resistance activities of the incumbent actors and promote the overall legitimacy of the policy program.

In line with the geography of sustainability transitions literature (Hansen and Coenen 2015), our study emphasizes that it is essential to regard technology decline in general as a multi-scalar phenomenon. It involves different interdependent scales on the regional, national and supranational level which all exhibit specific circumstances that affect the decline dynamic not only in that particular scale but also in other connected scales. Therefore, as argued by Miörner and Binz (2021), we have to consider how local changes to technological development on one scale can be re-scaled to an overarching scale, e.g., from the regional to the national level. Accordingly, breaking up important actor coalitions or development centers in the most influential regions or nations can trigger a destabilization of the prevailing regime. In the case of ICE, we suggested California and its Zero

Emission Vehicle regulation as an example of a place-based policy lever that could initiate a technological phase-out on the international level. Such multi-scalar interdependencies and feedback loops allow for a more targeted regulation approach and concentrated policy-mixes (Kivimaa and Kern 2016). This follows the idea of a multi-local and multi-scalar territorial innovation policy mix proposed by Jeannerat and Crevoisier (2022), which is concerned with transformative policies on the local territorial level and their interplay with the broader sustainability transformation on global markets.

6.8 Limitations

Considering our theoretical and methodological framework, we want to highlight the following important limitations of our paper. Firstly, to give a full picture of the adaptation process of a mature TIS, the analysis has to consider additional dimensions and indicators put forward by the TIS life cycle framework, e.g., considering the competition from other technologies, legitimacy issues or actual performance increases of combustion cars (Markard 2020; Markard et al. 2020; Markard et al. 2023). Moreover, although we briefly sketched the transnational linkages to expand the discussion of our results, further research has to integrate a complete spatial perspective into the TIS life cycle framework to give us a full understanding of territorial dynamics and feedback loops in the face of technology decline (see for example Wieczorek et al. 2015; Normann and Hanson 2018). This should also include transnational patenting activities of multinational car manufacturers across multi-territorial development clusters (Bohnsack et al. 2015).

Secondly, our research should not be taken as a forecast of the future development of the ICE-TIS. In our empirical study, we used patents as an indicator that only measures inventions that have yet to evolve into innovations successfully (Grupp 1997). Furthermore, we only focused on the most significant main paths in the citations network and pointed to other methods that capture a greater complexity (Verspagen 2007, Tu and Hsu 2016, Kim and Shin 2018). Additionally, we have to consider the accelerating technological and societal changes in the car industry, illustrated by the rise of electric vehicles pioneered by Tesla, the shift of traditional carmakers to a completely electric vehicle strategy, and aggravating ICE regulations around the globe, which arguably could influence the interpretation of our results moving forward (Wappelhorst 2020). For example, an accelerating shift of large car manufacturers to fully electric cars could potentially lead to a shift of the identified hybrid trajectory towards more heavy-duty focused applications like trucks or construction machinery.

Finally, as our study merely related the identified trajectories with historical developments taken from secondary literature, future research needs to establish a causal relationship between the changes in the context, the TIS structure, territorial particularities and the choice of the dominant technological trajectory. This could be accomplished using additional empirical analysis based on survey or interview data. Besides technological regulations, such work has to consider cultural differences and firm characteristics and investigate the linkages between the focal TIS and its context in greater detail (Markard 2020; Markard et al. 2023).

Notes

1 For a more extensive review see for example Berggren and Magnusson (2012), Dijk et al. (2013) or Pucher et al. (2017).
2 The database was accessed through www.patentinspiration.com.
3 For the EU we included patents from the EPO and the national patent offices of the member countries.
4 The policy program neglects emissions and pollutants occurring during the production process or the extraction of necessary resources.

References

Aghion, P. *et al.* (2016) Carbon taxes, path dependency, and directed technical change: Evidence from the auto industry. *Journal of Political Economy*, 124(1), 1–51.

Åhman, M. (2006) Government policy and the development of electric vehicles in Japan. *Energy Policy*, 34(4), 433–443.

Barberá-Tomás, D., Jiménez-Sáez, F. and Castelló-Molina, I. (2011) Mapping the importance of the real world: The validity of connectivity analysis of patent citations networks. *Research Policy*, 40(3), 473–486.

Batagelj, V. (2003) Efficient algorithms for citation network analysis (arXiv preprint cs/ 0309023). https:// arxiv.org / pdf/ cs/ 0309023.pdf.

Bathelt, H., Malmberg, A. and Maskell, P. (2004) Clusters and knowledge: Local buzz, global pipelines and the process of knowledge creation. *Progress in Human Geography*, 28(1), 31–56. doi:10.1191/0309132504ph469oa.

Berg, S., Wustmans, M. and Bröring, S. (2019) Identifying first signals of emerging dominance in a technological innovation system: A novel approach based on patents. *Technological Forecasting and Social Change*, 146, 706–722. doi:10.1016/j.techfore.2018.07.046.

Bergek, A. *et al.* (2008) Analyzing the functional dynamics of technological innovation systems: A scheme of analysis. *Research Policy*, 37(3), 407–429. doi:10.1016/j.respol.2007.12.003.

Bergek, A. *et al.* (2013) Technological discontinuities and the challenge for incumbent firms: Destruction, disruption or creative accumulation? *Research Policy*, 42(6–7), 1210–1224.

Bergek, A. *et al.* (2015) Technological innovation systems in contexts: Conceptualizing contextual structures and interaction dynamics. *Environmental Innovation and Societal Transitions*, 16, 51–64.

Berggren, C. and Magnusson, T. (2012) Reducing automotive emissions: The potentials of combustion engine technologies and the power of policy. *Energy Policy*, 41, 636–643.

Berggren, C., Magnusson, T. and Sushandoyo, D. (2015) Transition pathways revisited: Established firms as multi-level actors in the heavy vehicle industry. *Research Policy*, 44(5), 1017–1028.

Binz, C. and Truffer, B. (2017) Global innovation systems: A conceptual framework for innovation dynamics in transnational contexts. *Research Policy*, 46(7), 1284–1298.

Binz, C., Truffer, B. and Coenen, L. (2014) Why space matters in technological innovation systems: Mapping global knowledge dynamics of membrane bioreactor technology. *Research Policy*, 43(1), 138–155.

Bohnsack, R., Kolk, A. and Pinkse, J. (2015) Catching recurring waves: Low-emission vehicles, international policy developments and firm innovation strategies. *Technological Forecasting and Social Change*, 98, 71–87.

Borgstedt, P., Neyer, B. and Schewe, G. (2017) Paving the road to electric vehicles: A patent analysis of the automotive supply industry. *Journal of Cleaner Production*, 167, 75–87.

Breuer, C. and Zima, S. (2017) Geschichtlicher Rückblick. In van Basshuysen, R. and Schäfer, F. (eds) *Handbuch Verbrennungsmotor: Grundlagen, Komponenten, Systeme, Perspektiven*. Springer Fachmedien.

Cames, M. and Helmers, E. (2013) Critical evaluation of the European diesel car boom: Global comparison, environmental effects and various national strategies. *Environmental Sciences Europe*, 25(1), 15. doi:10.1186/2190-4715-25-15.

Carlsson, B. and Stankiewicz, R. (1991) On the nature, function and composition of technological systems. *Journal of Evolutionary Economics*, 1. doi:10.1007/BF01224915.

De Nooy, W., Mrvar, A. and Batagelj, V. (2018) *Exploratory Social Network Analysis with Pajek: Revised and Expanded Edition for Updated Software*. Cambridge University Press.

Dechezleprêtre, A., Neumayer, E. and Perkins, R. (2015) Environmental regulation and the cross-border diffusion of new technology: Evidence from automobile patents. *Research Policy*, 44(1), 244–257.

Dewald, U. and Fromhold-Eisebith, M. (2015) Trajectories of sustainability transitions in scale-transcending innovation systems: The case of photovoltaics. *Environmental Innovation and Societal Transitions*, 17, 110–125.

Díaz, S. *et al.* (2017) *Shifting Gears: The Effect of a Future Decline in Diesel Market Share on Tailpipe CO2 and NOx Emissions in Europe* (Whitepaper). Available at: https://theicct.org/sites/default/files/publications/Shifting-gears-EU-diesel-futures_ICCT-white-paper_06072017_vF.pdf.

Dijk, M. and Yarime, M. (2010) The emergence of hybrid-electric cars: Innovation path creation through co-evolution of supply and demand. *Technological Forecasting and Social Change*, 77(8), 1371–1390. doi:10.1016/j.techfore.2010.05.001.

Dijk, M., Orsato, R.J. and Kemp, R. (2013) The emergence of an electric mobility trajectory. *Energy Policy*, 52, 135–145.

Dosi, G. (1982) Technological paradigms and technological trajectories: A suggested interpretation of the determinants and directions of technical change. *Research Policy*, 11(3), 147–162.

European Automobile Manufacturers' Association (ACEA) (2018) *Vehicles in Use Europe 2018*. www.acea.auto/publication/report-vehicles-in-use-europe-2018/.

European Automobile Manufacturers' Association (ACEA) (2019) *The Automobile Industry Pocket Guide 2019–2020*. www.acea.auto/files/ACEA_Pocket_Guide_2019-2020.pdf.

European Automobile Manufacturers' Association (ACEA) (2021) *Vehicles in Use Report January 2021*. www.acea.auto/files/report-vehicles-in-use-europe-january-2021-1.pdf.

Fontana, R., Nuvolari, A. and Verspagen, B. (2009) Mapping technological trajectories as patent citation networks: An application to data communication standards. *Economics of Innovation and New Technology*, 18(4), 311–336.

Frenken, K., Hekkert, M. and Godfroij, P. (2004) R&D portfolios in environmentally friendly automotive propulsion: Variety, competition and policy implications. *Technological Forecasting and Social Change*, 71(5), 485–507.

Furr, N. and Snow, D. (2012) *Last Gasp or Crossing the Chasm? The Case of the Carburetor Technological Discontinuity*. Social Science and Technology Seminar, Fall, Stanford, CA. BYU working paper.

Grupp, H. (1997) *Messung und Erklärung des Technischen Wandels: Grundzüge einer empirischen Innovationsökonomik*. Springer.

Hansen, T. and Coenen, L. (2015) The geography of sustainability transitions: Review, synthesis and reflections on an emergent research field. *Environmental Innovation and Societal Transitions*, 17, 92–109. doi:10.1016/j.eist.2014.11.001.

Hascic, I. *et al.* (2009) Effects of environmental policy on the type of innovation: The case of automotive emission-control technologies. *OECD Journal: Economic Studies*, 2009(1), 1–18.

Hekkert, M.P. and Negro, S.O. (2009) Functions of innovation systems as a framework to understand sustainable technological change: Empirical evidence for earlier claims. *Technological Forecasting and Social Change*, 76(4), 584–594. doi:10.1016/j.techfore.2008.04.013.

Hekkert, M.P. *et al.* (2007) Functions of innovation systems: A new approach for analysing technological change. *Technological Forecasting and Social Change*, 74(4), 413–432. doi:10.1016/j.techfore.2006.03.002.

Hekkert, M.P. et al. (2011) *Technological Innovation System Analysis: A Manual for Analysts*. Utrecht University.

Hellsmark, H. *et al.* (2016) Innovation system strengths and weaknesses in progressing sustainable technology: The case of Swedish biorefinery development. *Journal of Cleaner Production*, 131, 702–715.

Hummon, N.P. and Doreian, P. (1989) Connectivity in a citation network: The development of DNA theory. *Social Networks*, 11(1), 39–63.

Islam, N. and Miyazaki, K. (2009) Nanotechnology innovation system: Understanding hidden dynamics of nanoscience fusion trajectories. *Technological Forecasting and Social Change*, 76(1), 128–140.

Isoaho, K. and Markard, J. (2020) The politics of technology decline: Discursive struggles over coal phase-out in the UK. *Review of Policy Research*, 37(3), 342–368. doi:10.1111/ropr.12370.

Jacobsson, S. and Johnson, A. (2000) The diffusion of renewable energy technology: An analytical framework and key issues for research. *Energy Policy*, 28(9), 625–640.

Jaffe, A.B. and Rassenfosse, G. (2017) Patent citation data in social science research: Overview and best practices. *Journal of the Association for Information Science and Technology*, 68(6), 1360–1374.

Jeannerat, H. and Crevoisier, O. (2022) From competitiveness to territorial value: Transformative territorial innovation policies and anchoring milieus. *European Planning Studies*. doi:10.1080/09654313.2022.2042208.

Kim, J. and Shin, J. (2018) Mapping extended technological trajectories: Integration of main path, derivative paths, and technology junctures. *Scientometrics*, 116(3), 1439–1459.

Kivimaa, P. and Kern, F. (2016) Creative destruction or mere niche support? Innovation policy mixes for sustainability transitions. *Research Policy*, 45(1), 205–217. doi:10.1016/j.respol.2015.09.008.

Köhler, J. *et al.* (2013) Leaving fossil fuels behind? An innovation system analysis of low carbon cars. *Journal of Cleaner Production*, 48, 176–186. doi:10.1016/j.jclepro.2012.09.042.

Liesenkötter, B. and Schewe, G. (2014) *E-Mobility: Zum Sailing-Ship-Effect in der Automobilindustrie*. Springer.

Liu, J.S., Lu, L.Y.Y. and Ho, M.H.-C. (2019) A few notes on main path analysis. *Scientometrics*, 119(1), 379–391.

Markard, J. (2020) The life cycle of technological innovation systems. *Technological Forecasting and Social Change*, 153, 119407.

Markard, J. and Truffer, B. (2008) Technological innovation systems and the multi-level perspective: Towards an integrated framework. *Research Policy*, 37(4), 596–615.

Markard, J. *et al.* (2020) Destined for decline? Examining nuclear energy from a technological innovation systems perspective. *Energy Research & Social Science*, 67, 101512.

Markard, J., Isoaho, K. and Widdel, L. (2023) Discourses around decline: Comparing the debates on coal phase-out in the UK, Germany and Finland. In Koretsky, Z. *et al.* (eds) *Technologies in Decline: Socio-Technical Approaches to Discontinuation and Destabilisation*. Routledge.

Martinelli, A. (2012) An emerging paradigm or just another trajectory? Understanding the nature of technological changes using engineering heuristics in the telecommunications switching industry. *Research Policy*, 41(2), 414–429.

Mina, A. *et al.* (2007) Mapping evolutionary trajectories: Applications to the growth and transformation of medical knowledge. *Research Policy*, 36(5), 789–806.

Miörner, J. and Binz, C. (2021) Towards a multi-scalar perspective on transition trajectories. *Environmental Innovation and Societal Transitions*, 40, 172–188. doi:10.1016/j.eist.2021.06.004.

Negro, S.O., Hekkert, M.P. and Smits, R.E. (2008) Stimulating renewable energy technologies by innovation policy. *Science and Public Policy*, 35(6), 403–416.

Noailly, J. and Shestalova, V. (2017) Knowledge spillovers from renewable energy technologies: Lessons from patent citations. *Environmental Innovation and Societal Transitions*, 22, 1–14.

Normann, H.E. and Hanson, J. (2018) The role of domestic markets in international technological innovation systems. *Industry and Innovation*, 25(5), 482–504.

Oltra, V. and Saint Jean, M. (2009) Variety of technological trajectories in low emission vehicles (LEVs): A patent data analysis, *Journal of Cleaner Production*, 17(2), 201–213.

Phirouzabadi, A.M. *et al.* (2020) The evolution of dynamic interactions between the knowledge development of powertrain systems. *Transport Policy*, 93, 1–16.

Pohl, H. and Yarime, M. (2012) Integrating innovation system and management concepts: The development of electric and hybrid electric vehicles in Japan. *Technological Forecasting and Social Change*, 79(8), 1431–1446.

Pucher, E. *et al.* (2017) Abgasemissionen. In van Basshuysen, R. and Schäfer, F. (eds) *Handbuch Verbrennungsmotor: Grundlagen, Komponenten, Systeme, Perspektiven*. Springer Fachmedien.

Putnam, J.D. (1996) *The Value of International Patent Rights* (PhD thesis, Department of Economics, Yale University).

Quitzow, R. (2015) Dynamics of a policy-driven market: The co-evolution of technological innovation systems for solar photovoltaics in China and Germany. *Environmental Innovation and Societal Transitions*, 17, 126–148. doi:10.1016/j.eist.2014.12.002.

Reitz, R.D. *et al.* (2020) IJER editorial: The future of the internal combustion engine. *International Journal of Engine Research*, 21(1), 3–10. doi:10.1177/1468087419877990.

Rohe, S. and Chlebna, C. (2021) A spatial perspective on the legitimacy of a technological innovation system: Regional differences in onshore wind energy. *Energy Policy*, 151, 112193. doi:10.1016/j.enpol.2021.112193.

Schäfer, F. *et al.* (2017) Hybridantriebe. In van Basshuysen, R. and Schäfer, F. (eds) *Handbuch Verbrennungsmotor: Grundlagen, Komponenten, Systeme, Perspektiven*. Springer Fachmedien.

Schmoch, U. and Grupp, H. (1990) *Wettbewerbsvorsprung durch Patentinformation-Handbuch für die Recherchepraxis*. TÜV Rheinland GmbH.

Seiffert, U. *et al.* (2017) Alternative Fahrzeugantriebe und APUs (auxiliary power units). In van Basshuysen, R. and Schäfer, F. (eds) *Handbuch Verbrennungsmotor: Grundlagen, Komponenten, Systeme, Perspektiven*. Springer Fachmedien.

Sushandoyo, D., Magnusson, T. and Berggren, C. (2012) "Sailing ship effects" in the global automotive industry? Competition between "new" and "old" technologies in the race for sustainable solutions. In Calabrese, G. (ed.) *The Greening of the Automotive Industry*. Palgrave Macmillan.

Trautmann, P. (2017) Filtration von Betriebsstoffen. In van Basshuysen, R. and Schäfer, F. (eds) *Handbuch Verbrennungsmotor: Grundlagen, Komponenten, Systeme, Perspektiven*. Springer Fachmedien.

Tu, Y.-N. and Hsu, S.-L. (2016) Constructing conceptual trajectory maps to trace the development of research fields. *Journal of the Association for Information Science and Technology*, 67(8), 2016–2031.

Verspagen, B. (2007) Mapping technological trajectories as patent citation networks: A study on the history of fuel cell research. *Advances in Complex Systems*, 10(1), 93–115.

Walrave, B. and Raven, R. (2016) Modelling the dynamics of technological innovation systems. *Research Policy*, 45(9), 1833–1844.

Wappelhorst, S. (2020) *The End of the Road? An Overview of Combustion-Engine Car Phase-Out Announcements across Europe*. Briefing. International Council on Clean Transportation.

https://theicct.org/wp-content/uploads/2021/06/Combustion-engine-phase-out-briefing-may11.2020.pdf.

Weiss, D. and Nemeczek, F. (2021) A text-based monitoring tool for the legitimacy and guidance of technological innovation systems. *Technology in Society*, 66, 101686. doi:10.1016/j.techsoc.2021.101686.

Weiss, D. and Scherer, P. (2022) Mapping the territorial adaptation of technological innovation systems—Trajectories of the internal combustion engine. *Sustainability*, 14(1), 113. doi:10.3390/su14010113.

Wieczorek, A.J. *et al.* (2015) Broadening the national focus in technological innovation system analysis: The case of offshore wind. *Environmental Innovation and Societal Transitions*, 14, 128–148.

7

THE ROLE OF ALTERNATIVE TECHNOLOGIES IN THE ENACTMENT OF (DIS)CONTINUITIES

Frédéric Goulet

It would almost seem that the decline of agricultural pesticides is now a foregone conclusion. Identified as problematic technologies for decades (Carson 1962; Gunter and Harris 1998), scientific evidence now abounds to demonstrate their effects on the health of agricultural workers (Evangelakaki et al. 2020) and rural populations (Dereumeaux et al. 2020), on populations of pollinating insects (Durant 2019) or on quality of water and soil (Pelosi et al. 2021). Collective movements in rural areas (Arancibia 2013) have also played an essential role in increasing the visibility of problems associated with the use of these technologies. In response to this evidence, public policies in many countries are becoming increasingly stringent in regulating their registration and use, reflecting a willingness to engage pesticides and their regime in a discontinuation dynamic (Stegmaier et al. 2014). In parallel, the burgeoning market for "pesticide-free" organic agriculture seems, even if it is still mostly a niche, to leave no doubt about the growing preference of consumers.

But clearly, all this is not enough. The global market for pesticides is also constantly growing, even in industrialised countries (Shattuck 2021). And the forecasts in this area are rather pleasing for the industry, which is constantly claiming the interest of these technologies to face major challenges such as world hunger or demographic growth (Fouilleux et al. 2017). Molecules are indeed regularly withdrawn from the market, but only to be replaced by molecules with similar functions, as is the case in the pharmaceutical sector (Kessel 2022). These elements contribute to making pesticides a set of technologies torn between a desired decline and a still promising future. Of course, pesticides are not the only technology in this situation, and it is probably one of the contributions of transition studies to have shown that changes rarely occur suddenly (Geels and Schot 2007). Work on technology life cycles is consistent with this (Taylor and Taylor 2012). It shows that decline is a long-term process (see Newman 2023), and that technologies that are set to decline can coexist for some time with emerging technologies that are set to replace them, and sometimes even hybridise (Pistorius and Utterback 1997).

DOI: 10.4324/9781003213642-7

It is precisely this relationship between incumbent and alternative technologies that I propose to address in this chapter. By alternative technologies, I mean technologies that fulfil the same purposes or functions as dominant technologies, but from different mechanisms, components or entities, which do not present the same problematic effects as dominant technologies. In the context of this book's reflection on the decline of technologies, I propose to address the following question: How does the emergence of alternative technologies contribute to the decline of incumbent technologies? The answer might seem intuitively obvious: a technology deeply embedded in an incumbent regime can only be interrupted when alternatives make it possible to replace it (see also discussion in Koretsky 2023). At least, if we stick to the case of pesticides, this is how politicians and public decision-makers tend to put things. In the high-profile case of glyphosate, for example, French President Emmanuel Macron declared in 2017 that glyphosate would be banned in France "as soon as alternatives have been found". The development of alternatives would thus be a decisive element, an obligatory point of passage (Callon 1986b) on the path to decline. But the reverse is not necessarily true: technological alternatives can exist in a niche for a long time, without managing to disrupt the existing regime and its dominant technologies. Michel Callon showed us in the 1980s how electric vehicle technology failed to impose itself against the combustion engine (Callon 1986a). And if we widen the focus beyond technologies, the replacement of one technology by another does not in any way presume the wider transformation of the socio-technical regime in which the problematic technologies are inserted. This is what Levain et al. have shown, still in the context of pesticides, with the ban on DDT (Levain et al. 2015).

The role of alternative technologies and their emergence in the decline of problematic technologies is therefore complex. In this chapter I propose to contribute to this reflection by focusing on a specific stage in the trajectory of alternative technologies: their expansion. By expansion I mean the stage in which alternative technologies are officially supported by public policies and increasingly important industrial actors. This is therefore not the stage of design or prototyping, but rather the stage of expanding existing technologies, even if they obviously continue to be the subject of R&D and innovation. In this spirit, I will focus on biopesticides, also known as biocontrol technologies (Box 7.1), which have been around for several decades but have undergone intensified development in recent years. I will focus on the cases of Argentina and Brazil. These two Latin American giants are distinguished first of all by their unwavering support for the agro-industrial sector, which is a major consumer of pesticides and which, thanks to the exports it generates, is essential to the fiscal revenues of the States. However, in recent years they have also seen the implementation of public policies to support innovation in biocontrol, and an undeniably dynamic market for these biological alternatives. By analysing how the state manages these emerging technologies, and how the landscape of the agricultural pesticide industry has been reconfigured to develop them, I show that the expansion of alternative technologies can support both the decline and the permanence of problematic technologies (in this case chemical pesticides). I thus draw attention to the interest of following the processes of expansion of alternative technologies to better understand the processes of decline and the dynamics of (dis)continuities.

7.1 Biocontrol: Disruptive technologies and historical trajectory

Biocontrol is a breakthrough alternative technology compared to synthetic pesticides. Both technologies aim to eliminate crop pests, mainly through insecticide or fungicide functions. However, the action of pesticides is based on molecules derived from chemical synthesis, whereas biocontrol is based on the use of living beings or substances of biological origin (see Box 7.1). Beyond this composition, the break between the two families of technologies occurs on various levels. First of all, the effects on pests are contrasting: chemical pesticides cause a rapid death of almost the entire pest population, whereas the effects of biocontrol products are only observable after a longer period of time, and with often less radical lethal effects. From a practical point of view, pesticides are most often applied by spraying solutions containing the active ingredients, either on the soil or on plants. Biocontrol technologies can also be sprayed, but are also often used by releasing them into cultivated fields. This is how, for example, the control of the corn borer works with trichogramma: trichogramma are tiny insects which, once present in infested plots, lay their eggs in those of the corn borer, a parasitic moth, and thus prevent them from hatching. Finally, pesticides and biocontrol differ logistically. Chemicals are stable materials that can be used under most climatic and biological conditions and transported or stored over long distances. This is not the case for biocontrol. These are living materials, which are by definition fragile, and which are moreover sensitive to the ecological conditions of their use. While some insects or micro-organisms may be effective in certain climates and against certain strains of pests, their effectiveness is not universal. In this sense, biocontrol and its use are based on more localised agricultural practices, adapted to agricultural production regions and their constraints.

BIOCONTROL TECHNOLOGIES

Biological control, or **biocontrol**, refers to a set of biological techniques used in plant protection to control plant pests (micro-organisms, insects, mites, nematodes, etc.). Biological control agents are generally divided into four main categories:

- Invertebrate beneficial macro-organisms, such as insects and mites.
- Micro-organisms (fungi, bacteria, viruses) used to protect crops against pests and diseases or to promote plant vitality.
- Chemical mediators, especially including insect pheromones, which help to control insect populations through sexual confusion methods or by attracting pests to traps.
- Natural substances from plants, animals or minerals.

Biological control technologies have been used and developed for a long time, since the end of the 19th century with the first work on auxiliary insects (Kogan 1998). Organic agriculture, which developed throughout the 20th century, is of course a particularly important sector for biocontrol. The same is true of the agroecological movements in the last quarter of the 20th century, particularly in Latin America, which support the development of these technologies and advocate a holistic approach to plant health, breaking with the agro-industrial model.

However, biocontrol is also developing in conventional agriculture, in crops such as grains, fruits and vegetables (Bonnaud and Anzalone 2021). Rather than competing with pesticides, it is rather an 'integrated pest management' approach (Flint and van den Bosch 1981) that has been gaining ground since the 1970s, mainly supported on the scientific front by entomologists. This integrated approach defends the idea that farmers should be able to use the full range of technical solutions available to protect their crops, without discrimination. Biological solutions, essentially based on macro-organisms, can thus coexist in certain plots with the use of pesticides. However, just as alternative agriculture such as organic agriculture cultivates its differentiation from conventional agriculture (Lehtimäki 2019), the chemical control and biological control industrial sectors were originally two very separate entities. From the 1970s onwards, there were large agrochemical companies developing molecules used throughout the world, and small and medium-sized companies, based locally, developing and marketing biocontrol products. The latter were only organised on a global scale very recently, compared to the agricultural chemical industry, with the creation of the International Biocontrol Manufacturer Association in 1995.

The development of biocontrol is therefore taking place in a fractured and polarised agricultural landscape with regard to pesticides, between alternative farmers who categorically reject chemical inputs, and a conventional sector that uses biological inputs without abandoning chemical control. For the latter, which is evolving under increasing pressure in industrialised countries, biocontrol has nevertheless recently become an increasingly attractive solution. It has been expanding rapidly since the end of the 20th century, as evidenced by the increasing number of studies examining the ways in which farmers adopt these technologies (Villemaine et al. 2021), or the ways in which they are authorised by the public authorities (Kvakkestad et al. 2020). But the modalities of this expansion quickly triggered warnings from the pioneers of biological control or agroecology. The latter called into question the idea of a pure and simple substitution of chemical inputs by biological inputs (Lockwood 1997), and favoured an in-depth transformation of industrial production systems (Rosset and Altieri 1997). As early as the late 1990s, they also warned against the growing investment of the input industry in biocontrol, calling for a reappropriation of these technologies by farmers to emancipate themselves from the industry (Altieri et al. 1997). But despite these warnings, the biotechnology sector, both public research and private companies (Schwindenhammer 2020), has become dominant in the field of biological control since the 1990s. Biological control research has had its ups and downs since its origins (Warner et al. 2011), but the turning point at this time is major. While innovations and uses of macro-organisms remain, a new wave of microbiology and biotechnology has profoundly reconfigured the research and development fronts of biocontrol. Bacteria or microscopic fungi selected and cultivated in the laboratory have become the new pillars of biological control. For example, in Brazil, the main South American biocontrol market, the registration figures for new products are unequivocal. Between 2000 and 2020, more than 60% of new biocontrol products registered were based on micro-organisms, compared to 17.6% on macro-organisms (the rest being chemical mediators and other natural substances). And this figure is only an average, as this proportion has increased drastically over this period, reaching 80% in 2020.

In the course of this expansion, biocontrol technologies have become an increasingly broad set of propositions. They constitute a diverse set of "configurations that work" (Rip and Kemp 1998), combining macro/micro-organisms and agricultural practices more or less compatible with the foundations, values and socio-technical systems of agroecology or industrial agriculture. But how has this expansion of biocontrol, and more precisely this microbiological and biotechnological turn, contributed to the evolution of the position of pesticides in the field of plant health? Has this development of biocontrol gone hand in hand with a desire to reduce the use of pesticides? Argentina and Brazil provide some answers to these questions. More specifically, the public policies implemented by the states to support biocontrol, as well as the reorganisation of the agricultural inputs sector, offer privileged observation sites for the transformations at work.

7.2 Public policies to manage technological coexistence

Biocontrol products have been used for several decades by farmers in Argentina and Brazil, as described in the previous section. They were first used by small farmers, often based on indigenous knowledge. They are promoted by many NGOs advocating agroecology to rural communities. They are also used by many agricultural producers using modern production techniques, whether in organic or conventional agriculture. This is the case, for example, of Argentinean fruit producers in Patagonia, major exporters to North American and European markets, who have developed the use of biocontrol in order to comply with the low pesticide residue levels required by these countries. But in parallel with these biocontrol developments, both Argentina and Brazil have developed an export sector of agricultural commodities, particularly soybeans, which has been extremely pesticide-intensive since the mid-1990s. The consumption of these chemical inputs has thus soared since this period, generating major local debates on how to reduce the associated damage. It is in this complex context that, from the mid-2010s onwards, Argentina and then Brazil have developed public policies aimed at supporting the development of biocontrol for all farmers, including those in the agro-industrial regime. The shaping of a new category, called bioinputs, played a central role in these processes.

7.2.1 The dilution of biocontrol in bioinputs

The emergence of a policy dedicated to the promotion of biocontrol occurred in 2013 in Argentina, and in 2019 in Brazil. Argentina is the first country in the region to develop such a policy, and is accompanied at the time by IICA (Inter-American Institute for Cooperation on Agriculture), an international organisation whose mission is to provide support for public agricultural policies. IICA notably encourages the construction of policies favouring sustainable agriculture or the development of a bioeconomy on a regional scale. It is within this framework that biocontrol is being promoted as an option that would enable Latin American agriculture to be transformed, while ensuring its competitiveness in world markets that are increasingly demanding with regard to pesticide residues.

But biocontrol is then captured by the Argentine government in a broader category, bioinputs, which also includes biofertilisers. The latter, which are alternatives to chemical fertilisers, refer to a heterogeneous set of technologies ranging from plant residue composts, to manures, or to advanced technologies such as bioinnoculants. Let us focus on the latter, as they played a key role in the way the category of bioinputs has been set up, and how biocontrol products were approached by the state. Bioinoculants are technologies based on micro-organisms, rhizobium bacteria. These bacteria exist naturally in soils and develop symbioses with the roots of plants of the legume family such as soybeans. This symbiosis allows the plants to take up nitrogen from the soil. This mechanism is important because nitrogen is an essential element for plant growth, and most fertiliser applications are aimed at this chemical element. The presence of rhizobia thus makes it possible to reduce the use of synthetic fertilisers necessary for plant growth. Laboratory work carried out since the 1980s has made it possible to isolate, select and improve these bacteria, and to fix them to the surface of seeds implanted in the soil. Plants inoculated in this way require less nitrogen fertiliser during their life cycle. The 1990s saw a massive expansion in soybean cultivation in Argentina and Brazil. The market for bioinnoculants literally boomed from this period onwards, becoming a flourishing industry for national biotechnology companies. In a few years, small national companies have become major entities, exporting to international markets and investing heavily in R&D, particularly in partnership with public microbiology laboratories.

In the mid-2010s, when the Argentine government wanted to encourage biocontrol, these bioinnoculants and their success story were a key reference in the field of biological inputs. But the link between bioinnoculants and biocontrol is made all the more easily since the latter has, as I have mentioned, taken a microbiological turn. While it was originally essentially a matter for entomologists, biocontrol is gradually becoming a research front for microbiologists and biotechnologists. This rapprochement is clearly taking place in Brazil at the end of the 2010s. Embrapa, the public agricultural research organisation, has traditionally been organised into thematic portfolios that structure its major research areas. Until 2019, there was a portfolio dedicated to biological control, and another dedicated to biological nitrogen fixation, i.e., bioinnoculants. That same year, the two portfolios were merged into a single "biological inputs" portfolio. As a microbiologist specialising in biocontrol points out, this merger seemed obvious in the end, given that both research groups work on a microscopic scale and depend on common banks of micro-organisms:

> EMBRAPA has collections of microorganisms, so we are organizing ourselves as a biological resources centre and the activities are very similar (between N fixation and biological control)… It was easier to work together to exchange more information and so on, we more or less have the same line of work, the same skeleton of work, we have the collection.
>
> *(Interview, Brasília, 11/04/2019)*

Supported by a transformation of the scientific field, the notion of bioinput is therefore making its way onto the political agenda, in an integrated approach to biological inputs combining bioinnoculants and biocontrol. This framing makes it possible to support the development of biocontrol while relying on the existence and reputation of bioinnoculants, for which an industry and public policy support are already well developed. This approach allows biological alternatives to pesticides, and more broadly to chemical inputs, to be considered as compatible—and in no way as a threat or explicit hindrance—with the industrial agricultural regime and its technologies. Bioinoculants, a product of biotechnology in the same way as GMOs, which are essential to the agroindustry, are one of the pillars of this model. In this sense, the way in which the State takes charge of biocontrol contributes to diluting the latter within a set of technologies that is not defined, on the contrary, by its alternative character to the intensive agricultural regime using pesticides.

7.2.2 Institutional location and political framing

In both Argentina and Brazil, with the support of IICA, the concept of bioinputs gained increasing visibility in the mid- to late 2010s. Argentina set up the Cabua (Argentinean Commission of Agricultural Bioinputs), an institutional space that brings together the various stakeholders in organic inputs, and Brazil developed the National Bioinputs Programme. The institutional location of these bioinput initiatives confirms the compatibility of this process with the technologies and institutions of the agro-industrial regime. As mentioned earlier, biocontrol has long been promoted within the Argentine Ministry of Agriculture, but by the department in charge of family farming and agroecology, which is radically opposed to the agroindustrial model. In Brazil, the same is true. Biocontrol is historically the preserve of the Planapo (National Plan for Agroecology and Organic Production), developed and coordinated by the MDA, the ministry in charge of agrarian development, small producers and agroecology. However, it is not the historical promoters of biocontrol and bioinputs who have been mandated to ensure their expansion. In fact, Cabua was created in Argentina within the Biotechnology

Division, and is dependent on Conabia, the Argentine biotechnology cenacle. From this point of view, Argentina is making a choice that is in continuity with the precursory action of IICA, which promoted bioinsumos from its Biotechnology Area. In Brazil, the National Bioinputs Programme is being created within the Secretariat for Innovation, Rural Development, and Irrigation, an entity created in 2019 within the Ministry of Agriculture. This positioning comes after the Ministry and the services working in favour of agroecology and family farming were dissolved following the impeachment of Dilma Roussef and the arrival of the right wing in power in 2016.

The anchoring of biocontrol in the institutional landscape in Argentina and Brazil thus contributes to making these technologies compatible with the dominant agroindustrial regime. The officials in charge of these commissions or programmes clearly defend this compatibility, stating that it is not a matter of the state defending a clean break from pesticides. Rather, it is a question of a slow transition, based on a diversification of available technologies, which will coexist to ensure plant health without compromising production. The Argentinian head of the Biotechnology Division thus mentions the fact that adopting a more radical point of view, in terms of technological breakthrough, could even be counterproductive for biocontrol:

> We must avoid thinking of it as a clash of technologies, as a substitution of technologies. We prefer to talk about the fact that bio-inputs can complement classical phytosanitary products and so on. Because if you say that it is a panacea and that chemicals are the disaster of the world … Manichaeism can work against you … we must avoid the duality between bioinputs and chemicals. Don't make (biocontrol) an exaggerated banner because you will gain many enemies who will find counterexamples of your Manichaeism.
>
> *(Interview, Buenos Aires, 11/10/2017)*

The chosen position is thus to "promote the use" of biological alternatives in a more neutral way, not to take sides with organic over chemical, and above all to dissociate technological questions from "political" or "ideological" questions. The latter would indeed risk making the promoters of biocontrol lose their objectivity, as says an official from the Environment Division of the Argentinean Ministry of Agriculture:

> We are not in favour of bio-inputs, we are not in favour of agrochemicals, we are in favour of sustainability. We are in favour of sustainability. We are in favour of continuing production… If you mix up the discussions, it generates a debate that does not end up moving forward. It ends up being a technical, ideological, political dispute, where models are confronted that in reality do not necessarily have to be confronted. But from the point of view of the development of the nation, they have to be compatible because there is room for everyone.
>
> *(Interview, Buenos Aires, 12/12/2016)*

In Brazil, a similar tone is used, seeking to bring the promotion of biocontrol and bioinputs back to higher, non-political or partisan issues. The Secretary of State for Family Agriculture in the Bolsonaro government thus justifies the National Bioinputs Programme as being institutionally supported under the banner of "innovation", understood here as being associated with more economic issues:

> It will be led by the innovation secretariat and, in our opinion, it is a relevant theme economically and in generating income for many communities, small producers.
>
> *(Communication event, Brasília, 09/05/2019)*

The first coordinator of the national programme also insists on this relationship with the notion of innovation, which acts as an umbrella term (Rip and Vos 2013) allowing for the inclusion of diverse and sometimes even divergent dynamics:

> Bio-inputs enters this logic of innovation, of seeing what possibilities there would be for these inputs that are still considered as if they were alternative inputs, and how these inputs now have the potential to actually innovate in the agricultural sector, gaining scale …, thus involving all these possibilities, all these potentials.
>
> *(Interview, Buenos Aires, 18/05/2015)*

Another umbrella term commonly used to situate bioinput initiatives is bioeconomy. At the time of the launch of the National Bioinputs Programme in 2020, the Brazilian Minister of Agriculture made bioinputs one of the flagship initiatives of this bioeconomy:[1]

> (The bioinputs programme) means we are entering the often-cited bioeconomy. The bioeconomy Brazilian agriculture is actually joining is based on what we expect from the agriculture of the future.
>
> *(Interview, Brasília, 11/04/2019)*

This desire to bring together and to depoliticise the different technological options available also applies to the public targeted by the development of bioinputs. While biocontrol has often been promoted in favour of agroecology and small-scale producers, as part of a more general alternative project to the agro-industrial regime, this time the project is intended to be inclusive. An Argentinean official, referring to the initiatives in favour of bioinputs, points out:

> The idea here is to make public policies that integrate the small with the big. Not that one of us is left out, not that we are left out, because the reality of the country is also like that.
>
> *(Interview, Buenos Aires, 11/10/2017)*

The president of a national network of large agricultural producers defending the use and production of biocontrol goes further regarding this inclusive position. He insists on the challenge of opening up to producers who are not necessarily opposed to the use of pesticides, but rather bearers of a certain pragmatism focused on the effectiveness of products. This would mean taking an interest in:

> conventional producers, that don't have, let's say, an ideological issue behind "I don't want pesticides" … No. These are people that have to be convinced of the efficiency of the process, of the success of the process.
>
> *(Online workshop, 11/08/2020)*

Through semantic work on the category of bioinputs, but also through positions taken on the coexistence or even compatibility between pesticides and biocontrol, the latter is placed in a specific position. It is presented as a viable technological alternative for ensuring the health of cultivated plants, capable of performing functions similar to those of pesticides. But at the same time, officials, supported by farmers, refuse to condemn pesticides and to use the expansion of biocontrol as a lever to displace pesticides. This tension is reflected, as we have seen, in a choice of classification and categorisation (Bowker and Leigh Star 1999). The trajectory of biocontrol has been reoriented so that it joins that of bioinoculants, which are resolutely inscribed in the world of biotechnologies and in the regime of the pesticide-consuming agroindustry. Through this framing, biocontrol finds itself in a "technical package" that includes the technology it is supposed to help develop. It is installed in a logic of coexistence, which would make it possible to respond to societal warnings about the dangers of pesticides, without hindering the agro-industrial regime. No government would dare to condemn the latter, given that the foreign currency it generates is essential to state budgets (Richardson 2009). But to understand the form taken by this expansion of biocontrol, it is important to look beyond the state and public policy. It is also important to look at the shifts that have taken place on the side of the pesticide and biocontrol industry.

7.3 Industrial mergers and technological equivalence

Having shown how biocontrol has become part of a broader public policy category—bioinputs—let me now return more specifically to this set of biological alternatives to pesticides. In the early 2010s, when biocontrol began to attract public policy interest, the organisation of the industrial sector that designs and produces these technologies was still in its infancy. While the bioinoculant sector was already well organised—in Brazil, the National Association of Inoculant Producers and Importers (ANPII) was created in 2000—the Association of Biocontrol Companies (ABC Bio) was only created in 2007. However, these companies were not new. Brazilian companies emerged mainly in the 1980s and 1990s in the Campinas region of the State of São Paulo, southern Brazil. They were originally small and medium-sized companies, founded by entrepreneurs closely linked to the scientific and university sector, which rapidly became connected to the demand from the sugar cane sector. One of the leaders of ABC Bio says:

> When this biological control started, it came much more from the university, by post-graduate students, who ended up developing a project with the university. And then they created their own company. This was very common … these are companies that came with some knowledge from the university, or from the Biological Institute of Campinas. The Biological Institute of Campinas provided some strains—what we call—microorganisms; selections, and some companies … and it was a sector that developed a lot in the state of São Paulo by providing solutions for the sugarcane crop. Which today is one of the crops that most uses biological control in Brazil.
>
> *(Interview, São Paulo, 15/06/2018)*

The ABCBio was originally formed around these companies, until the 2010s came to mark a turning point. It was at this time that large multinational companies in the agrochemical sector, in the same way as at the international level, began to take an interest in biocontrol and to invest in this sector through the acquisition of local companies. In Brazil, they invest in these companies in the south of the country and quickly join as members of ABCBio. The association's executive secretary mentioned this approach in 2018 and the change it generated in the association:

> I had several meetings with the agrochemical industries. Because I saw, outside Brazil, that they were already operating in this segment. So, I had several meetings with Bayer, BASF … which are traditionally agrochemical companies, but which were already carrying out research. And they were enthusiastic about this group of companies (ABCBio) and from there, it got stronger and we started to have a more operational team; bigger … They are now investing heavily in this technology. So, we have a Bayer, which is strongly active in the association; an FMC; a BASF; an Arysta.
>
> *(Interview, São Paulo, 15/06/2018)*

This rapprochement thus gave the association new capacity for action, which ended with the significant funding that the large multinationals could bring. It is also the developments made possible by these investments that finally convinced the historical companies of ABCBio, which initially did not like the arrival in their association of the agrochemical companies, of the benefit:

> So much so that some directors wanted to put in the statute the non-permission of companies that have agrochemicals as well, to participate in the board. But this ended up softening over time. Even today there are still some who are very reticent. But, on the other hand, they are seeing that innovation and development is being introduced … and it is a natural thing for the market.
>
> *(Interview, São Paulo, 15/06/2018)*

And, in fact, the association's position has become conciliatory vis-à-vis agrochemicals, with a discourse defending the complementarity between biological and chemical inputs:

> And we have no confrontation, for example. As we are still a very small segment in Brazil, the agrochemical industries, we practically act together. We don't have, for example, conflicts of interest. Mainly because today, these agrochemical companies are also acting in this segment. The idea is that the companies have options for the farmer, who will increasingly need agrochemicals, but will also need biological inputs. So, the idea is that these industries, they end up providing the solution for the farmer; a complete portfolio. Not just high toxicity products.
>
> *(Interview, São Paulo, 15/06/2018)*

This rapprochement of companies producing chemical and biological inputs within a single entity was concretely recorded at the institutional level in 2019. In Brasilia, at a ceremony organised in the presence of the Ministers of Agriculture and the Environment, CropLife Brasil was officially launched, a union bringing together industries producing four types of agricultural inputs: pesticides, biocontrol, seeds and biotechnologies. As in public policy, biocontrol is included in a set of technologies that combine technologies that are symbols of agribusiness, such as biotechnology. But above all, it is associated with synthetic pesticides, i.e., the technologies to which it was supposed to represent an alternative. On its website, Croplife Brasil refers to this association by linking it to the same banner term used in public policy, namely innovation. With this new organisation, the following would come together:

> In a single platform the experience and track record of associations that have led discussions on innovation in agriculture for decades.
>
> *(https://croplifebrasil.org/sobre-croplife/)*

And the executive secretary of ABC Bio, who has become the "Biological Director" at Croplife, again mentions the "non-competitive" place of biocontrol in this group of technologies and industrial players. The emergence of an alternative technology would not necessarily accompany or cause the decline of the dominant technology:

> We represent an efficient tool that can, when used properly, help reduce excessive pesticide consumption. But one technology is not going to replace the other.
>
> *(Interview, São Paulo, 15/06/2018)*

The argument of coexistence and complementarity between technologies is therefore put forward by the chemical and biological input industry as well as by public administration. This idea is obviously reflected in the composition of the new industrial alliances, such as CropLife Brasil, and in the positions taken by their leaders defending a unified approach to plant protection technologies. But the close connection between the two types of technologies and the two industrial sectors can also be seen in the individual trajectories of some individuals, such as entrepreneurs in the biocontrol sector. This is the case, for example, with one of the managers of an Argentinean company marketing biocontrol solutions, who returned to Buenos Aires in the early 2010s after working for many years in the pesticide industry in Brazil and the US. This shift from chemical to organic did not prevent him from maintaining activities in the pesticide sector, and from developing certain convictions about the importance of using both types of technology. He points out:

> I listen to all the bells. Then I decide what is best for me … Besides, I work as a consultant for many agrochemical companies. I don't think there is going to be a total disappearance of chemicals.
>
> *(Interview, Buenos Aires, 26/04/2017)*

The idea of a massive decline of pesticides and a generalised replacement by biocontrol is thus rejected as a utopia by these industrialists, pioneers of biocontrol without defining themselves as enemies of the pesticide industry. This utopia of substitution or decline is, in their eyes, carried by actors who basically do not understand the technical elements of the plant health issue. The secretary of ABC Bio explains:

> Our segment is seen positively as the solution to all problems. And this is very positive. However, when they talk about our segment, about bioinputs, there is a very big rejection with respect to agrochemicals, saying—"No. Bioinputs are going to be the solution to the problems"—and, in fact, we are one more tool within the integrated pest management.
>
> *(Interview, São Paulo, 15/06/2018)*

Biocontrol and pesticides are both said to have virtues, as well as drawbacks. Those involved in the industrial alliance between biocontrol and pesticides insist on this point, thus making an equivalence between biological and chemical inputs. While pesticides are denounced for their environmental or health effects, some industrial players do not hesitate to point out the risks that biocontrol can generate, especially products based on micro-organisms. An Argentinean consultant supporting companies in the registration of chemical or biological inputs with the public regulator mentions this similarity between the two types of inputs:

> There are a lot of things that are natural, that are organic, in my vision, doesn't mean they are less toxic. Or that they can't bring some kind of trouble. So, I agree that perhaps the current regulation is too chemical or too complex and

> does not fit in with certain studies, but to say well, that this is not the case… we have to look for different methods of identity, we have to evaluate it … all are compounds that are applied. It's all an artificial thing you're going to apply to a crop. It's not water, it's not sun. Bioinputs also have their risks.
>
> *(Interview, Buenos Aires, 04/09/2017)*

The risks pointed out are obviously not of the same order as those induced by chemical molecules. It is the risk of dispersing micro-organisms such as bacteria or viruses into nature that is pointed out, when the selection or reproduction process is not controlled. This would be the case in particular when farmers venture to produce certain micro-organisms by themselves. But it is also the risk of poor agricultural practices, of farmers misusing biocontrol technologies, that is pointed out. This last risk is precisely the same as the one regularly identified by the pesticide industry to defend their products. Faced with accusations against their molecules and their dangerousness, the industry often argues that it is the misuse of products by farmers—overdosing, treatments in bad weather conditions—that is to be condemned, rather than the technologies themselves. This link between the product and its use is even recognised in the legislation of certain countries. In France, for example, it is not the commercial products alone that are approved, but the commercial products and their conditions of use (Pellissier 2021). As a result of these arguments and movements in the chemical and biological input industry, biocontrol is thus associated with and equated to the dominant technologies that are pesticides. While alternative technologies undoubtedly offer a breakthrough and a credible alternative to problematic technologies, they are not necessarily the levers of their decline. They can be involved in both the discontinuity and continuity of the regime developed around problematic technologies.

7.4 Conclusion

How does the emergence of alternative technologies contribute to the decline of incumbent technologies? This is the question I asked in the introduction to this chapter. To answer it, I looked at the Argentine and Brazilian cases of the development of biocontrol technologies, alternatives to chemical pesticides. More specifically, I considered the expansion stage of these technologies, i.e., a period in which existing technologies start to receive increasing interest from both public and private stakeholders. While they were a niche for decades, their market has been growing strongly for the last ten years, and they are the subject of dedicated public policies. Following on from the work done in both countries (Goulet 2021; Goulet and Hubert 2020), I have shown here that the development of these alternative technologies plays an ambiguous role in the decline of the technologies in place. The dilution of these alternative technologies into larger ensembles, including technologies essential to the functioning of the incumbent technological regime and even including the incriminated technologies, contributes to the idea of coexistence between technologies. The scientific, political and industrial spheres are at the heart of this dynamic, by incorporating these alternative technologies into their activity portfolios, and by defending a vision of coexistence or even complementarity between technologies. Alternative technologies are thus placed at the service of both technological continuity and discontinuity. This position obviously does not hide the fact that certain actors, historical defenders of alternative technologies and promoters of an explicit discontinuity, do not find their way in the direction taken by this expansion. This is the case, for example, of the promoters of agroecology or organic farming, for whom only the firm and definitive withdrawal of pesticides constitutes an acceptable outcome. But in our case study, marked by the joint support of the state and the input industries for the agro-industrial system based on the use of pesticides, the space for strategies of counter-framing alternative technologies remains slim. This raises the question of what would happen in other contexts, such as those of industrialised countries where society and consumers are becoming more aware of the risks associated with pesticides. Analysis of the development of biocontrol in France (Aulagnier and Goulet 2017), intervening in the framework of an ambitious pesticide reduction policy, seems however to show us that things are not different in the end. While political speeches in favour of a frank break are certainly formulated, biological alternatives remain, as in South America, one option among others to guarantee high levels of agricultural productivity in the long term.

Therefore, in the same way as in other sectors (Bergek et al. 2013; Kim and Park 2018; Goulet and Vinck 2022), agricultural technologies that have become problematic and are set to decline are thus associated with and attached to alternative technologies, within technological mixes. But once this conclusion has been reached, it is the question of the future of these mixes that should be of concern, and that of the balances or competitions that arise or shift over a longer period, beyond the expansion phase. In the case of the combination of bioinputs and pesticides, an Argentinean public servant does not hesitate to delegate the evolution of these balances to "natural" forces, which are related to the market or to the evolution of the efficiency of technologies:

> The technological paradigm shift will necessarily take place gradually, that is, these products will be introduced and show their potential, they will be adopted to the extent that farmers want to use them … If later one model proves to be more efficient, naturally one will win out over the other.
>
> *(Interview, Buenos Aires, 11/10/2017)*

Based on a short period marking the expansion of alternative technologies, this chapter has provided elements for a better understanding of the role of these alternatives in the decline of problematic technologies. The challenge now is to better characterise how this relationship can evolve over time, without relying on naturalistic explanations, and how they can be governed (see Stegmaier 2023). The three spheres considered here—political, scientific and industrial—offer important places to observe and consider these developments and their drivers.

Note

1 Bioinputs are currently handled by IICA within a "Bioeconomy and Productive Development Programme".

References

Altieri, M.A., Rosset, P.M. and Nicholls, C.I. (1997) Biological control and agricultural modernization: Towards resolution of some contradictions. *Agriculture and Human Values*, 14, 303–310.

Arancibia, F. (2013) Challenging the bioeconomy: The dynamics of collective action in Argentina. *Technology in Society*, 35, 79–92.

Aulagnier, A. and Goulet, F. (2017) Des technologies problématiques et de leurs alternatives: Le cas des pesticides agricoles en France. *Sociologie du Travail*, 59. https://doi.org/10.4000/sdt.840

Bergek, A., Berggren, C., Magnusson, T. and Hobday, M. (2013) Technological discontinuities and the challenge for incumbent firms: Destruction, disruption or creative accumulation? *Research Policy*, 42, 1210–1224.

Bonnaud, L. and Anzalone, G. (2021) A perfect match? The co-creation of the tomato and beneficial insects markets. *Journal of Rural Studies*, 83, 11–20.

Bowker, G. and Leigh Star, S. (1999) *Sorting Things Out: Classification and its Consequences*. MIT Press.

Callon, M. (1986a) The sociology of an actor-network: The case of the electric vehicle. In Callon, M., Law, J., Rip, A. (eds) *Mapping the Dynamics of Science and Technology: Sociology of Science in the Real World*. Palgrave Macmillan.

Callon, M. (1986b) Some elements of a sociology of translation: Domestication of the scallops and the fishermen of the St Brieuc Bay. In Law, J. (ed.) *Power, Action and Belief: A New Sociology of Knowledge?*Routledge.

Carson, R. (1962) *Silent Spring*. Houghton Mifflin.

Dereumeaux, C., Fillol, C., Quenel, P. and Denys, S. (2020) Pesticide exposures for residents living close to agricultural lands: A review. *Environment International*, 134, 105210.

Durant, J.L. (2019) Where have all the flowers gone? Honey bee declines and exclusions from floral resources. *Journal of Rural Studies*, 65, 161–171.

Evangelakaki, G., Karelakis, C. and Galanopoulos, K. (2020) Farmers' health and social insurance perceptions – A case study from a remote rural region in Greece. *Journal of Rural Studies*, 80, 337–349.

Flint, M.-L. and van den Bosch, R. (1981) *Introduction to Integrated Pest Management*. Plenum Press.

Fouilleux, E., Bricas, N. and Alpha, A. (2017) "Feeding 9 billion people": Global food security debates and the productionist trap. *Journal of European Public Policy*, 24, 1658–1677.

Geels, F.W. and Schot, J. (2007) Typology of sociotechnical transition pathways. *Research Policy*, 36, 399–417.

Goulet, F. (2021) Characterizing alignments in sociotechnical transitions: Lessons from agricultural bio-inputs in Brazil. *Technology in Society*, 65, 101580.

Goulet, F. and Hubert, M. (2020) Making a place for alternative technologies: The case of agricultural bio-inputs in Argentina. *Review of Policy Research*, 37, 535–555.

Goulet, F. and Vinck, D. (eds) (2022) *Faire sans, Faire avec Moins. Les Nouveaux Horizons de L'innovation*. Presses des Mines.

Gunter, V.J. and Harris, C.K. (1998) Noisy winter: The DDT controversy in the years before silent spring. *Rural Sociology*, 63, 179–198.

Kessel, N. (2022) Marchés pharmaceutiques et retrait de médicament. In Goulet, F. and Vinck, D. (eds) *Faire sans, Faire avec Moins. Les Nouveaux Horizons de L'innovation*. Presses des Mines.

Kim, J. and Park, S. (2018) A contingent approach to energy mix policy. *Energy Policy*, 123, 749–758.

Kogan, M. (1998) Integrated pest management: Historical perspectives and contemporary developments. *Annual Review of Entomology*, 43, 243–270.

Koretsky, Z. (2023) Dynamics of technological decline as socio-material unravelling. In Koretsky, Z. *et al.* (eds) *Technologies in Decline: Socio-Technical Approaches to Discontinuation and Destabilisation*. Routledge.

Kvakkestad, V., Sundbye, A., Gwynn, R. and Klingen, I. (2020) Authorization of microbial plant protection products in the Scandinavian countries: A comparative analysis. *Environmental Science & Policy*, 106, 115–124.

Lehtimäki, T. (2019) Making a difference: Constructing relations between organic and conventional agriculture in Finland in the emergence of organic agriculture. *Sociologia Ruralis*, 59, 113–136.

Levain, A., Joly, P.-B., Barbier, M., Cardon, V., Dedieu, F. and Pellissier, F. (2015) *Continuous discontinuation – The DDT Ban revisited*. International Sustainability Transitions Conference: "Sustainability transitions and wider transformative change, historical roots and future pathways". University of Sussex, Brighton, UK.

Lockwood, J.A. (1997) Competing values and moral imperatives: An overview of ethical issues in biological control. *Agriculture and Human Values*, 14, 205–210.

Newman, P. (2023) The end of the world's leaded petrol era: Reflections on the final four decades of a century-long campaign. In Koretsky, Z. *et al.* (eds) *Technologies in Decline: Socio-Technical Approaches to Discontinuation and Destabilisation*. Routledge.

Pellissier, F. (2021) *Tuer les Pestes pour Protéger les Cultures: Sociohistoire de L'administration des "Pesticides" en France* (PhD, Université Paris-Est).

Pelosi, C., Bertrand, C., Daniele, G., Coeurdassier, M., Benoit, P., Nelieu, S., Lafay, F., Bretagnolle, V., Gaba, S., Vulliet, E. and Fritsch, C. (2021) Residues of currently used pesticides in soils and earthworms: A silent threat? *Agriculture, Ecosystems & Environment*, 305, 107167.

Pistorius, C.W.I. and Utterback, J.M. (1997) Multi-mode interaction among technologies. *Research Policy*, 26, 67–84.

Richardson, N. (2009) Export-oriented populism: Commodities and coalitions in Argentina. *Studies in Comparative International Development*, 44, 228–255.

Rip, A. and Kemp, R. (1998) Technological change. In Rayner, S. and Malone, E.L. (eds) *Human Choice and Climate Change*. Battelle Press.

Rip, A. and Vos, J.P. (2013) Umbrella terms as mediators in the governance of emerging science and technology. *Science, Technology & Innovation Studies*, 9, 39–59.

Rosset, P.M. and Altieri, M.A. (1997) Agroecology versus input substitution: A fundamental contradiction of sustainable agriculture. *Society & Natural Resources*, 10, 283–295.

Schwindenhammer, S. (2020) The rise, regulation and risks of genetically modified insect technology in global agriculture. *Science, Technology and Society*, 25, 124–141.

Shattuck, A. (2021) Generic, growing, green? The changing political economy of the global pesticide complex. *The Journal of Peasant Studies*, 48, 231–253.

Stegmaier, P. (2023) Conceptual aspects of discontinuation governance: An exploration. In Koretsky, Z. *et al.* (eds) *Technologies in Decline: Socio-Technical Approaches to Discontinuation and Destabilisation*. Routledge.

Stegmaier, P., Kuhlmann, S. and Visser, V.R. (2014) The discontinuation of socio-technical systems as a governance problem. In Borrás, S. and Edler, J. (eds.) *The Governance of Socio-Technical Systems*. Edward Elgar.

Taylor, M. and Taylor, A. (2012) The technology life cycle: Conceptualization and managerial implications. *International Journal of Production Economics*, 140, 541–553.

Villemaine, R., Compagnone, C. and Falconnet, C. (2021) The social construction of alternatives to pesticide use: A study of biocontrol in Burgundian viticulture. *Sociologia Ruralis*, 61, 74–95.

Warner, K.D., Daane, K.M., Getz, C.M., Maurano, S.P., Calderon, S. and Powers, K.A. (2011) The decline of public interest agricultural science and the dubious future of crop biological control in California. *Agriculture and Human Values*, 28, 483–496.

8

CARING FOR DECLINE

The case of 16mm film artworks of Tacita Dean

Dirk van de Leemput and Harro van Lente

8.1 Introduction[1]

Typically, when a technology is seen as becoming obsolete, users, developers, maintainers and other relevant groupings will reassess their relationship to the technology, and may decide, for practical or strategic reasons, to follow the "next" technology—hence reinforcing the processes of decline (see Turnheim 2023). Often, however, pockets of resistance will remain, working against the tide of decline and seeking to keep the technology working. Such dedication includes caring for the constitutive networks of supply, skills and valuation.

In this chapter we analyze how care works for technologies in decline. We look at museums, as these are sites where in general, decline is resisted. After all, a key societal task of museums is to preserve and present important cultural expressions. (e.g., ICOM n.d.; Domínguez Rubio 2020) By caring for art, museums work to counteract the processes of decline, aging and obsolescence. In many cases, for instance in the case of time-based media art, this includes caring for technology, too. Museums, therefore, are a good site to trace and analyze care activities for declining technologies.

We will proceed in three steps. In the first step, we will review the notion of care, drawing from feminist theories and maintenance and repair studies. To summarize the insights, we develop a framework of care activities, based on basic distinctions in objects and strategies of care. First, what benefits from care: are activities targeted at a specific artwork, a range of artworks or a technology? Second, what strategies are employed: are care activities organized around an (art) object, a network of care or an ecosystem?

In the second step, we will give an account of the care activities related to the artworks of the contemporary artist Tacita Dean, in particular her *Disappearance at Sea* (1996) at Tate, London. For Dean 16mm film is her primary artistic medium

DOI: 10.4324/9781003213642-8

and as such it is fundamental to the identity of her work (Smith 2012). Once 16mm was a prominent and ubiquitous technology used for the production of moving images; today it is relegated to the realms of aficionados and museums. We trace the various concerns, actor groupings and their care activities related to the works of Tacita Dean. This account is based on an extensive document study on the history of the 16mm film sector, on field work at Tate, London, over five weeks and on interviews with technicians and conservators at Tate and elsewhere by the first author in late 2019 and early 2020. [2] We focus our analysis on external technicians and Tate's time-based media conservators; the latter are responsible for the conservation of time-based media artworks, that is, works of art that incorporate audio, film, video, 35mm slides or computer-based elements (Laurenson 2004).

Our third and last step is to reflect on how care is part of technological decline. While our insights are drawn from the domain of museums, our findings bring the general insight that the decline of technologies comes with additional care efforts. Our analysis aligns with earlier research that stresses the importance of invisible work that comes with any maintenance or conservation (e.g., de la Bellacasa 2017; Domínguez Rubio 2020; Jackson 2019; Miller 2021; Star & Ruhleder 1996; Tronto 1993; Vinsel & Russell 2020).

8.2 Decline and care

In recent years, there has been a surge in scholarly interest in maintenance and repair (Vinck 2019; Denis, Mongili & Pontille 2015, Russell & Vinsel 2018). In his bestseller *The Shock of the Old*, the historian David Edgerton (2019) argues that technology-in-use includes much more than invention and innovation, it also includes maintenance. A key observation is that the societal presence of technologies requires maintenance and repair, otherwise technologies would fade away. In a sense, decline is the normal state of a technology, counteracted by practices of maintenance and repair (Denis, Mongili & Pontille 2015). Maintenance and repair connect the "two radically different forces and realities" of "an almost-always-falling apart world" and "a world in constant process of fixing and reinvention" (Jackson 2014: 222).

Maintenance is often thought of in terms of care, such as a "care of things" (Denis & Pontille 2015: 341) or "material care" (Jackson 2019: 427). Thinking about maintenance as care draws attention to the attentive relation of a maintainer to an object, or "a kind of patient attending, a slow and attentive being with by which the trajectory of others is secured and maintained through time" (Jackson 2019: 428). Above all, "[t]his care of things reverses the traditional view of the role of artifacts in society in that it concentrates on the material fragility of things and the constant necessity of taking care of them" (Denis & Pontille 2015: 341). Joan Tronto defines care as:

> everything that we do to maintain, continue and repair "our world" so that we can live in it as well as possible. That world includes our bodies, ourselves

and our environment, all that we seek to interweave in a complex life sustaining web.

(*Tronto 1993: 103*)

This definition stresses the broad range of care, from health care to the care of things, while it is at the same time open as to what activities can count as care.

Care has been on the agenda in feminist studies for several decades. The earliest feminist-inspired studies focused on care in the more obvious places, such as the house, in health care or farming. These are places where women often perform most of the care directed towards patients or others in a household. The studies stress both the importance of care work as well as that it tends to be hidden (Mol, Moser & Pols 2015). Care, as well as maintenance and repair, is typically directed at more than one thing (Mol 2002). Graham & Thrift (2007: 4) noted that "it becomes increasingly difficult to define what the 'thing' is that is being maintained and repaired. Is it the thing itself, or the negotiated order that surrounds it, or some 'larger' entity?" At the same time, maintenance and repair is not only directed at the material order, but also the social order (Henke 2000).

For the purpose of this chapter, we distinguish two dimensions of care, relating respectively to goals—*what* is intended to be sustained?—and strategies—*how* is this goal to be achieved? When museums care for, say, the 16mm projector in an artwork, the goal of care may be a specific artwork of 16mm film, the possibility of 16mm film art or the fate of 16mm film technology in general. In our case study, we will trace these different goals alongside each other. The second dimension is how these care activities are performed and where the care work is taking place. Here we distinguish between three strategies: 1) work at the level of an object, 2) work at the level of a network of care on which an object depends,[3] or 3) work at the level of an "ecosystem", the broader environment allowing a technology to function. We will use these two dimensions as a heuristic in our exploration of how museums care for technologies in decline.

As a preliminary note we would like to stress that we are very aware that the care of artworks has long been the subject of discussion within museum and conservation studies, as well as in the practice of art conservation. An important issue has been the acknowledgement that artworks can change, and that this is not necessarily a bad thing (Depocas, Ippolito & Jones 2003; Laurenson 2006). There is however a strong understanding that Tacita Dean's film-based artworks should be experienced as film projection and should not be converted to other media (Jennings 2007). For the sake of our analysis, we will be agnostic as to what constitutes the best way to care for art.

8.3 Tacita Dean and 16mm film

The artwork *Disappearance at Sea*, made by the British artist Tacita Dean, was shot in 1996 in and around St. Abb's lighthouse in Berwick upon Tweed. It was shot on 16mm film—at that time already a fading technology—and it was first shown on the Berwick Ramparts Project in the town of Berwick. The work, produced in an edition of three and an artist's proof,[4] was rewarded with a nomination for the Turner Prize, a prestigious prize for British visual artists.

Tate Britain, organizer of the award and the exhibition, bought an edition of *Disappearance at Sea* in 1998 directly after the Turner prize exhibition. The acquisition prompted Tate to develop practices and policies for the conservation of artworks incorporating film technology. The work was displayed at Tate in 1998–1999 ("Turner Prize Show"), in 2001 ("Tacita Dean: Recent Films and Other Works"), in 2005–2007 (part of a collection display) and in 2013 ("Looking at the View"). In 2019–2020 the work went on a tour to three smaller museums in England. During the 2001 exhibition at Tate, Ken Graham, the art handling technician at Tate, installed *Disappearance at Sea* and some other works by Tacita Dean. Since then, he has worked as her technical assistant and also writes the installation instructions that Tate now follows. All later installations at Tate were installed by the museum's staff in cooperation with Ken Graham, by then running his own company KSO-AV. The 2019–2020 travelling exhibition was installed by David Leister, under the guidance of Ken Graham.

Today, Tate has a large set of its own 16mm projectors. Most of them are EIKI's in diverse models—the most common brand of 16mm projectors used within the art world. Dean's technician Ken Graham explains their popularity because "they are easier to maintain, easier to get parts for, because they were such a popular model … Everything is bare and accessible" (Interview, Ken Graham, 2–8–2019). In addition, Tate also owns an important array of accessories, such as lenses and loopers. Tate has over the years developed quite a large collection of artworks on film. In 2001 the museum invested in cold storage for their films. Since 1996 Tate has worked with several film labs to make exhibition prints of *Disappearance at Sea*, including Colour Film Services and Metrocolour (both 1998), Henderson (2001), Soho Images (2005–2006), ARRI (2013) and Fotokem (2013 and 2019). Tate has followed Tacita Dean in her choice of labs (Dean 2018).

In the following, we will first sketch the general rise and decline of 16mm film technologies. Then we will highlight the care work for 16mm film at Tate, UK. We present how film installations at Tate are being maintained and how their technicians are trained (section 3.2). We also follow the care work in the broader networks of care: the maintenance and repair of projectors by technicians external to Tate (section 3.3). Finally, we address the care work for 16mm in the broadest sense, in terms of the "ecosystem" to be sustained (section 3.4). Here, Tacita Dean plays an important role, too, as she cares for the social order of film technology in general.

8.3.1 The rise and fall of 16mm film

16mm film is a physical, analogue film format that dates back to 1923. Through the years the format has undergone some changes, but the basic features remain. It is a photographic film (with either positive, negative or reversal emulsion) 16mm wide, with standardized perforations on at least one side. The film is shot in a camera, processed and printed in a laboratory, and viewed with the use of a projector. 16mm film thus depends on large infrastructures for its production, reproduction and display.

When projected, the film wears in the process. In a projector the film is constantly moved forward one frame, stopped for a split second while light falls through it and then moved forward one frame again. In this process the film deteriorates from two sources: light and movement. The light that falls through the film slowly makes the film fade away. In museums and galleries, film is usually projected for eight hours a day and five to seven days a week. More aggressive is the wear by projection movements. The film comes in constant contact with parts of the projector and is quite aggressively pulled through its system. Within a museum or gallery context the film usually runs on a looping system that allows the film to be shown continuously. In this looper, the film is also moving against itself, causing even more friction. In a museum, a print needs to be replaced every two or three weeks, depending on the length of the film. Likewise, the portable projectors designed for incidental industry or classroom use wear down and need maintenance and repair every now and then.

16mm film was first introduced in 1923 by Kodak as a relatively cheap and safe technology for making home movies. Until then, motion picture film was almost exclusively made in 35mm width. The 35mm width made the material expensive, nitrate cellulose was highly flammable and thus unsafe, and the development of film always required a two-step process: first developing the negative in-camera film, followed by printing and developing on positive stock. This made the use of motion picture film expensive and dangerous for amateurs. Kodak tackled these issues by presenting a package containing camera, projector and film in a new format: 16mm. The format was smaller and thus cheaper, the acetate celluloid base was almost non-flammable and safer to use, and the film used a reversal emulsion. These properties of being cheaper and safe, but also more portable, led to an increased interest in the format from different fields. As an amateur format it was soon to be replaced by the even cheaper standard or double 8 format (Kattelle 1986; Matthews & Tarkington 1955).

In 1935 Kodak introduced Kodachrome, a color reversal film. In 1938 a reproduction method for that film stock was introduced. This led to a surge of interest in professional usage of 16mm film. Hollywood and other filmmakers however still regarded 16mm as a substandard format at this time, not suitable for serious movie making (Norris 2016). In the 1950s, the format became more and more popular with the television industry, for news broadcasts, documentary and drama. By the 1970s there was a big infrastructure in place for the production of 16mm reversal prints for use in television broadcasting.

The 16mm format stayed popular far into the 1990s, when it was mostly replaced by video technologies. It became more common to use digital intermediates in the processing of 16mm and 35mm film, while digital capture became more common near the year 2000. Around 2010 many cinemas converted to digital projection after distribution companies actively promoted the installation of digital projectors in cinemas. This led to a decrease in the demand for film printing on 35mm, which also meant that this possibility declined for 16mm film. In the 21st century, the availability of film stocks and film projectors also declined significantly (Read 2006; Cleveland & Pritchard 2015, Eisloeffel 2013; Lucas 2011).

8.3.2 Care work for objects

One morning during the fieldwork of the first author, he joined two young time-based media technicians, Kate and Fernando,[5] in their round to maintain some of the artworks on display at Tate Modern. The work *Black Is* (1965) by the Italian-American artist Aldo Tambellini, an abstract 16mm film installation, had been on display for a while and needed its weekly maintenance, which includes changing the lamp in the projector. These are excerpts from the field notes:

> With a screwdriver Kate unscrews the two screws on the bottom of the protective acrylic housing to reach the projector. She complains that they are wiggly and too loose in the MDF-construction of the plinth. It dropped a few weeks ago, so the door is damaged. They do not turn on the light and work in the semi-dark gallery lights. Kate tries to remove the casing of the projector but has some trouble finding how to open it. Their colleagues exchanged the projector a few weeks ago and she does not know this one. She goes away to find a tool but comes back without and then tries again. Now she feels that the casing should be turned to the top. She removes another protective metal cover and can remove the light bulb from its house. While she tries to open the projector, she says things like "I think …" and "it should …"
>
> Fernando in the meanwhile has taken the new lamp from the box and hands it over to Kate. She puts on a white glove and says: "The lamp has a longer life when you do not touch it with your bare hands". Fernando answers: "That is good knowledge. I did not know". Kate asks Fernando to put the housing back on. He is struggling a bit with the outer housing, Kate says there should be a "satisfying click". After trying for a while, Fernando manages to find the right way of putting the housing back on. While they are working, Kate asks who replaced the film. She thinks it is not correctly routed through the projector. The work nevertheless still shows on the screen. When they turn it back on, the focus is lost. Kate adjusts the focus control of the projector until the focus seems to be good. This is hard to see since the projection is a hand painted strip of film with no conventional image. They decide to wait until the text comes on. After some waiting, they decide that the lines look OK and that this means that the focus is good enough.
>
> *(Field notes, 26–7–2019)*

Although there is a lot going on in this scene, it is above all a clear example of care work for objects. The two junior technicians are at work on the gallery floor, they improvise and interact with the equipment to ensure the projector and installation are working properly. The weekly routine maintains the projector, but it also maintains a community of experts. It trains new people in the basic workings of a projector and its parts. This vignette also shows how care for a material order requires an interaction with materials as well as the activation of knowledge and experience (Harper 1987). The practice of training and learning in repair and maintenance is crucial for good repair. In a similar way, Lejeune (2019) shows how repair of locomotives depends on knowledge gathered from cooperation, lunch talks and support circles, where others stand in a circle around those doing the repair work, as jazz musicians stand around someone doing a solo. Likewise, Orr (2016) showed that the repair of photocopiers depends on how repair people improvise their procedures and learn from working and interacting with machines.

A second example is about how technician David Leister works on the 16mm film projector that was used in the Tambellini display mentioned above. It was now stored with a label: "Ex-Tambellini. Needs lots of TLC", an abbreviation of "tender love and care".

> When he starts the 110 volt projector it makes some noise. He turns the fly-wheel, and then the motor slowly runs. He then holds his hand against the spindle that comes out of the motor end. The spindle stops. "I should not be able to hold it back" he says. David looks around for a new motor, but finds only old ones, or 240 volt motors. "I can turn it into a 240 volt … I did it before at my studio". David looks around the workshop to see what he can use. He finds a brand new motor in a box, but decides not to use it. He also looks at two projectors that have a sticker "for spares" on them, but those two already have no motor left. He takes another box and puts it on the table. He also takes a transformer from the shelf. He carefully looks at all the wirings in the connector plugs of the two parts and the chassis of the projector. First he is afraid the plugs will not connect. He studies the plugs and the wires for a long time and then tries to plug them. They fit. He says "I Frankenstein this projector". Then he starts making funny noises, imitating short cuts and explosions. He seems to enjoy it and makes fun of himself as a Dr. Frankenstein, while at the same time he is worried about the thing he is making. Again he carefully examines all wirings and connections and decides it is safe to try it.
>
> *(Field notes, 2–8–2019)*

Despite the joke, this scene indicates how a relatively standard procedure of exchanging parts is mixed with problematizing the procedure. In his reference to Dr. Frankenstein technician David Leister shows that he is aware that he is breaking some order that should not be broken. The story of Frankenstein (Shelley 1993) is often used as an icon of science and technology gone bad (Halpern et al. 2016; Hammond 2004; Latour 1996; Stubber & Kirkman 2016). The novel is often interpreted as a story about an evil scientist who breaks the natural order of things and thereby creates a monster. Another reading of the text is to focus on care (de la Bellacasa 2017: 43; Halpern et al. 2016; Latour 2005). The lesson in these interpretations is that the "evil" was not in Victor Frankenstein creating his monster, but in him abandoning his creature. In de la Bellacasa's (2017: 43) words: "This version of caring for technology carries well the double significance of care as an everyday labor of maintenance that is also an ethical obligation: we must take care of things in order to remain responsible for their becomings".

In the scene above, the technician thoughtfully converted his projector and took many measures to care for it in the long term: replacing power plugs, labeling and filing the changes. Like in the Frankenstein novel, there is the desire to continue something, which is only possible by altering the existing order. In the novel, Victor Frankenstein embarked on his mission to create life after the death of his mother. He had a mission to sustain the life of his fellow humans and could only do so by assembling human body parts (Shelley 1993: 34–45). In our case, David Leister has a desire to continue the possibility of exhibiting 16mm film art and his only possibility to extend the "life" of the projector is by combining parts of projectors that were not designed to be together.

Apart from maintenance in its diverse forms, buying parts, transferring parts and making parts are common strategies to care for objects. When Ken Graham started his business two decades ago, in close cooperation with Tacita Dean, he began collecting projectors and developing loopers. Today, buying and rescuing all kinds of equipment is still one of Ken Graham's activities, even if it is just buying equipment to use for spare parts.

> I [also] buy equipment that I don't need, just to know that it is safe and is not gonna end up in a skip somewhere. You know, I had a friend who worked at [a film production company] in west London. He called me up and said: how quick can you get a van here? They are filling up a skip with 16mm. So I got a friend and a van and we went over there. Pulled up to the dock, where they were gonna throw stuff in the skip, we would just put it in the van instead. It was like six projectors and loads of reels and splicers and all kinds of stuff they were just gonna throw away. So we rescued it … I will buy a lot of knackered projectors, just because I know there are parts in there that I can use … There was a dealer, a guy who had the EIKI franchise in the UK, but he died a few years back and all of his stuff disappeared … So it is difficult to get the parts, but there are transferrable parts, you can modify stuff. You can, I am an engineer, I can make my own parts.
>
> *(Interview Ken Graham, 2–8–2019)*

Indeed, another strategy is to remake parts and for this the network of technicians is important, too.

> You know I have been at Ruud's funeral last week [Ruud Molleman was a film technician working for Dutch museums]. We were talking about this very thing, about which parts were still available, which you had to make, who was making them. So, just at that funeral I was asked by two guys, how I got

> drawings of parts that we need and if we can get a need for a bulk of them enough, then we can get them made. And then they will always be there.
>
> *(Interview Ken Graham, 2–8–2019)*

The remaking of parts replaces the supply of spare parts that was once organized by manufacturers. Networks and networking events—even a respected technician's funeral—help to support the continued presence of 16mm film technology.

8.3.3 Sustaining networks of care

With the decline of 16mm film technology, the network of care around that technology gradually transforms. When 16mm film was in general use, producers provided maintenance for their projectors and film labs were abundant and equipped with skilled technicians. Now these facilities have largely disappeared, other networks have emerged. The vignette above of the maintenance of a Tambellini installation also illustrates how junior technicians learn and employ new skills on the job with regard to film technology. They find out by trial and error how small adjustments and maintenance in this projector work.

Indeed, there are many efforts to sustain the "network of care" of film art in general. These are mostly related to learning and communicating, such as training on the job, the care for employees and specialists, communicating with film labs and film technicians. It also includes the sustenance of a broader network of technicians, as was referred to in the quote about the funeral of a Dutch professional being attended by technicians from various places. Tate is one of the hubs for technicians in the film world, as many people get their (continued) training at the time-based media conservation department and then work at other galleries or start their own business. Meanwhile, Tate is keen to keep experienced employees in the museum, to maintain the internal institutional knowledge about 16mm film and other technologies. Moreover, junior technicians at Tate learn how to communicate with film labs and expert communities.

In sustaining networks of care, the naming of things is crucial. An example at Tate is the internal Wiki that the team built about time-based media conservation and the technologies they used. The Wiki had an extensive page about 16mm film, where the team collected information about the technical details of the technology. It also contains a long list of terms that are used in film labs to name processes, film prints and equipment. The list differentiates between names used in the US and the UK and between black and white and color processes. According to one of the team members, the Wiki is important to record the knowledge needed to communicate with others in the networks of care. He wished the others spent more time filling and maintaining the Wiki (Field notes, 21–8–2019). The maintenance of endangered languages is a common topic in linguistic research and is generally associated with social networks. The coherence of social networks determines how languages are maintained (Sallabank 2010). Likewise, the jargon used in film labs can be considered an endangered language as well. The number of film labs has sharply diminished, as has the number of people who work in them. The associated language is only spoken within film labs, and by some film enthusiasts, but not by a broader audience.

Another effort by the conservators to sustain the network of care for a technology in decline is to support specialized businesses.

> We are increasingly trying to use our freelancers with specialist skills as much as possible for various different projects to provide a source of income and ongoing relationship … They want to work on their terms, on the thing they are really good at.
>
> *(Interview Chris King, 19–2–2021)*

8.3.4 Caring for an ecosystem

The prospects of a technology in decline also depend on an ecosystem of expertise, resources and use. In the case of 16mm film, the artist Tacita Dean herself is very active in the care of such an ecosystem. Although she also works with other mediums (sound, blackboards, installations, drawing, photography, painting), she is best known for her film works. Film is her dearest medium. She has become an activist for the continuation of the possibility of using film for art, but also in general.

Dean's first intervention is her film installation *Kodak* (2006), in response to the end of the production of 16mm black and white stock by Kodak, and features their factory in Chalon-sur-Saône. "The idea of the film was to use its obsolete stock on itself. The point is that it's a medium that's just about to be exhausted" (Tacita Dean, quoted in Manchester, 2009). In the same manner her later installation FILM (2011)—a commission in Tate's Turbine hall—is a homage to film and analogue technology (Smith 2012). Both works are targeted at a broad audience, showing what film can do and be.

In 2011, when she noticed that the Soho Film Lab had stopped processing 16mm film, she embarked on a long project to fight for the coexistence of film next to digital formats. She became involved in Savefilm.org, an organization "to protect and safeguard the medium of film, the knowledge and practice of film-making and the projection of film print"[6] (Savefilm.org, n.d. a). One of the aims was to have film protected under UNESCO's Cultural Diversity Convention. When most cinemas had switched to digital, Savefilm.org started a campaign to keep film technology alive, supported by many artists and cinema directors, producers, actors and cinematographers, including big names like Wes Anderson, Cate Blanchett, Danny de Vito, Ken Loach, Steve McQueen, Martin Scorsese and Steven Spielberg. Other film organizations, like the International Federation of Film Archives (FIAF) and the Association of Moving Image Archivists (AMIA) also decided to support Savefilm.org's efforts (Savefilm.org n.d. b; Filmadvocacy.org 2013). The petition highlights the beauty and the importance of film and the threat that it will be fully replaced by digital technologies:

> Film and digital are different mediums, in that they differ materially and methodologically in their artistic rather than technological use, and so make different cinema and different art. They have their own unique disciplines, image structures and visual qualities. Their co-existence is essential to keep diversity and richness in our moving-image vocabulary. The ascendance of one does not have to mean the capitulation of the other, unless we allow this to happen.
>
> (*Savefilm.org, n.d. a)*

The petition emphasizes that especially the projection of film is endangered. When left to the market, film will not survive, it claims. This will also end precious knowledge and skills in photochemical industry. Hence the call on UNESCO "to protect and safeguard the medium of film, the knowledge and practice of film-making and the projection of film print" (Savefilm.org, n.d. a).

In March 2015, when Dean was artist in residence at the J. Paul Getty Museum in Los Angeles, she organized an event in collaboration with Christopher Nolan on "reframing the future of film". The morning session was a private roundtable with key figures from the film industry, including representatives of Kodak, film labs, studios, cinemas, archives and museums. According to Dean this was the first time most of these people actually sat together and met (The Kodakery 2015). It was followed by a public session in the afternoon of which Scott Foundas (2015) recalls "the most euphoric moment of the day" when Kodak CEO Jeff Clarke announced that Kodak would continue to produce celluloid film, and contradicted prevalent concerns that Kodak might be exiting the motion-picture film business. Today Kodak is still making 16mm film, but the amount of available stocks continues to decline. For the future, the ecosystem of this technology in decline stays vulnerable and requires continuous care.

8.4 Conclusion: Care and decline

In this chapter, we have outlined the care activities for a declining technology, 16mm film. We noted how 16mm film technology became widely used in the 20th century and that at the turn of the century its use dropped rapidly, giving way to video technologies. In spite of its decline, 16mm film technology is still used, for instance in collections of time-based media art. In this chapter we followed how art museums face the task of keeping this obsolete technology working; we studied how technicians in and around Tate care for 16mm film technology. We highlighted the artwork *Disappearance at Sea* of the British artist Tacita Dean and traced the various care efforts.

In our review of the notions of care and maintenance we delineated a heuristic of following care goals (what) and strategies (how). For this chapter we identified the goals of sustaining specific artworks, 16mm film art in general and 16mm film technology. Strategies of care can include work at the level of objects, networks of care and ecosystems. Accordingly, we followed the care for 16mm film projectors and their parts, such as lamps, belts and loopers. We observed how technicians have to improvise with the materials and the procedures and how, by doing so, they train and sustain dedicated skills and knowledge. The care for technologies in decline also includes a version of the Frankenstein story: putting parts together to render it a new life. Subsequently, we traced how the skills and expertise are, in turn, sustained by a "network of care" in which the technicians, artists and the museum at large take part. Technicians depend on such networks of care, and make sure they keep the networks with their fellow professionals elsewhere alive, by an economy of giving and taking. We finally reported the efforts to care for the ecosystem of 16mm film and other analogue film. Tacita Dean herself has been active here, as well as many well-known actors and movie directors. Clearly, care for decline is multifaceted.

A few additional observations stand out in our exploration of care for a technology in decline. In most cases, care efforts are *invisible* to museum visitors, and often within the organization of the museum, too. Second, the *heterogeneity* of care work and of caretakers is striking: technicians and conservators come from different backgrounds, ranging from engineering to art schools, and work in different organizations. Their care efforts, organized in different practices, follow different strategies, too. An important issue for all of them is their network outside the museum: typically, they see themselves as working across two worlds, providing expertise and resources.

The heterogeneity also brings different goals people have in mind when they care. Those employed by a museum have the aims of the museum in mind when they care for 16mm; their care is aimed towards single film-based artworks and the possibility of film art in general. Others, such as the independent technicians Ken Graham and David Leister, are embedded in both the art world and the engineering and technical communities. In addition to museums, they also support cinemas and film heritage institutes. They operate at the boundaries of different institutions, they bridge these worlds and have built international networks to source their materials and to share knowledge. These people aim to continue both film art and film technology. Tacita Dean, finally, is a special case whose authority seems to come from many sources. She is famous because of her film-based artworks, and she continues to care for them. At the same time she has become an activist for film art and the prospects for film in general. She displays a large spectrum of activities and relations, which results in a broad spectrum of care practices.

To conclude, a technology in decline may incite many efforts to continue its societal presence. These efforts reveal, on the one hand, what a "working" technology entails when it is not in decline: a myriad of suppliers, technical skills and linguistic conventions. Care for decline highlights the "lived working of the networks" that tend to be ignored in technology and innovation studies with their focus on newness (Sormani, Bovet & Strebel 2019: 2). On the other hand, the efforts to sustain a technology in decline bring about new networks and ecosystems of care, with new boundaries and new authorities.

Notes

1 This publication builds on the fieldwork conducted by the first author as part of his PhD project. This is a collaborative doctorate between Maastricht University and Tate, supervised by the second author, Vivian van Saaze and Pip Laurenson. The research is part of the project "Reshaping the Collectible: When Artworks Live in the Museum". The project is supported by a grant from the Andrew W. Mellon fund.
2 Tate is a family of four museums and holds the United Kingdom's national collection of British art, and international modern and contemporary art.
3 The term network of care is inspired by Annet Dekker's (2018) analysis of the networks that form around and sustain works of net art.
4 The editions are now owned by Tate, De Pont (Tilburg, The Netherlands) and FRAC Bretagne (France).
5 These are fictive names.
6 An earlier version of the campaign talked about "declaring it a world heritage" (Savefilm. org, n.d. c)

References

Cleveland, D. and Pritchard, B. (2015) *How Films Were Made and Shown: Some Aspects of the Technical Side of Motion Picture Film, 1895–2015*. David Cleveland.

de la Bellacasa, M.P. (2017) *Matters of Care Speculative Ethics in More Than Human Worlds*. University of Minnesota.

Dean, T. (2018) *Tacita Dean. [Volume 2], Complete works & filmography, 1992–2018*. Royal Academy of Arts.

Dekker, A. (2018) *Collecting and Conserving Net Art. Moving Beyond Conventional Methods*. doi:10.4324/9781351208635.

Denis, J., and Pontille, D. (2015) Material ordering and the care of things. *Science, Technology, & Human Values*, 40(3), 338–367. www.jstor.org/stable/43671239.

Denis, J., Mongili, A. and Pontille, D. (2015) Maintenance and repair in science and technology studies. *Tecnoscienza: Italian Journal of Science & Technology Studies*, 6(2), 5–15.

Depocas, A., Ippolito, J. and Jones, C. (2003) *The Variable Media Approach: Permanence through Change [L'Approache des Médias Variable: La Permanence par le Changement]*. Guggenheim Museum; Daniel Langlois Foundation for Art, Science and Technology.

Domínguez Rubio, F. (2020) *Still Life: Ecologies of the Modern Imagination at the Art Museum*. The University of Chicago Press.

Edgerton, D. (2019) *The Shock of the Old: Technology and Global History since 1900*. 2nd edn. Profile Books.

Eisloeffel, P. (2013) A brief history of the 16mm film format. *MAC Newsletter*, 41(1), 32–33.

Filmadvocacy.org (2013) *Motion Picture Film, World Heritage and Freedom of Expression*. www.filmadvocacy.org/2013/09/03/motion-picture-film-world-heritage-and-freedom-of-expression/.

Foundas, S. (2015) Christopher Nolan rallies the troops to save celluloid film. *Variety*. https://variety.com/2015/film/columns/christopher-nolan-rallies-the-troops-to-save-celluloid-film-1201450536/.

Graham, S., and Thrift, N. (2007) Out of order: Understanding repair and maintenance. *Theory, Culture & Society*, 24(3), 1–25. doi:10.1177/0263276407075954.

Halpern, M.K., Sadowski, J., Eschrich, J., Finn, E., and Guston, D.H. (2016) Stitching together creativity and responsibility: Interpreting Frankenstein across disciplines. *Bulletin of Science, Technology & Society*, 36(1), 49–57. doi:10.1177/0270467616646637.

Hammond, K. (2004) Monsters of modernity: Frankenstein and modern environmentalism. *Cultural Geographies*, 11(2), 181–198. www.jstor.org/stable/44250971.

Harper, D.A. (1987) *Working Knowledge: Skill and Community in a Small Shop*. University of Chicago Press.

Henke, C.R. (2000) The mechanics of workplace order: Toward a sociology of repair. *Berkeley Journal of Sociology*, 44, 55–81. www.jstor.com/stable/41035546.

ICOM (n.d.) *Museum Definition*. https://icom.museum/en/resources/standards-guidelines/museum-definition/.

Jackson, S. (2014) Rethinking repair. In Tarleton, G. (ed.) *Media Technologies: Essays on Communication, Materiality, and Society*. MIT Press.

Jackson, S. (2019) Material Care. In Gold, M.K. and Klein, L.F. (eds) *Debates in the Digital Humanities 2019*. University of Minnesota Press.

Jennings, K. (2007) Tacita Dean, disappearance at sea, 1996. In Scholte, T. and 't Hoen, P. (eds) *www.inside-installations.org*. ICN/SBMK.

Kattelle, A.D. (1986) The evolution of amateur motion picture equipment 1895–1965. *Journal of Film and Video*, 38(3–4), 47–57. www.jstor.org/stable/20687736.

The Kodakery (2015) *Episode 1: Tacita Dean*. Podcast, 22 December. https://soundcloud.com/the-kodakery/episode-1-tacita-dean.

Latour, B. (1996) *Aramis, or, The Love of Technology* (C. Porter, Trans.). Harvard University Press.

Latour, B. (2005) Victor Frankenstein's real sin – What is the right use of technology? A conversation between philosophers. *Domus*, 878, 28.

Laurenson, P. (2004) Developing strategies for the conservation of installations incorporating time-based media: Gary Hill's between cinema and a hard place. *Tate Papers*, 1. www.tate.org.uk/research/publications/tate-papers/01/developing-strategies-for-the-conservation-of-installations-incorporating-time-based-media-gary-hills-between-cinema-and-a-hard-place.

Laurenson, P. (2006) Authenticity, change and loss in the conservation of time-based media installations. *Tate Papers*, 6. www.tate.org.uk/download/file/fid/7401.

Law, J. (2015) Care and killing: Tensions in veterinary practice. In Mol, A., Moser, I. and Pols, J. (eds) *Care in Practice: On Tinkering in Clinics, Homes and Farms*. Transcript Verlag.

Lejeune, C. (2019) Interruptions, lunch talks, and support circles: An ethnography of collective repair in steam locomotive restoration. In Strebel, I., Bovet, A. and Sormani, P. (eds) *Repair Work Ethnographies: Revisiting Breakdown, Relocating Materiality*. Springer & Palgrave Macmillan.

Lucas, R.C. (2011) *Crafting Digital Cinema: Cinematographers in Contemporary Hollywood* (Unpublished PhD thesis, University of Texas). https://repositories.lib.utexas.edu/bitstream/handle/2152/ETD-UT-2011-08-4147/LUCAS-DISSERTATION.pdf?sequence=1&isAllowed=y.

Manchester, E. (2009) *Tacita Dean. Kodak. 2006*. www.tate.org.uk/art/artworks/dean-kodak-t12407.

Matthews, G.E. and Tarkington, R.G. (1955) Early history of amateur motion-picture film. *SMPTE Journal*, 64(3). doi:10.5594/J14538.

Miller, Z. (2021) Practitioner (in)visibility in the conservation of contemporary art. *Journal of the American Institute for Conservation*, 60(2–3), 197–209. doi:10.1080/01971360.2021.1951550.

Mol, A. (2002) *The Body Multiple: Ontology in Medical Practice*. Duke University Press.

Mol, A., Moser, I. and Pols, J. (2015) *Care in Practice: On Tinkering in Clinics, Homes and Farms*. doi:10.14361/transcript.9783839414477.

Norris, P. (2016) Kodachrome and the rise of 16mm professional film production in America, 1938–1950. *Film History*, 28(4), 58–99. doi:10.2979/filmhistory.28.4.03.

Orr, J.E. (2016) *Talking about Machines: An Ethnography of a Modern Job*. EBSCOhost. doi:10.7591/9781501707407.

Read, P. (2006) A short history of cinema film post-production. In Polzer, J. (ed.) *Zur Geschichte des Filmkopierwerks*. Joachim Polzer.
Russell, A.L. and Vinsel, L. (2018) After innovation, turn to maintenance. *Technology and Culture*, 59(1), 1–25. doi:10.1353/tech.2018.0004.
Sallabank, J. (2010) The role of social networks in endangered language maintenance and revitalization: The case of Guernesiais in the Channel Islands. *Anthropological Linguistics*, 52(2), 184–205. www.jstor.org/stable/41330796.
Savefilm.org (n.d. a) *Join the Campaign*. www.savefilm.org/savefilm-org/.
Savefilm.org (n.d. b) *Supporters*. www.savefilm.org/supporters/.
Savefilm.org (n.d. c) *Savefilm.org*. Archived version of 24 June 2014. http://web.archive.org/web/20140124200805/http:/www.savefilm.org:80/.
Sexton, J. (2003) "Televerite" hits Britain: Documentary, drama and the growth of 16mm filmmaking in British television. *Screen: The Journal of the Society for Education in Film and Television*, 44(4), 429.
Shelley, M.W. (1993) *Frankenstein or The Modern Prometheus*. Wordsworth Classics.
Smith, C. (2012) "The last ray of the dying sun": Tacita Dean's commitment to analogue media as demonstrated through FLOH and FILM. *Necsus*, 1(2), 269–298.
Sormani, P., Bovet, A. and Strebel, I. (2019) Introduction: When things break down. In Strebel, I., Bovet, A. and Sormani, P. (eds) *Repair Work Ethnographies: Revisiting Breakdown, Relocating Materiality*. Springer & Palgrave Macmillan.
Star, S.L. and Ruhleder, K. (1996) Steps toward an ecology of infrastructure: Design and access for large information spaces. *Information Systems Research*, 7(1), 111–134. doi:10.1287/isre.7.1.111.
Stubber, C. and Kirkman, M. (2016) The persistence of the Frankenstein myth: Organ transplantation and surrogate motherhood. *Soundings: An Interdisciplinary Journal*, 99(1), 29–53. doi:10.5325/soundings.99.1.0029.
Tronto, J. (1993) *Moral Boundaries: A Political Argument for an Ethic of Care*. Routledge.
Turnheim, B. (2023) Destabilisation, decline and phase-out in transitions research. In Koretsky, Z. *et al.* (eds) *Technologies in Decline: Socio-Technical Approaches to Discontinuation and Destabilisation*. Routledge.
Van Saaze, V. (2013) *Installation Art and the Museum: Presentation and Conservation of Changing Artworks*. Amsterdam University Press.
Vinck, D. (2019) Maintenance and repair work. *Engineering Studies*, 11(2), 153–167. doi:10.1080/19378629.2019.1655566.
Vinsel, L. and Russell, A.L. (2020) *The Innovation Delusion: How our Obsession with the New Has Disrupted the Work that Matters Most*. Currency.
Zimmermann, P.R. (1995) *Reel Families: A Social History of Amateur Film*. Indiana University Press.

PART III

Governance explorations

9

IMPLEMENTING EXNOVATION?

Ambitions and governance complexity in the case of the Brussels Low Emission Zone

Ela Callorda Fossati, Bonno Pel, Solène Sureau, Tom Bauler and Wouter Achten

9.1 Introduction

It is becoming common wisdom that the development of sustainable innovations needs to be accompanied with deliberate actions and policies aiming at the destabilisation, decline and phase-out of unsustainable technologies and practices (Kivimaa & Kern 2016; Rogge & Johnstone 2017; Ayling & Gunningham 2017). The pursuit of transition S-curves (cultivation, diffusion and institutionalisation of sustainable technologies and practices) has been extended into more encompassing strategies towards sustainable "X-curves" (Hebinck et al. 2022). These strategies aim for accelerated sustainability transitions (Markard et al. 2020) through the active unmaking of outdated and unsustainable socio-technical structures.

One of the concepts that expresses this shift in transition strategies is "exnovation" (Arnold et al. 2015; David 2017, 2018; David & Gross 2019, Heyen et al. 2017). This concept is coming from science, and more precisely, it is elaborated by researchers in transitions governance who flag the need for a paradigm shift to redefine current challenges in sustainability transitions.[1] Exnovation has been defined as "the purposive termination of existing (infra)structures, technologies, products, and practices" (Heyen et al. 2017: 326). Thus, the exnovation concept expresses how transitions research is moving beyond its "comfort zone" of cultivating disruptive innovation (Heyen et al. 2017), engaging more with intentionality directed towards the closing of certain (carbon-intensive or otherwise unsustainable) technologies and broader socio-technical configurations. Also, exnovation is more than a mere reversal of innovation. As a concept introduced in transitions governance research, it engages with the identification of broader and relevant policy mixes (Kivimaa & Kern 2016) and policy intervention points (Kanger et al. 2020) for transformative system change.

In this chapter, we propose the following working definition of *exnovation*: it refers to the active unmaking of mature and new technologies and broader socio-

DOI: 10.4324/9781003213642-9

technical configurations that raise critical sustainability problems. This definition calls attention to at least three features: 1) the need to deal with critical and increasingly acute sustainability problems; 2) the need for active, purposive intervention into decline processes; and 3) the uncertainty about appropriate policy goals, instruments and vertical/horizontal modes of governance.

The first point distinguishes the exnovation concept from the literature grounded in historical cases of technological decline (Turnheim & Geels 2012; Normann 2019; Koretsky & van Lente 2020) where sustainability issues tend to be missing or take a less central place. It also distinguishes our definition of exnovation from the one given by Heyen and colleagues, which they do not anchor in sustainability issues even if they are considered of "particular interest" (2017: 326). By referring to "critical sustainability problems", we understand the term "critical" as put by Latour (2021: 41): It "no longer merely reflects a subjective and intellectual quality [i.e., *l'esprit critique*] but a perilous and terribly objective situation [i.e., *habitabilité de la zone critique*]". Second, the exnovation concept involves the direct allusion to *in*novation, for which policies and purposive intervention repertoires exist and have gained a great deal of visibility in society. Other concepts, such as "discontinuation" (Hoffmann et al. 2017; Stegmaier et al. 2021), that also address purposive intervention into decline processes from a transitions governance research perspective do not imply this direct allusion to innovation and innovation governance. The third point contrasts with common terms for purposive intervention into decline processes such as "phase-out", "ban" or "moratoria" that point to top-down governance approaches and insular targets. Relatedly, it is worth noting that the "what question" behind our definition of the exnovation concept goes beyond already mature unsustainable technologies and established socio-technical regimes, contrary to the underlying assumption of large parts of the literature (including the discontinuation of socio-technical regimes literature or the definition of exnovation given by Heyen et al.). Active unmaking concerns not just "the actually existing". Indeed, the futurisation of politics for sustainability is a key element in the conceptualisation of exnovation (David & Gross 2019). For example, considering the *Energiewende*, the authors highlight how nuclear power was introduced by the German government as a "new technology", "a modernising option" and how, soon after its emergence, large parts of German civil society objected to it because they considered it too risky and unsustainable.

Empirically, this chapter explores how exnovation is pursued in the Low Emission Zone (LEZ) policy in the Brussels Capital Region (BCR or Brussels).[2] The Brussels LEZ is one of the few schemes (amongst more than 200 LEZ[3] existing in Europe) involving (since recently) a planned full access ban on internal combustion engines (ICE) vehicles (i.e., with official announcements and "roadmaps"). We propose to discuss a set of exnovation challenges, reflecting in particular on the lessons drawn from a metropolitan-level perspective at which governance complexity manifests particularly heavily.

The chapter is structured as follows. After a discussion of exnovation in complex governance settings (section 9.2) and a clarification of methodology (section 9.3), we present the LEZ as the centrepiece of a longer-term technologically centred exnovation policy agenda. The empirical analysis highlights three key challenges, pertaining to dynamics in the science–policy interface (section 9.4), distributed decision making (section 9.5), and the engagement with widening exnovation futures beyond one-dimensional strategies (section 9.6). The conclusion teases out the broader implications for exnovation research and practice (section 9.7).

9.2 Exnovation and governance complexity: Setting the scene of the LEZ case

There is already a certain stock of knowledge on how technologies decline (see Koretsky, 2023; Turnheim, 2023; Stegmaier, 2023). Still, it remains difficult to move from general, retrospective insights to an understanding of current challenges when pursuing exnovation through policy interventions. The arguments to engage more with system innovation *in-the-making* (Smith et al. 2010; Garud & Gehman 2012) apply even more to processes of unmaking: these processes make it painfully felt how transitions tend to be confusing mixtures of purposive actions and spontaneously emerging processes (Smith et al. 2005; Rotmans 2005). Governance actors need to find out what their scope for agency is in these processes—often learning the hard way.

Exnovation is in our case, the LEZ (cf. Box 9.1), a highly sensitive matter. It starts from the formidable forces of continuity in the automotive sector (Wells & Nieuwenhuis 2012). These have impeded the substitution of the ICE by cleaner technological options. The Dieselgate episode, in which several major car manufacturers turned out to have manipulated the emission profiles of their cars, marks a classical crisis in environmental governance: governments, betting on responsive regulation, and ambitious innovation programmes, saw their trust in the private sector betrayed. Meanwhile, actors from industry eventually felt themselves compelled to game the testing procedures, as they considered the technological performance requirements impossible to meet. The subsequent introduction of the LEZs forms a new phase in this exnovation process. Accordingly, this bold and rigorous measure should not be confused for the straightforward implementation of an exnovation policy. It is a move in the exnovation game—and neither the opening of it nor the checkmate to end it. In the words of Stegmaier et al. (2021: 4): "It is utterly hard to predict how many withdrawals from withdrawals and continuations will occur until a socio-technical regime is all over and past".

The pursuit of exnovation though policy interventions happens in complex governance settings. This complexity manifests particularly heavily in metropolitan-level policymaking. For instance, the Dieselgate episode showed how German cities were constrained to act upon judicial decisions to ban diesel cars for the associated health hazards, these court decisions being inextricably connected with the broader context of global car markets and EU-level negotiations over regulations (Gross & Sonnberger 2020). Whilst metropolitan administrations worldwide are launching ambitious sustainable development programmes, they can also be considered pivotal actors in the governance of multi-level exnovation processes.[9]

BOX 9.1 CHARACTERISTICS OF THE BRUSSELS LOW EMISSION ZONE

An LEZ is a demarcated urban zone certain vehicles are not allowed to access based on emission standards.[4] Traditionally, the aim of these zones has been to counter persistent and acute traffic related air quality problems in urban areas. These zones involve top-down regulation: governments decide and control the implementation of this policy instrument. The EU Directive on Air Quality has played a key role in their dissemination across European cities. However, LEZ are implemented in different ways, there is not a single or standard policy design. They vary notably according to the criteria highlighted in the box below. More importantly, they are becoming key instruments for the phase-out of ICE vehicles in urban areas.

The BCR introduced an LEZ (1.0) on 1 January 2018. It has been in operation since then, with an exception for the period of the first Covid-19 lockdown.[5] The measure forms part of the Air Climate Energy Plan adopted by the Brussels Government in 2016.[6] At the time of the LEZ introduction, the calendar for the gradual banning of certain vehicles was settled until 2025. Since June 2021, the calendar extending the LEZ (2.0) has been officially laid out.[7] It is this extension that is expected to lead to the phase-out of ICE vehicles by 2035.

The terminology LEZ "1.0" and "2.0" to refer to the different phases of the policy was used by a representative of an administration during one of the interviews we conducted for this research. However, this expression is not officially used. LEZ 1.0 corresponds to the first phase where the policy was envisaged as a lever to fight air pollution without the objective of technological phase-out. In this sense, LEZ 1.0 was compatible with incremental innovation in ICE vehicles. LEZ 2.0, which could be dated from June 2021 when it became an exnovation policy with phase-out targets, has a dedicated calendar and various associated support measures. Moreover, unlike LEZ 1.0, which was justified based on air quality problems, LEZ 2.0 incorporates the problem of (direct) CO_2 emissions to a greater extent into its justification.

Main characteristics of the Brussels LEZ

Area covered	A large area is covered: the whole region, i.e., around 160 km^2 (except for a few sections of the ring road) The perimeter refers to the administrative jurisdiction of the BCR (it is not defined in relation to the spatial concentration of air pollution or metropolitan-based considerations)
Types of vehicles	Cars, minibuses, buses, coaches, vans Vehicles registered in Belgium and abroad Motorbikes, scooters, and trucks were not covered by the first phase but are included in the new roadmap as of 2025

Access criteria	Based on the type of fuel (diesel, petrol, GPL/CNG) and European toxic emissions standards (Euro standards) Progressive scheme initially planned until 2025, and now full ban planned for 2035 On 1 January 2021, diesel vehicles without Euro standards and with Euro standards 1/2/3 were excluded, as well as petrol vehicles without Euro standards and with Euro standard 1 (see the official LEZ website[8])
Time schedule	Permanent (24 hours a day, 7 days a week, as with most LEZ)
Control technology and penalties	Camera surveillance system (ANPR) Non-compliance with the LEZ results in a fine of 350 euros. However, a new fine can only be imposed for a new infraction if it takes place at least three months after the previous one
Exceptions	Several derogations exist (e.g., vehicles adapted for the transport of disabled persons, oldtimers' vehicles, fair/market vehicles, etc., cf. official LEZ website) A one-day pass exists (at 35 euros per day and limited to eight days per year). The terms have been reviewed since 1 January 2021: the pass can be purchased prior to accessing the LEZ, or the day after
Support measures	Prime Bruxell'air for modal shift and Prime LEZ for electric vehicles for small and medium-sized enterprises. Non-take-up is high for both primes which are currently under reform Infrastructure: Parking at the edges of the zone (for transit/dissuasion) and deployment of charging infrastructure for electric vehicles

Our analysis of the Brussels LEZ is particularly insightful regarding three challenges: 1) the dynamics of the science–policy interface; 2) distributed decision-making; and 3) the engagement with widening exnovation futures beyond one-dimensional (technologically centred) strategies. These three exnovation challenges are not identified deductively, from some theoretical framework. Neither were they found purely inductively in the development of the LEZ case. They follow roughly from the three disciplinary lines of inquiry defined by the research project.[10]

1. The dynamics of the science–policy interface. Exnovation comes with accusations directed towards technologies, practices, and those held responsible for them (Gross & Sonnberger 2020). The articulation of problems and the setting of political agendas come with an extensive negotiation of proof and of attributable responsibility. Historical cases on the destabilisation of smoking/cigarettes (Normann 2019) and on ban of the production/use of DDT and glyphosate (Tosun et al. 2019) already speak volumes about this. The assessment of impacts may not be a prominent theme in (innovation-oriented)

transitions theory, but it tends to gain in political importance when looking at the practice of purposive interventions into decline processes. This is especially true with sustainability issues brought to the forefront of the scene and evidence-based policy forged as a strong ideal. In other terms, exnovation revolves around evidence. And this reliance on evidence and data generates considerable governance complexity: *When is there enough evidence to implement exnovation policies?* Trencher et al. (2019) describe for example how exnovation policies evoke "discursive resistance", i.e., active resistance, delegitimisation and counter-evidence from actors who have come to depend on the technologies and structures targeted for phase-out. Accordingly, heated processes of reality construction arise in which decision-makers, experts and the public are collectively negotiating and exploring (Jasanoff 2004) the expected impacts of exnovation policies.

2. Distributed decision-making. The issues already arise in the phase of problem articulation: Which less disruptive and easier to enforce measures can be taken by others, elsewhere, or at a later time? *Who should be implementing exnovation policies?* As underlined by Loorbach (2007), the very calls for systemic sustainability transitions follow from the experience with sluggish and fragmented governance structures that fail to organise collective action in the face of sustainability problems. The governance of transitions revolves indeed around the capacity to seize "windows of opportunity" (Geels 2005) in socio-technical regimes. As matured institutional settings, they tend to be inert, yet they also give rise to tensions, institutional contradictions, and institutional voids: certain issues fall through the cracks. These tensions can be leveraged by institutional entrepreneurs to initiate radical policy proposals (Fünfschilling 2019). Importantly, there is little scope for unilateral control or straightforward implementation of exnovation policies by governmental actors. Contemporary multi-level governance structures behave as complex systems (Teisman et al. 2009): discretions, resources and political power tend to be dispersed across local, national, and supranational governance levels, and they are also differentiated functionally by administrative departments. Moreover, private sector and civic actors play important parts in governance networks, through their resources, their knowledge, and their powers to legitimate proposals. Governmental actors need to establish their position in governance networks (Johnstone & Newell 2018).
3. Exnovation politics and the engagement with widening exnovation futures beyond one-dimensional strategies. A third exnovation challenge resides in the pressures toward one-dimensional exnovation strategies and the attempts to move towards wider and more ambitious visions. Radical attempts at systemic change tend to be watered down and captured by incumbent actors. This is true when system change is examined from an innovation perspective (Smith 2007), but also when it comes to exnovation (van Oers et al. 2021). Transition champions may envision the wholesale termination of unsustainable systems, but governance actors tend to have greater concerns for continuity and incremental

approaches. Exnovation policies tend to target only particular regime elements. In the LEZ case, the policy targets the use of a mature technology: fossil fuel propulsion technologies and de facto ICE vehicles. *The issue arises whether the other dimensions of sociotechnical configurations are targeted*, which levers can be pulled, which instruments seem effective, and which policy discourses will be politically acceptable. In our case, this issue of political acceptability is of well-known importance, given the cultural entrenchment of the automobile (Geels 2005; Urry 2005). This challenge of arriving at coherent, comprehensive exnovation policies clearly shows in urban and regional-level studies. The case of underground parking in Maastricht shows how the municipal government is tied to long-term contracts with the companies exploiting the underground parking. Regarding the possible roll-back of the car-dominated inner city, at least this hand is tied: the underground parking is locked-in, which hinders a possible roll-back of the car-dominated inner city (Stanković et al. 2020). Graaf et al. (2021) similarly point how the policy mixes towards "regime destabilization" of Kivimaa & Kern (2016) are seldom rolled out in comprehensive fashion: many cities are complementing their innovation-oriented policies with restrictions on traffic circulation, yet few of them also include bans/regulations of mobility services based on vehicle characteristics or specific speed limit reductions. Certain kinds of exnovation policies hit upon political taboos: the financial advantages of the company car are acquired rights that are difficult to withdraw (May et al. 2019). These examples remind us that there are serious barriers towards comprehensive exnovation policies.

9.3 Methodology: Inquiry into the LEZ as an exnovation case

The inquiry into the LEZ has been organised around "exnovation arenas". This involves two main clarifications about case selection and case development.

A first important consideration is that diverse Brussels actors participated in the selection of the LEZ as a case of exnovation. The selection of the "LEZ case" has been informed by exchanges with actors from administrations, businesses, trade unions, and civil society.[11] The case selection was shaped by a salient event: by June 2018, a first declaration of principle was made regarding the phase-out of diesel vehicles. The future of the recently introduced LEZ appeared as a matter of great concern, since 2019 would bring regional elections and consequently the negotiation of the next general policy statement of the Brussels government (Gouvernement de la Région de Bruxelles-Capitale 2019). In that context, for Brussels stakeholders the LEZ resonated with the exnovation concept that was outlined by the researchers who mentioned both the limits of cultivating disruptive innovation and the X curves. In that sense, Brussels stakeholders recognised the LEZ as a case of exnovation even if they did not use or know the term. The terms more commonly used in the Brussels context are "phase-out", "ban", "exclusion" (e.g., of Diesel EURO 4 in 2022), "low/zero emission zone" and "low emissions mobility".[12]

A second key to our methodological approach is the idea that exnovation is redefined, negotiated, and eventually possibly managed and resolved by networks of actors who have a stake in these issues. The transition management framework underlines how the persistence of current sustainability problems is strongly due to the closedness of governance networks (Rotmans 2005). Broader and more diverse actor arenas need to be actively created and managed to get processes of system innovation and sustainability going. Hence the transitions governance instrument of the "transition arenas": staying away from the established channels of political bargaining. Involving innovation-minded and institutionally independent individuals, these arenas are geared towards the development of alternative visions and the experimentation with radical alternatives (Loorbach 2007). Exnovation arenas are similar examples of not-yet established actor networks around not yet fully defined policy issues. They are similarly defined and shaped by actors with an acute awareness of particular exnovation challenges. They reflect the developments and uncertainties of a transition process "in the making"—which in this case explicitly involves "unmaking". Otherwise, our exnovation arenas have not been envisioned as transition governance instruments, and they are not deliberately aiming to bring together transformation-seeking frontrunners. Exnovation arenas involve both regime and niche actors. They also address the central role that governmental actors can play.

Specifically, the case analysis combines four data collection methods: documentary analysis, interviews, observation of what we called "official exnovation arenas", and workshops. Through the documentary analysis, we aimed to shed light on the state of play for the elaboration and implementation of the LEZ policy. We analysed the Brussels strategy for mobility (i.e., Good Move plan) as well as other various documents related to the implementation of the LEZ (e.g., Air Climate and Energy Plan). The desk study also included a monitoring of press articles. We conducted 12 individual semi-structured interviews (April–September 2020) with different categories of stakeholders: Brussels administrations, business federations, trade unions, researchers, and a civil society organisation working on air quality issues. The case study development also relied on direct observation of the official exnovation arenas (September 2020 to February 2021). Indeed, a series of official and institutional discussion processes exist. In 2020/2021, these were articulated around working groups on ex-ante evaluations of the phasing-out of ICE vehicles. We joined the steering committee of this study as observers and followed the progress of its "impact assessment" component.[13] The discussions within the steering committee and the review of the first drafts of the study all contributed to the analysis. Finally, we organised two workshops (February and May 2021) on exnovation scenarios for urban mobility transitions which brought together around 30 participants (mainly beyond the experts of the official exnovation arenas—e.g., trade unionists, institutional entrepreneurs, representative of more peripheral administrations, including social administrations ...). The objective of these workshops was to experiment with the confrontation of viewpoints on the future of the LEZ and its inclusion in a wider societal debate on exnovation for urban (mobility) transitions.

9.4 The dynamics of the science–policy interface

The LEZ case is emblematic of the ideal of evidence-based policy. The challenge is to know when there is enough evidence to implement exnovation policies.

For the Brussels governments the main and particularly acute problem targeted by the LEZ is air pollution.[14] Yet, the LEZ came into being almost 30 years after environmental epidemiology studies had developed a new paradigm around the phenomenon of "air pollution". This paradigm focused on the concentration of pollutants in ambient air (rather than just the pollution escaping from industrial plants) and was able to unravel complex causalities and estimate attributable cases of mortality and morbidity (Boutaric & Lascoumes 2008). Today, this paradigm is established and the health benefits of the reduction of traffic-related air pollution (avoided deaths, illnesses and costs) are the primary justification for the introduction of the LEZ in Brussels (and elsewhere). However, issues such as the thresholds of harmful concentrations or the quality of air pollution measurements (what exactly is measured, how it is measured, by whom and where) are still debated in Brussels.[15]

Apart from the evidence-based overall rationale, there remains the crucial issue of policy effectiveness. Will the policy be effective in reducing air pollution? This issue engages with the specific criteria that are selected for the implementation. According to the official roadmap (BE 2021), the phase-out is conceived based on the engine types (diesel, petrol, natural gas, hybrid) and Euro Standards. At the core of a controversy over the LEZ effectiveness there is the policy assumption that vehicles with lower Euro Standards (or older vehicles) have a relatively worse environmental performance and should be phased out first.[16] Reports from academia and NGOs[17] showed that the former assumption (the older it is, the more polluting it is) is not valid for NOx emissions from diesel vehicles. Since Dieselgate this result is familiar to the local associations that fight for better air quality and to some extent to the general public. This scandal reveals that for several years real NOx emission levels of diesel vehicles were much higher than what car manufacturers declared and that was allowed by the Euro Standards. It is only from 2019 that new vehicles started to be roughly compliant with NOx limit values (Bernard et al. 2021). In that context, Hooftman et al. (2018: 15) conclude that "it would make more sense banning diesel vehicles entirely in LEZs". However, with the current incremental calendar—phasing-out vehicles of one Euro Standard every one to three years—significant reductions in NOx will happen in Brussels in 2025 only, as shown by the ex-ante assessment conducted by regional authorities (BE et al. 2019a). And diesel cars will eventually be banned from 2030 in Brussels, i.e., five to six years after London and Paris.

Euro Standards are not only not consistently stronger for the different pollutants measured, but they are also incomplete: not all air pollutants of concern are included, such as ultrafine particles or non-exhaust PMs.[18] Standards are in the end the results of political negotiations at the EU level and represent contested proxies of the environmental performance of vehicles. Due to the failure of Euro Standards on NOx especially (Hooftman et al. 2018: 15) "basing LEZ access requirements on Euro Standards proves to be problematic". Civil society actors in Brussels who are engaged in campaigns for a better air quality, contrary to the petrol and to a lesser extent the car lobbies, support the LEZ rationale. However, they are concerned about its implementation based on a not sufficiently critical approach that results in a slow phase-out calendar.

Moreover, the LEZ case elicits how challenging an evidence-based exnovation policy is when current scientific evidence is inconclusive about the appropriate substitute technology. On some aspects, there does not seem to be a scientific consensus yet about impacts of various alternative technologies. Related knowledge is constantly evolving, as are the practices of manufacturers. For example, the common view was that petrol vehicles emit more greenhouse gas emissions than diesel vehicles (e.g., EEA 2018), and are thus not a good alternative to diesel vehicles. This view has been however challenged by recent measurements in real use conditions (ICCT 2019), and calculations on the whole life cycle (European Commission cited by T&E 2017). On some other aspects where knowledge is consolidated and where assessments find some impacts are displaced to other geographical areas, there seems to be a lack of debate about what should be preserved at the expense of what (e.g., air quality in European urban areas versus freshwater quality in mining areas or air quality in production areas)[19] or what could change the terms of the debate. The displacement of impacts[20] is often neglected in the Brussels LEZ debate, which revolves around evidence of local environmental impacts (and particularly local air pollutant emissions).

The evaluation of an exnovation policy is a multifaceted and complex matter that challenges common understandings of policy effectiveness. Certainly, much of the debate focuses on considerations of effectiveness in relation to air pollution (main policy aim) and the appropriate technological substitute. However, other topics are emerging. Effectiveness in a broader sense refers to the potential destabilising effects of the LEZ on the automobility regime (which is not a stated policy aim, but rather a possible side-effect). Ex-post evaluations reveal no evidence of the LEZ reducing traffic or the fleet size; rather it tends to accelerate the "natural" renewal of the fleet with banned vehicles being mostly replaced with recent petrol vehicles and hybrid vehicles (BE et al. 2019b, 2020). Additionally, little is known about the socio-economic impacts of the LEZ. A preliminary study suggested that low-income households—if at all—tend to own mainly one old car, whilst for high-income households it is mainly the second car that is considered for replacement (Transport & Mobility Leuven 2011). This (poor) state of knowledge is affecting the elaboration of the "support measures" component of the exnovation policy.

The LEZs may thus be widely considered as one of the most effective tools to fight against air pollution (Hoen et al. 2021)—yet a closer look at the Brussels LEZ case suggests the need for a broader understanding of the effectiveness of exnovation policies and the presence of impacts that remain contested issues.

9.5 Distributed decision making

The LEZ case highlights the complexity of multi-level governance and administrative fragmentation. The key issue is then who should be implementing exnovation policies.

Generally speaking, much is at stake at the EU level, for example with the EU regulations on the automotive industry and on air quality. Regarding the latter, Belgium and the BCR have been put on notice repeatedly by the European Commission for non-compliance with pollutant concentration limits. These infringements have influenced the decision of the BCR to adopt and pursue the implementation of the LEZ.[21] From a legal perspective, the question relates to the division of powers between EU and Member States, and within a Member State to the institutional organisation of the Member State (Feltkamp & Dootalieva 2021).

Importantly, at the Belgian federal government level, to date, there has been no announcement of a phase-out decision for the sale of new ICE vehicles.[22] The federal government agreement (30 September 2020) simply states that "in order to improve air quality, reduce the climate impact of our transport and encourage innovation in the automobile sector, the government will work, in consultation with the federated entities, on phasing-out the sale of vehicles that do not meet the zero-emission standard".[23] At the federal level exnovation is envisioned as a *technological substitution* process where regime actors are likely to maintain a strategic position. This is also apparent in the conditions formulated to announce the phase-out, that include: a sufficiently developed market for zero emission cars at an affordable cost, the availability of life-cycle analyses, and the development of the infrastructure necessary for the electrification of cars. The only date introduced in the federal government agreement specifically concerns the "company car regime".[24] The existence of the latter is not contested by the federal government, but rather is envisaged as a lever for the electrification of the (general) car fleet (i.e., a lever for a "sufficiently developed market" since all new company cars will have to be "carbon neutral" by 2026). This suggests that "good candidates" in a first phase of exnovation (i.e., the process of technological substitution) include actors who seek to regain legitimacy (company car regime).

The Belgian automobile federations are not categorically opposed to this vision of exnovation as a technological substitution process, which is formulated by the federal government.[25] If the opposition is not categorical, it is because this vision would allow carmakers and intermediaries "to continue with what is essential": to sell cars and to delay the phasing-out with a view towards developing the assembly lines and distribution networks for electric cars. The federations support the pursuit of "greening the car fleet" and the pioneering role that the company car regime should play in this process, even if at the same time they participate in some sort of delegitimation work, emphasise the obstacles and call for a less tight calendar (for the "greening" of the company car regime—negotiated at federal level—and for the LEZ—negotiated at the federated level of the BCR). In sum, the federal state that has important powers related to exnovation policies such as the regulation of sales (product standards) and taxation schemes (company car regime), is held back by the political taboos surrounding car ownership and company cars. It is in this context that the extension of LEZ ("2.0") has been elaborated and entered the adoption processes at the BCR parliament—without a well-defined phase-out policy backup at the highest levels of government and clear challenges of legal transposition and translation.

In the BCR the official phase-out announcement came in 2019 with the new environmentally minded regional government (Gouvernement de la Région de Bruxelles-Capitale 2019). This announcement is different in nature from the one that would be possible at the federal level. With the Brussels LEZ, the problem is formulated in terms of a ban on use of (or on access for) the existing and future vehicle fleet rather than in relation to a ban on the sales of new cars. Regarding this aspect, it is also important to focus on the so-called "residual car fleet",[26] which refers to the future car fleet, i.e., the projected stock of cars having access to circulation. It is this residual fleet that is conceived by the BCR in terms of technological substitution. In other words, not all banned ICE vehicles should be replaced by low/zero emission vehicles. This suggests that the scope of technological substitution is limited by the pursuit of another regional policy objective: to discourage private car ownership and its use.[27] This nuance in the vision of exnovation as technological substitution can be found at the BCR level, but not at the federal level. However, the notion of "residual vehicle fleet" remains unclear: what is its size and who is included/excluded? For example, how does the federal decision to maintain the company car regime, and even consider it as a lever for the electrification transition, interact with the more balanced objectives of the BCR?

Finally, the continuity at the level of the policy instrument (from LEZ 1.0 to LEZ 2.0, cf. Box 9.1 in section 9.2) raises a series of questions around the stakeholders involved in an exnovation policy and the identification of competences required. For example, the concerns around the development of the electric alternative (uncertainties regarding technological and environmental performance, market maturity, targeted fleets, recharging infrastructure, transition of the electricity sector) were virtually absent from the LEZ 1.0 problematisation, whereas they are critical elements when the LEZ 2.0 becomes a "proper" exnovation policy. By whom and how will these concerns be integrated in the policy design and execution? Also, some issues were addressed by LEZ 1.0 but insufficiently, such as the incentives for transport behavioural change (non-take-up of the primes by individuals and companies). While the issue of non-take-up is well known for social policy administrations, it is less well known to the Brussels environment agency, which is piloting the LEZ (1.0 and 2.0), or to the other regional agencies involved in the governance bodies.

9.6 Widening exnovation futures beyond one-dimensional strategies

Since socio-technical configurations are complex multidimensional and embedded constructs, the politics of exnovation policies has much to do with the scope of "active unmaking". The LEZ has become a phase-out policy that targets a single dimension of the automobile regime: its technological core, i.e., fossil fuel propulsion technologies. The main solution envisioned by the LEZ lies on the *technological substitution* of that core, i.e., here, to accelerate the electric mobility transition. Beyond this vision, we have identified through the exnovation arenas four other visions of exnovation. These visions that problematise the unmaking of other regime/system dimensions (including industrial organisation, user practices and behaviour, urban infrastructure and space, sufficient provisioning) are less prominent—but could however play an important role in the development of more encompassing exnovation policies for a sustainable urban mobility transition.

A first alternative vision of exnovation problematises certain characteristics of the cars, beyond their motorisation. The SUVs symbolise the problem of "unreasonable" cars: they represent cars which are highly consuming of resources (energy, materials, urban space, household budgets) and give an image of a luxury good, or at least divergence from the car democratisation movement that characterised the evolution of the regime until the 1970s. For environmental NGOs in Brussels, exnovation is envisioned as a process of *technological contestation*, where the technological substitute solution provided by the regime (electrification of cars) hides an endogenous problem (i.e., the generalised upgrading[28] of the industry), which is likely to interfere with and even block the sustainability benefits expected from the former.

Technological contestation is informed by recent upwards trends in transport-related CO2 emissions that are partly explained by the industry's move towards upgrading.[29] The proponents of this vision call for exnovative actions to be taken on those variables (size, weight, mass) which are "undeniable determinants" of CO2 emissions. These variables are not targeted by regulation and the move towards upgrading is even said to correspond to an adaptation strategy of the carmakers facing regulation (Pardi, 2020). The issue of car lines is more fundamental than it first appears. It ultimately concerns the macro conditions for an effective demand for new greener cars. Would the industry be able to reconfigure in the absence of this move towards upgrading? Can innovation in this area (lightweighting[30]) reasonably be expected to come from regime actors or is it more the domain of "low tech" niches?

An NGO that supports cyclists endorses this vision. For them, the first phase of exnovation policies (e.g., a ban on car advertising) should target "any vehicle whose weight, power and speed are excessive and whose front face design is dangerous for other road users".[31] The ultimate goal of these organisations is that product standards should be set for ending these unreasonable vehicles, so that they are no longer produced. While resistance persists in the EU political agenda to regulate in accordance with this vision, there are some developments at the regional policy level. The issue of "increasingly bigger vehicles" is mentioned in the Good Move Plan as a safety issue ("zero accidents", "zero death"). More recently, the issue is associated with the debates on fairer pricing measures (i.e., it appears as a possible corrective parameter for the design of a "smart kilometre tax" that is currently discussed in the BCR). In any case, regional public authorities are weakly committed to this exnovation vision, compared to the vision of "technological substitution". This is paradoxical according to the NGOs, since measures such as the LEZ, by inducing the renewal of the car fleet, enhance the legitimacy of public action on the orientation of the market for new cars.[32]

A second alternative vision of exnovation problematises the *divestment of the most collectively inefficient mobility behaviours*, i.e., those related to the private owned car paradigm and solo driving, rather than characteristics linked to propulsion technologies or car lines. Under this vision, unmaking focuses on user practices and the cultural dimensions of the dominant regime. Niches and alternatives are to be found in the "new mobilities", carsharing, ridesharing, and mobility as a service. This is linked to the mainstream vision (technological substitution) by the notion of "captive fleet". Cars that are used most intensively (e.g., carsharing) are "good candidates" for a first phase of the ICE vehicles phase-out. Duties on environmental performance seem to be particularly binding for the captive fleets.

This vision, which is quite openly embraced by regional authorities, is often coupled with pricing measures. The idea that a higher total cost of cars will discourage car ownership is however contested since it raises issues of social inequality.[33] Motorised working classes are proportionally more affected than motorised wealthy classes while their dependence on cars tends to be stronger (constraints on housing and employment location, work at non-standard hours, complex travel arrangements to access services, etc.). Also, it is not a given that alternatives such as carsharing can internalise these kinds of constraints. It is in this context that a debate on more sophisticated taxes (i.e., the "smart kilometre tax" based on car use rather than just on ownership, as well as on other parameters with environmental and social impact) is being raised in the BCR.

Given the damaging effects of car mobility and the uncertainties linked to its zero emissions, zero accidents and zero private car ownership futures, a third alternative and more radical exnovation vision engages with the *dismantling of car mobility* and a post-car future. The contestation of infrastructures and of urban space devoted to cars are the crucial dimensions highlighted here. These advantages conferred to the regime prevent niches and alternatives such as those belonging to soft mobility (i.e., human powered mobility) to fully develop. This lock-in appears to be more prevalent in the mobility domain than in other domains. Indeed, daily mobility and car use are little sensitive to the pro-environmental values and attitudes since individuals tend to face certain spatial-temporal problems that can only be addressed by the car (Demoli et al. 2020). While we find that some spatial interventions (e.g., traffic plans, cycle roads, pedestrian areas) are dealing with the spatial competition involved in the deployment of alternative modes of transport, the very idea of dismantling car mobility remains a political taboo.[34]

A fourth vision problematises *transport needs*, i.e., unnecessary and constrained daily mobility. The main idea is that part of our daily transport demand can be avoided and that this would improve life quality. According to participants of our first workshop (cf. section 9.3 on methodology), the problem is "deeply structural" and "spatially embedded": it is about unmaking urban sprawl, high real estate prices in the city centre, shopping malls in the suburbs, the dismantling of proximity public services, the culture and organisation of work based on the workplace, etc. The automobility regime and the transport system are understood in their complex interactions with other systems, in particular the urban system (Kivimaa et al. 2015).

By raising the question of compressibility of transport needs and departing from *any* idea of substitution, this vision engages with sufficiency perspectives on sustainability debates. Indeed, this vision is in line with developments in more mature debates on exnovation, particularly in the field of energy transitions where the issue of demand reduction has reached an influential role. In the academic debate and in the discourse of local environmental NGOs, this vision is mediated by concepts such as the "right to the city" (Marcuse 2014) or "transport justice" (Gössling 2016). In the Brussels policy debate, this vision is reflected in the notion of "city of proximity" or "ten minutes city".[35]

These are not per se alternative visions of the LEZ. Rather, by problematising various regime/system dimensions, they represent alternative visions of exnovation for sustainable urban mobility transitions. In this way, the exnovation arenas helped to put into perspective the LEZ policy and its limited exnovative scope. Also, these alternative visions are not equally engaged with system change. While the second alternative (divestment of the most collectively inefficient mobility behaviours) seems compatible with ongoing transition pathways of the automobility regime, the three others may involve exnovation scenarios more radically different from the present.

9.7 Conclusion: From LEZ to comprehensive exnovation policy mixes

This chapter has presented exnovation as a new concept in transitions governance research. Purposeful unmaking targeting critical sustainability problems is clearly much more complex than the implementation of a "flipside of innovation". Considering a metropolitan-level perspective at which governance complexity manifests particularly heavily, we have addressed the question of how exnovation is pursued in the Brussels LEZ policy. More specifically, three exnovation challenges have been discussed in this chapter.

First, the case brings out the challenge that arises from the science–policy interface. A key question is when is there enough evidence to implement exnovation policies. Pursuing exnovation through policy interventions requires extensive "proof of harmfulness" and justification by impact assessments—it is well-known from cases on smoking or ozone-depleting substances how scientific ambiguity and the agency of "merchants of doubt" can delay closure. The LEZ case shows well how the need for impact assessment extends well beyond the judgement of whether a technology is harmful: this exnovation policy involved a range of effects, beyond the effects targeted. The types, magnitude and distribution of impacts are clearly the subject of intense debate on social acceptability, fairness and proportionality. These debates are politically particularly sensitive, as bans are top-down policy instruments that are as such prone to meeting resistance. Most of all perhaps, the LEZ reminds us that exnovation policies entail heavy burdens of proof on policy effectiveness. A key issue at the science–policy interface was the choice of priorities and phasing. Even if agreeing that a gradual, progressive phase-out trajectory is needed, what should be exnovated first? And what can come first in terms of implementation programmes and political agenda-setting? Exnovation policies are controversial, especially against the background of past policies with disappointing results in terms of environmental performance and policy effectiveness. Meanwhile, real achievements in environmental performance can reduce the political urgency of exnovation policies.

A second challenge concerns distributed decision making and the key question of who should be implementing exnovation policies. If it is possible to implement an exnovation policy at a city-region scale, problems of coherence quickly arise with "law as a brake" (Soininen et al. 2021): when the initiative is not pursued at higher governance levels, is not pursued in border regions, and involves a policy instrument not originally designed with an exnovation mindset. Section 9.5 clarifies how various other measures (a ban on sales or withdrawal of the company car regime) are not available at the BCR level and how the status quo at the highest levels limits the pursuit of regional objectives. Thus, a second point of attention presents the LEZ case as a clear example of a territory-bounded measure to address larger problems (metropolitan and global-level problems). It raises questions for regional governments about whether they are in the position to wait for exnovation to be taken up on the national/transnational level, instead of running ahead of the troops. Alternatively, regional governments may decide to join arms with cities in other countries that are similarly desperate to get certain technologies/practices exnovated or connect sectoral government structures in radically different ways.

Third and finally, our study has highlighted the fact that the LEZ represents a rather narrow vision of exnovation, largely relying on technological substitution. More encompassing and more radical visions exist, challenging the scope of exnovation policies beyond one-dimensional strategies targeted at technologies and artefacts. This range of moves towards wider exnovation futures (section 9.6) is arguably relevant well beyond engines, vehicles and urban mobility.

Notes

1 Originally the concept emerges in the field of organisational studies (Kimberley 1981) with a particular focus on health organisations and on how they remove past innovations to provide space for new innovations. The concept of exnovation has in fact been "borrowed" to that field and is being reframed in the field of sustainability transitions (and in the strand of transitions governance in particular) with rather a view on change of socio-technical systems. The bridging of the two research fields remains largely unexplored.
2 The BCR, which is one of the three administrative regions of Belgium, forms part of the metropolitan area of Brussels. These regions (federated entities) have gained a high degree of autonomy in recent decades.
3 www.ademe.fr/zones-a-faibles-emissions-low-emission-zones-lez-a-travers-leurope.
4 https://urbanaccessregulations.eu/low-emission-zones-main/what-are-low-emission-zones.

5 https://lacapitale.sudinfo.be/570236/article/2020-05-29/la-lez-redemarre-le-1er-juillet-bruxelles.
6 http://document.environnement.brussels/opac_css/elecfile/PLAN_AIR_CLIMAT_ENERGIE_FR_DEF.pdf.
7 www.lalibre.be/belgique/mobilite/2021/06/25/fin-des-moteurs-thermiques-a-bruxelles-selon-bruxelles-environnement-une-centaine-de-deces-pourraient-etre-evites-chaque-annee-QN4UW4HXCFHFVAY4KTSOHJQEOY/.
8 www.lez.brussels/mytax/practical?tab=Agenda.
9 Cities and city-level governance are prominent topics of investigation in (innovation-oriented) sustainability transitions literature. We can mention, for example, the urban anchoring of the work in grassroots innovations (Wolfram 2018), the territorial dimension of strategic niche management (Turnheim and Geels 2019), or the work on the geography of transitions. With the remarkable exception of Graaf et al. (2021), there is no such strong interest for urban contexts in the exnovation literature.
10 The first challenge benefits from insights of sustainability impact assessment while the second challenge captures insights from legal analysis. The third challenge joins more conventional analyses in the field of sustainability transitions around the politics of transitions policies.

More information on the research project can be found here: https://exnovation.brussels/en/homepage/.
11 The mentors of the project are the regional agency for environment and energy (Bruxelles Environnement: BE), the regional agency for business and entrepreneurship (Hub.brussels), the main regional socioeconomic deliberative body (Brupartners) and a citizen environmental local movement (BRAL).
12 More specifically, it is the notion of "sortie des véhicules à moteur thermique" that is mainly used in the early debates (preparatory work in 2019 and 2020). However, since the publication of the official roadmap in 2021 (that includes the phase-out calendar), it is the notion of "low emissions mobility" that is being put forward. For the BCR, implementation comes with a preference for the language of opportunities and innovation.
13 The results and the official roadmap were published in June 2021 (for an overview, cf. BE 2021).
14 This statement concerns the "LEZ 1.0" but should be contrasted for the "LEZ 2.0" which places greater emphasis on the problem of direct CO2 emissions.
15 www.rtbf.be/article/la-region-bruxelloise-condamnee-pour-ne-pas-avoir-lutte-assez-contre-la-pollution-de-l-air-10685986.
16 It's interesting to note that this assumption assimilates exnovation implementation with the reverse of innovation.
17 E.g., Hooftman et al. (2018) and Bernard et al. (2021).
18 A Brussels researcher active in urban mobility politics pointed out the example of innovations such as filters to reduce PMs that can result in the emission of even smaller and dangerous particles that are not on the radar of the Euro Standards. See also Transport & Environment (2017).
19 The electric car reduces CO2 emissions globally and air pollution locally but will bring new and significant pollution elsewhere (EEA 2018; BE 2021).
20 In environmental assessment, "impact displacements" result from a change in the life cycle of a product or a technology; impacts can be displaced between life-cycle stages or between impact categories.
21 www.rtbf.be/info/belgique/detail_qualite-de-l-air-la-commission-demande-des-explications-a-la-belgique-qui-ne-replit-pas-ses-obligations-2?id=10700971.
22 National governments' announcements on the phase-out of ICE vehicles are recent (since 2017), they are part of official national policy documents such as climate plans or transport plans, and most of them have not yet resulted in implementation plans or changes in regulations. Pioneers include Norway (announcement in 2017 for 2025), Holland (announcement in 2019 for 2030), Denmark (announcement in 2018 for 2030/

2035) and the UK (announcement in 2020 for 2030/2035). Beyond national policies, the phase-out of ICE vehicles is now on the EU agenda following a request from Denmark which stipulated that member countries should be free to ban sales of fossil-fuelled cars by 2030. www.transportenvironment.org/news/end-fossil-fuel-car-eu-agenda.

23 www.belgium.be/fr/la_belgique/pouvoirs_publics/autorites_federales/gouvernement_federal/politique/accord_de_gouvernement.

24 In Belgium, company cars benefit from a favourable tax regime that reduces the taxation of the organisation that provides the car and of the beneficiary compared to a cash salary. The phenomenon concerns 11.5% of the total car fleet and covers 23% of the kilometres driven by Belgian cars (May et al. 2019).

25 www.traxio.be/fr/articles/toutes-les-nouvelles-voitures-de-societe-a-zero-emission-reponse-de-traxio-febiac-renta/#/.

26 http://document.environnement.brussels/doc_num.php?explnum_id=9807.

27 Objective "D3" of the Good Move Plan.

28 This means that the market shares of SUV-type vehicles tend to increase in a way that it is no longer only the premium carmakers (e.g., BMW or Mercedes-Benz) who produce this (upgraded) type of vehicle.

29 https://exnovation.brussels/blog-article/missions-co2-vehicules-neufs-baissent-1ere-fois-depuis-5-ans/.

30 The notion of "lightweighting" can be confusing, as the problem is the increase in weight linked to the models of cars, rather than weight linked to the kind of materials, which tend to decrease. In the exnovation arenas, this confusion was found in the arguments of car lobbies.

31 www.gracq.org/actualites-du-gracq/sante-climat-et-securite-reclament-linterdiction-de-la-publicite-pour-les.

32 https://lemap.be/il-ne-faut-plus-laisser-les-constructeurs-nous-mener-en-automobile.

33 www.ieb.be/Diminuer-l-usage-de-la-voiture-en-ville-a-tout-prix-social.

34 www.sudinfo.be/id221463/article/2020-07-14/elke-van-den-brandt-ministre-bruxelloise-de-la-mobilite-je-ne-suis-pas-une-anti.

35 www.bruxelles.be/lavilleendevenir.

References

Arnold, A., David, M., Hanke, G. and Sonnberger, M. (eds) (2015) *Innovation-Exnovation: Über Prozesse des Abschaffens und Erneuerns in der Nachhaltigkeitstransformation*. Metropolis-Verlag.

Ayling, J. and Gunningham, N. (2017) Non-state governance and climate policy: The fossil fuel divestment movement. *Climate Policy*, 17(2), 131–149. https://doi.org/10.1080/14693062.2015.1094729.

Bernard, Y., Dallmann, T., Lee, K., Rintanen, I. and Tietge, U. (2021) *Evaluation of Real-World Vehicle Emissions in Brussels (The Real Urban Emissions (TRUE) Publication)*. ICCT (International Council on Clean Transportation). https://theicct.org/publications/true-brussels-emissions-nov21.

Boutaric, F. and Lascoumes, P. (2008) L'épidémiologie environnementale entre science et politique: Les enjeux de la pollution atmosphérique en France. *Sciences Sociales et Santé*, 26(4), 5–38.

Brauers, H., Oei, P.Y. and Walk, P. (2020) Comparing coal phase-out pathways: The United Kingdom's and Germany's diverging transitions. *Environmental Innovation and Societal Transitions*, 37, 238–253.

Bruxelles Environnement (BE) (2021) *Low Emission Mobility Brussels: En Route vers une Mobilité Basses Émissions*. Roadmap (first version). https://environnement.brussels/sites/default/files/user_files/roadmap1.4_lowemissionmobility_fr_final_clean.pdf; https://environnement.brussels/thematiques/mobilite/strategie-low-emission-mobility.

Bruxelles Environnement (BE), Bruxelles Prévention et Sécurité, Bruxelles Mobilité, CIRB, & Bruxelles Fiscalité (2019a) Effets attendus de la Zone de basses émissions sur le parc automobile et la qualité de l'air en région bruxelloise. https://lez.brussels/mytax/fr/practical?tab=Impact.

Bruxelles Environnement (BE), Bruxelles Prévention et Sécurité, Bruxelles Mobilité, CIRB, & Bruxelles Fiscalité (2019b) *Evaluation de la Zone de Basses Emissions – Rapport 2018.* https://lez.brussels/mytax/fr/practical?tab=Impact.

Bruxelles Environnement (BE), Bruxelles Prévention et Sécurité, Bruxelles Mobilité, CIRB, & Bruxelles Fiscalité (2020) *Evaluation de la Zone de Basses Emissions – Rapport 2019.* https://lez.brussels/mytax/fr/practical?tab=Impact.

David, M. (2017) Moving beyond the heuristic of creative destruction: Targeting exnovation with policy mixes for energy transitions. *Energy Research and Social Science*, 33, 138–146. https://doi.org/10.1016/j.erss.2017.09.023.

David, M. (2018) The role of organized publics in articulating the exnovation of fossil-fuel technologies for intra-and intergenerational energy justice in energy transitions. *Applied Energy*, 228, 339–350.

David, M. and Gross, M. (2019) Futurizing politics and the sustainability of real-world experiments: What role for innovation and exnovation in the German energy transition? *Sustainability Science*, 14(4), 991–1000.

Demoli, Y., Sorin, M. and Villaereal, A. (2020) Conversion écologique vs dépendance automobile: Une analyse des dissonances entre attitudes environnementales et usages de l'automobile auprès de ménages populaires en zone périurbaine et rurale, *Flux*, 119–120, 41–58.

European Environment Agency (EEA) (2018) *Electric Vehicles from Life Cycle and Circular Economy Perspectives.* TERM report 13/2018. www.eea.europa.eu/publications/electric-vehicles-from-life-cycle.

Feltkamp, R. and Dootalieva, A. (2021) *Exnovation in the Brussels Capital Region – What Effective Governance Power Does the Brussels Capital Region Have?* Working paper. https://researchportal.vub.be/en/publications/exnovation-in-the-brussels-capital-region-what-effective-governan

Fünfschilling, L. (2019) An institutional perspective on sustainability transitions. In Boons, F. and McMeekin, A. (eds) *Handbook of Sustainable Innovation.* Edward Elgar Publishing.

Garud, R. and Gehman, J. (2012) Metatheoretical perspectives on sustainability journeys: Evolutionary, relational and durational. *Research Policy*, 41(6), 980–995.

Geels, F.W. (2005) Processes and patterns in transitions and system innovations: Refining the co-evolutionary multi-level perspective. *Technological Forecasting and Social Change*, 72(6), 681–696.

Geels, F.W. (2019) Socio-technical transitions to sustainability: A review of criticisms and elaborations of the multi-level perspective. *Current Opinion in Environmental Sustainability*, 39, 187–201.

Gössling, S. (2016) Urban transport justice. *Journal of Transport Geography*, 54, 1-9. https://doi.org/10.1016/j.jtrangeo.2016.05.002.

Gouvernement de la Région de Bruxelles-Capitale (2019) Déclaration de politique générale commune au Gouvernement de la Région de Bruxelles-Capitale et au Collège réuni de la Commission communautaire commune, législature 2019–2024. 20 July. www.parlement.brussels/wp-content/uploads/2019/07/07-20-D%C3%A9claration-gouvernementale-parlement-bruxellois-2019.pdf.

Graaf, L., Werland, S., Lah, O., Martin, E., Mejia, A., Muñoz Barriga, M.R., Nguyen, H.T.T., Teko, E. and Shrestha, S. (2021) The other side of the (policy) coin: Analyzing exnovation policies for the urban mobility transition in eight cities around the globe. *Sustainability*, 13, 9045. https://doi.org/10.3390/su13169045.

Gross, M. and Sonnberger, M. (2020) How the diesel engine became a "dirty" actant: Compression ignitions and actor networks of blame. *Energy Research & Social Science*, 61, 101359.

Hebinck, A., Diercks, G., von Wirth, T., Beers, P.J., Barsties, L., Buchel, S., Greer, R., van Steenbergen, F. and Loorbach, D. (2022) An actionable understanding of societal transitions: The X-curve framework. *Sustainability Science*. https://doi.org/10.1007/s11625-021-01084-w.

Heyen, D.A., Hermwille, L. and Wehnert, T. (2017) Out of the comfort zone! Governing the exnovation of unsustainable technologies and practices. *Gaia*, 26(4), 326–331.

Hoen, A., Hilster, D., Király, J., de Vries, J. and de Bruyn, S. (2021) *Air Pollution and Transport Policies at City Level. Module 2: Policy Perspectives*. CE Delft. https://epha.org/clean-airhealthy-cities-a-silver-lining-for-a-healthy-recovery/.

Hoffmann, S., Weyer, J. and Longen, J. (2017) Discontinuation of the automobility regime? An integrated approach to multi-level governance. *Transportation Research Part A: Policy and Practice*, 103, 391–408.

Hooftman, N., Messagie, M., van Mierlo, J. and Coosemans, T. (2018) A review of the European passenger car regulations – Real driving emissions vs local air quality. *Renewable and Sustainable Energy Reviews*, 86, 1–21. https://doi.org/10.1016/j.rser.2018.01.012.

Hopkins, D. (2017) Destabilising automobility? The emergent mobilities of generation Y. *Ambio*, 46(3), 371–383.

Hubert, M. (2008) Expo '58 and "the car as king": What future for Brussels's major urban road infrastructure? *Brussels Studies*. https://doi.org/10.4000/brussels.621.

International Council on Clean Transportation (ICCT) (2019) Gasoline versus diesel: Comparing CO2 emission levels of a modern medium size car model under laboratory and on-road testing conditions. https://theicct.org/publications/gasoline-vs-diesel-comparing-co2-emission-levels.

Jasanoff, S. (ed.) (2004) *States of Knowledge: The Co-Production of Science and the Social Order*. Routledge.

Johnstone, P. and Newell, P. (2018) Sustainability transitions and the state. *Environmental Innovation and Societal Transitions*, 27, 72–82.

Kanger, L., Sovacool, B.K. and Noorkõiv, M. (2020) Six policy intervention points for sustainability transitions: A conceptual framework and a systematic literature review. *Research Policy*, 49(7), 104072. https://doi.org/10.1016/j.respol.2020.104072.

Kemp, R. and Rotmans, J. (2004) Managing the transition to sustainable mobility. In Elzen, B., Geels, F. and Green, K. (eds) *System Innovation and the Transition to Sustainability: Theory, Evidence and Policy*. Edward Elgar.

Kemp, R., Schot, J. and Hoogma, R. (1998) Regime shifts to sustainability through processes of niche formation: The approach of strategic niche management. *Technology Analysis and Strategic Management*, 10, 175–196.

Kimberley, J.R. (1981) Managerial innovation. In Nystrom, P. and Starbuck, W. (eds) *Handbook of Organizational Design*, Vol. 1. Oxford University Press.

Kivimaa, P. and Kern, F. (2016) Creative destruction or mere niche support? Innovation policy mixes for sustainability transitions, *Research Policy*, 45(1), 205–217.

Kivimaa, P., Mäkinen, K. and Helminen, V. (2015) Path creation for urban mobility transition: Linking aspects of urban form to transport policy analysis. *Management of Environmental Quality*, 26(4), 485–504.

Koretsky, Z. (2023) Dynamics of technological decline as socio-material unravelling. In Koretsky, Z. *et al.* (eds) *Technologies in Decline: Socio-Technical Approaches to Discontinuation and Destabilisation*. Routledge.

Koretsky, Z. and van Lente, H. (2020) Technology phase-out as unravelling of socio-technical configurations: Cloud seeding case. *Environmental Innovation and Societal Transitions*, 37, 302–317. https://doi.org/10.1016/j.eist.2020.10.002.

Latour B. (2021) *Où Suis-Je? Leçons du Confinement à L'usage des Terrestres*. Editions La Découverte.

Loorbach, D. (2007) *Transition Management: New Mode of Governance for Sustainable Development*. http://hdl.handle.net/1765/10200.

Marcuse, P. (2014) Reading the right to the city. *Analysis of Urban Change*, 18(1), 4–9. https://doi.org/10.1080/13604813.2014.878110.

Markard, J., Geels, F.W. and Raven, R. (2020) Challenges in the acceleration of sustainability transitions. *Environmental Research Letters*, 15(8), 081001.

May, X., Ermans, T. and Hooftman, N. (2019) Les voitures de société: Diagnostics et enjeux d'un régime fiscal. In Vandenbroucke, A. *et al.* (eds) *Voitures de Société et Mobilité Durable: Diagnostic et Enjeux*. Éditions de l'Université de Bruxelles.

Niko, S., Romppanen, S., Huhta, K. and Belinskij, A. (2021) A brake or an accelerator? The role of law in sustainability transitions. *Environmental Innovation and Societal Transitions*, 41, 71–73. https://doi.org/10.1016/j.eist.2021.09.012.

Normann, H.E. (2019) Conditions for the deliberate destabilisation of established industries: Lessons from U.S. tobacco control policy and the closure of Dutch coal mines. *Environmental Innovation and Societal Transitions*, 33, 102–114.

Pardi, T. (2020) *Everything Must Change for Everything to Stay the Same? Prospects and Contradictions of the Electrification of the European Automotive Industry*. https://sase.confex.com/sase/2020/meetingapp.cgi/Paper/15984.

Rogge, K.S. and Johnstone, P. (2017) Exploring the role of phase-out policies for low-carbon energy transitions: The case of the German Energiewende. *Energy Research & Social Science*, 33, 128–137.

Rotmans, J. (2005) Societal innovation: Between dream and reality lies complexity. Inaugural lecture Erasmus Universiteit Rotterdam.

Smith, A. (2007) Translating sustainabilities between green niches and socio-technical regimes. *Technology Analysis & Strategic Management*, 19(4), 427–450.

Smith, A., Stirling, A. and Berkhout, F. (2005) The governance of sustainable socio-technical transitions. *Research Policy*, 34(10), 1491–1510.

Smith, A., Voß, J.P. and Grin, J. (2010) Innovation studies and sustainability transitions: The allure of the multi-level perspective and its challenges. *Research Policy*, 39(4), 435–448.`

Soininen, N., Romppanen, S., Huhta, K. and Belinskij, A. (2021) A brake or an accelerator? The role of law in sustainability transitions. *Environmental Innovation and Societal Transitions*. https://doi.org/10.1016/j.eist.2021.09.012

Stanković, J., Dijk, M. and Hommels, A. (2020) Upscaling, obduracy, and underground parking in Maastricht (1965–present): Is there a way out? *Journal of Urban History*, 47(6), 1225–1250.

Stegmaier, P. (2023) Conceptual aspects of discontinuation governance: An exploration. In Koretsky, Z. *et al.* (eds) *Technologies in Decline: Socio-Technical Approaches to Discontinuation and Destabilisation*. Routledge.

Stegmaier, P., Visser, V.R. and Kuhlmann, S. (2021) The incandescent light bulb phase-out: Exploring patterns of framing the governance of discontinuing a socio-technical regime. *Energy, Sustainability and Society*, 11(1), 1–22.

Switzer, A., Bertolini, L. and Grin, J. (2013) Transitions of mobility systems in urban regions: A heuristic framework. *Journal of Environmental Policy & Planning*, 15(2), 141–160.

Teisman, G., van Buuren, A. and Gerrits, L.M. (eds) (2009) *Managing Complex Governance Systems*. Routledge.

Tosun, J., Lelieveldt, H. and Wing, T.S. (2019) A case of "muddling through"? The politics of renewing glyphosate authorization in the European Union. *Sustainability*, 11(2), 440. https://doi.org/10.3390/su11020440.

Transport & Environment (T&E) (2017) Diesel: The true (dirty) story. www.transportenvironment.org/publications/diesel-true-dirty-story.

Transport & Mobility Leuven (2011) Studie betreffende de relevantie van het invoeren van Lage-emissiezones in het Brussels Hoofdstedelijk Gewest en van hun milieu-, socio-economische en mobiliteitsimpact. Brussels Instituut voor Milieubeheer. https://document.environnement.brussels/opac_css/index.php?lvl=notice_display&id=6833.

Trencher, G., Healy, N., Hasegawa, K. and Asuka, J. (2019) Discursive resistance to phasing out coal-fired electricity: Narratives in Japan's coal regime. *Energy Policy*, 132, 782–796.

Turnheim, B. (2023) Destabilisation, decline and phase-out in transitions research. In Koretsky, Z. *et al.* (eds) *Technologies in Decline: Socio-Technical Approaches to Discontinuation and Destabilisation*. Routledge.

Turnheim, B. and Geels, F.W. (2012) Regime destabilisation as the flipside of energy transitions: Lessons from the history of the British coal industry (1913–1997). *Energy Policy*, Special Section: "Past and Prospective Energy Transitions – Insights from History", 50 (November), 35–49.

Turnheim, B. and Geels, F.W. (2019) Incumbent actors, guided search paths, and landmark projects in infra-system transitions: Re-thinking strategic niche management with a case study of French tramway diffusion (1971–2016). *Research Policy*, 48(6), 1412–1428. https://doi.org/10.1016/j.respol.2019.02.002.

Urry, J. (2004) The "system" of automobility. *Theory of Culture and Society*, 21(4–5), 25–39.

Van Oers, L, Feola, G., Moors, E. and Runhaar, H. (2021) The politics of deliberate destabilisation for sustainability transitions. *Environmental Innovation and Societal Transitions*, 40, 159–171. https://doi.org/10.1016/j.eist.2021.06.003.

Wells, P. and Nieuwenhuis, P. (2012) Transition failure: Understanding continuity in the automotive industry. *Technological Forecasting and Social Change*, 79(9), 1681–1692.

Wolfram, M. (2018) Cities shaping grassroots niches for sustainability transitions: Conceptual reflections and an exploratory case study. *Journal of Cleaner Production*, 173, 11–23. https://doi.org/10.1016/j.jclepro.2016.08.044.

10

PHASE-OUT AS A POLICY APPROACH TO ADDRESS SUSTAINABILITY CHALLENGES

A systematic review

Adrian Rinscheid, Gregory Trencher and Daniel Rosenbloom

10.1 Introduction

In 1921, an American engineer discovered that tetraethyl lead, when added to gasoline, works as an effective antiknock agent, inhibiting early ignition of internal combustion engines and thereby improving engine performance.[1] Leaded gasoline soon became the technological norm worldwide. This is despite the fact that its problematic effects on human health and ecosystems were known since the 1920s (Markowitz & Rosner 2002). The 1970s then saw an accumulation of strong epidemiological evidence about the adverse health and environmental consequences of lead. This prompted jurisdictions around the world to *phase out* the production and use of leaded gasoline over the ensuing decades (Newman 2023). Under this policy approach, governments announced schedules for the stepwise reduction of production and use, setting a final date by which a complete termination was to be achieved (Lovei 1998). These phase-outs proved a protracted affair. Indeed, the global phase-out of tetraethyl lead in automotive applications was only completed when the last service stations in Algeria ceased the sale of leaded gasoline in July 2021.

Today, the use of phase-out as a policy approach has proliferated around the world to target a diversity of elements with undesirable consequences for humans, society and the natural environment. Increasingly, these interventions are becoming an object of academic research. Following Rosenbloom and Rinscheid (2020), we define phase-out as a policy intervention that deliberately seeks to terminate one or several socio-technical elements (e.g., technologies, substances, processes or practices) in a gradual or stepwise process (see also Turnheim 2023; Stegmaier 2023). Phase-out interventions often specify a time horizon for the termination of targeted elements, sometimes also defining intermediate milestones (Trencher et al. in review). The gradual process of decline, which distinguishes phase-out from

DOI: 10.4324/9781003213642-10

abrupt bans, gives societal actors time to adapt and can help to minimize undesired disruptions to industries and supply chains.

State authority, sometimes delegated to international institutions, is central as a driver of phase-outs (Johnstone & Newell 2018). Yet private actors may also play an important role in their implementation, just as civil society and political actors will often advocate for their initiation and ongoing implementation. Phase-outs may be pursued via different policy instruments or combinations thereof. They are most readily associated with command-and-control styles of government intervention, which belong to the traditional repertoire of environmental regulation. Still other policy instruments may be used, such as market-based instruments or removal of support, such as subsidies (Geels et al. 2017).

While theoretical engagement with phase-out has, to date, been relatively dispersed (Rosenbloom & Rinscheid 2020), this approach can be situated among several that directly relate to *decline* (Koretsky 2023). Phase-out may be understood as a form of *discontinuation*, which refers to purposeful strategies to abandon or dismantle existing socio-technical systems by using a dedicated discontinuation policy approach (Stegmaier 2023). It may also be situated as an intentional driver of systemic change encompassed by the concept of *destabilization* (Turnheim 2023; Kivimaa & Kern 2016). Still others view phase-outs as the result rather than driver of broader forces of destabilization (Normann 2019). In contrast to other decline-related concepts, phase-out is more strongly embedded within the language and practice of policymaking, emerging as one of the most prominent practical efforts to address sustainability challenges over the past decades (Rinscheid et al. 2021).

So far, little effort has been undertaken to document where scholarly conversations about phase-out take place, what empirical contexts these conversations relate to, and what specific insights the literature has generated for research and policy. To address this gap, this chapter systematically explores the body of scientific knowledge on phase-out related to environmental sustainability. Offering an in-depth examination of the context of scientific production and science–policy interactions on phase-out, we first provide an overview of the evolution of this scientific work and subsequently trace the following six dimensions: 1) targets of phase-out interventions as explored in the literature, 2) academic communities that contribute to phase-out research, 3) sustainability challenges phase-out research seeks to address, 4) policy instruments at the center of scholarly work, 5) functions ascribed to phase-out interventions, and 6) societal motivations for enacting phase-outs.

We approach these dimensions through a systematic review. This method provides a transparent and replicable methodological framework to summarize and synthesize characteristics and trends within a defined body of scientific work (Haddaway et al. 2015; Peñasco et al. 2021). Systematic reviews are not only suited for charting a path through unexplored terrain, but also for identifying blind spots and distilling directions for further scholarly engagement with a topic. While our investigation allows us to provide an initial topography of sustainability-related phase-out research, this should be further refined by future work.

10.2 Methodological approach

Our review focuses on the scientific literature that discusses phase-out in the context of environmental sustainability. To identify relevant publications, we developed a search string consisting of three segments: first, the term "phase-out" and its variants; second, terms related to policy or governance; and third, terms related to sustainability challenges (see Table 10.1). This search string was used to query the titles, abstracts and keywords of English-language scholarly articles and reviews indexed in the Scopus database. We did not apply any restrictions with respect to academic fields, as we were interested in the ways phase-out is evoked across the entire body of scientific work. We selected 1972 as the starting point, corresponding with the Stockholm United Nations Conference on the Human Environment. This is widely regarded as kickstarting global awareness of human-driven environmental degradation and subsequent countermeasures through policy and technological development (Ashford & Hall 2019). The first publication identified based on our search strategy dates from 1976. We conducted our search on January 1st, 2022. It yielded 1,097 hits in total. Technically, the resulting dataset includes each academic output (published in English) in which at least one term from each segment of the search string is mentioned in the title, abstract or keywords.

Informed by accounts on systematic reviews (e.g., Grant et al. 2009; Haddaway et al. 2015), the ensuing research process was structured as follows. First, we assessed the *relevance* of publications against our research objective. Accordingly, we excluded studies in which use of the term "phase-out" did not reflect a deliberate process of downscaling the production or use of a socio-technical element in the context of environmental sustainability. Consequently, we excluded publications that 1) discuss phase-outs for health reasons solely (e.g., the phase-out of free formula milk for infants in developing countries), 2) discuss the phase-out of policies, institutions or programs not explicitly related to environmental sustainability (e.g.,

TABLE 10.1 Search string used to identify relevant literature

1st segment	*2nd segment*	*3rd segment*
'Phase-out' and variants thereof	Terms capturing policy instruments and governance approaches	Terms capturing sustainability challenges
"phase out" OR "phase-out" OR "phasing out" OR "phasing-out" OR "phased-out" OR "phased out"	*action OR agreement OR ban* OR commit* OR decision OR effort OR framework OR govern* OR incentive OR initiative OR instrument OR law OR legislat* OR management OR mandat* OR measure OR mechanism OR plan OR polic* OR program* OR regulation OR rule OR scheme OR strateg* OR treaty*	*environment OR sustainab* OR climate*

the phase-out of financial incentives for renewable energies), and 3) focus on individual behavior (e.g., the phase-out of "inappropriate" activities in China's national forest parks). After removing 229 irrelevant publications, our analyses drew on 868 publications.

Second, we developed a manual to guide the coding of relevant publications (see Table 10.2). We inductively developed the respective coding scheme for four of our properties of interest: *targets, sustainability challenges, functions* and *motivations.* Concretely, we iteratively developed an initial scheme based on a number of test runs before consistently coding the entire dataset using the final coding manual. For *policy instruments*, we drew on an established classification of environmental policy instruments (OECD 2001; Persson 2006) and adapted it in line with our specific research purpose and descriptions of policy instruments in the reviewed works. With respect to *academic communities*, we used the subject area codes

TABLE 10.2 Examined properties, guiding research questions, and examples

Property	*Associated research question*	*Examples*	*Origin of coding framework*
Phase-out target	What socio-technical element is targeted by the phase-out?	• Flame retardants • Lead • Dark-colored roofs	Inductively developed
Academic community	What academic communities have contributed to debates about phase-out?	• Environmental science • Engineering • Chemistry	Based on 'subject area' codes taken from Scopus
Sustainability challenge	What sustainability challenge is addressed by the phase-out?	• Atmospheric ozone depletion • Aquatic ecosystems • Climate change	Inductively developed
Policy instrument	What policy instruments are discussed to enact the phase-out?	• Command-and-control • Economic instruments • Voluntary approaches	Adapted from OECD (2001)
Functions	What functions are associated with the phase-out?	• Signaling • Substituting • Transforming	Inductively developed
Motivations	What motivations underscore the phase-out?	• Economic benefits • Ethics and equity • Public demand	Inductively developed

provided by Scopus. We then coded all relevant publications based on their titles, abstracts and keywords.

Third, we analyzed the data in line with the research questions introduced in Table 10.2. To identify and render visible the main characteristics and trends of scholarly work on phase-out, we used descriptive statistics and graphical illustrations. While the purpose of our work is mostly exploratory, we also offer some analytical reflections and sketch directions for further research.

10.3 Findings

10.3.1 Emergence, take-off and acceleration of phase-out research

Figure 10.1 depicts the volume of new publications over time and thereby illustrates the growth trend of phase-out research. Inspired by conceptual work on the phases of sustainability transitions (Rotmans et al. 2001), we distinguish three periods of scholarly engagement so far. The first period (emergence) is characterized by a limited number of works overall and a relatively low number of new scientific publications each year. The period starts with the first identified publication in 1976, which dealt with a policy introduced by the United States' Environmental

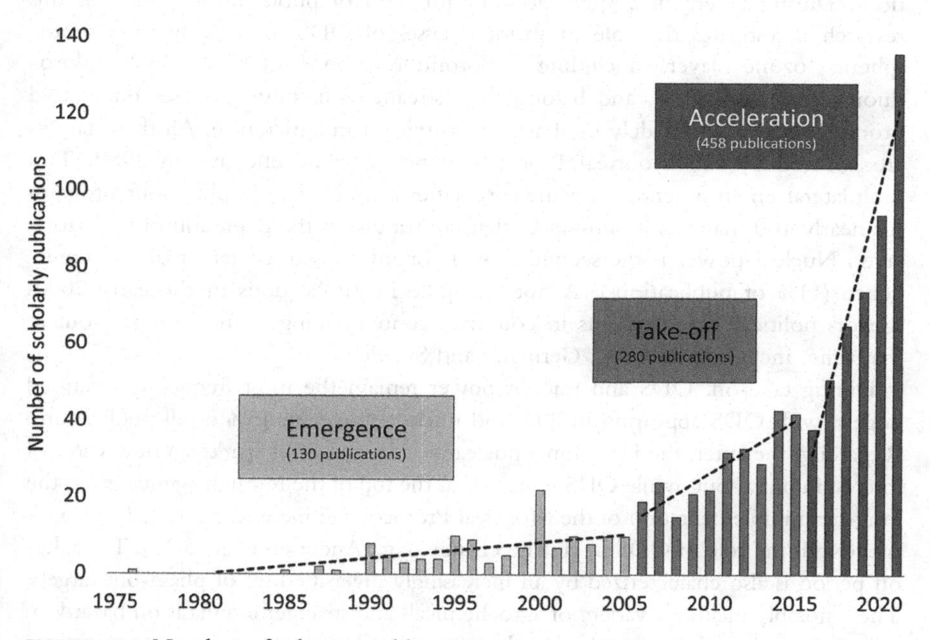

FIGURE 10.1 Number of relevant publications on phase-out in the context of climate, environment or sustainability identified on Scopus (1972–2021)

Linear trends indicate average annual increase (Δ) of the number of publications for three periods: Period 1 ("Emergence"; 1976–2005; Δ = 0.42), Period 2 ("Take-off"; 2006–2015; Δ = 2.76), Period 3 ("Acceleration"; 2016–2021; Δ = 17.94).

Protection Agency to phase out the dumping of industrial and municipal waste at sea by 1981 (Dewling & Anderson 1976). Phase-out was rarely invoked over the following years. It however appeared more frequently from 1990 to 2005, with three to ten scholarly publications each year, the exception being a spike of 22 publications in 2000. During this period, ozone depleting substances (ODS) and nuclear power were the most frequently discussed targets of phase-out (see Figure 10.2). Stronger and more sustained interest in phase-out set in after 2005, with 28 publications appearing on average each year between 2006 and 2015. During this take-off period, the average annual growth trend of the phase-out literature amounted to 13%. This considerably surpasses the average growth rate of all publications in the scientific enterprise as a whole, which has been around 5% since the 1950s (Bornmann et al. 2021). Finally, the period 2016 to 2021 reflects the acceleration phase. It is marked by a sharp increase in academic output on phase-out and an even higher annual growth rate of 29%.

Prominence of phase-out targets over time

Figure 10.2 provides a more detailed overview of the three periods of scholarly engagement with phase-out and identifies the main targets of phase-out interventions. During emergence, ODS account for 42% of publications. Much of this research investigates the role of various classes of ODS in depleting the stratospheric ozone layer, including chlorofluorocarbons (CFCs), hydrochlorofluorocarbons (HCFCs), and halons. One stream of literature focuses on methyl bromide, which was widely used as a soil fumigant in agriculture. All these targets are covered by the Montréal Protocol, which became effective in 1989. This multilateral environmental agreement specifies a multi-decade phase-out program for nearly 100 man-made substances that contribute to the depletion of the ozone layer. Nuclear power is the second most frequently discussed target during emergence (11% of publications). A notable spike in publications in the early 2000s mirrors political developments in countries contemplating a nuclear phase-out at that time, including Belgium, Germany and Sweden.

During take-off, ODS and nuclear power remain the most frequently discussed targets, with ODS appearing in 20% and nuclear power in 15% of all publications. Regarding the latter, the Fukushima nuclear accident in 2011 sparked a new wave of interest in phase-out, while ODS remained at the top of the research agenda given the long-term implementation of the Montréal Protocol and increasing scholarly engagement with the role of ODS in climate change (e.g., Andersen et al. 2013). The take-off period is also characterized by an increasingly diversified set of phase-out targets. These notably include a variety of agrochemicals and stronger emphasis on hazardous substances such as flame retardants and perfluorinated compounds.

While the take-off period provided an early fertile ground for scholarly engagement with phase-out in the context of decarbonization, in particular, discussions of fossil fuel-related phase-out targets have intensified since 2016. Although the Paris Agreement does not contain any direct reference to phase-out,

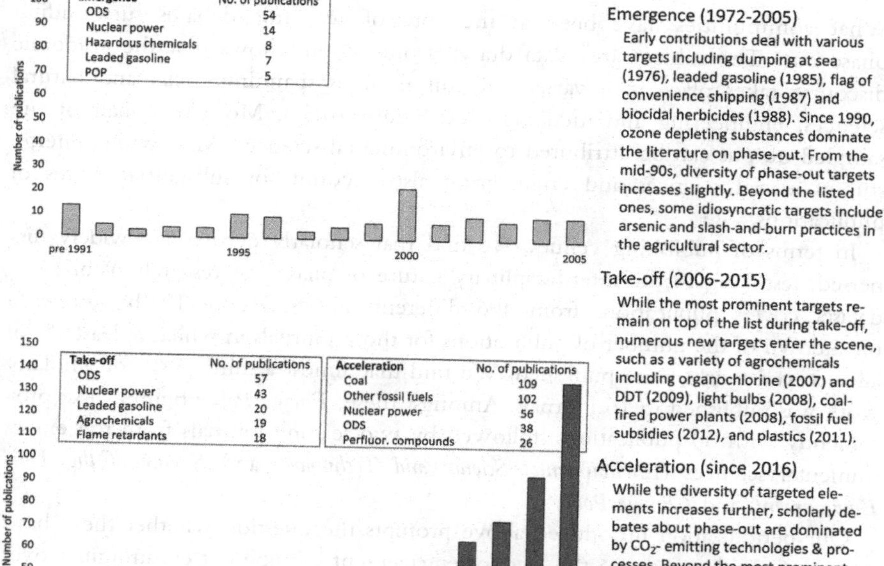

FIGURE 10.2 Overview of most prominent (top five) targets discussed in three periods (1976–2021)

the turn towards nationally determined contributions appears to have sparked strong academic interest in the domestic phase-out of carbon-intensive arrangements. Targets especially include coal mining and power generation and other fossil fuel based technologies such as gas heaters, internal combustion engines and conventional kerosene-powered aircraft. While nuclear power and ODS remain among the top targets, their relative prominence decreases to 12% and 7%, respectively. On the other hand, the acceleration period is characterized by a constantly rising number of new phase-out targets. These include waste incineration, mercury-based technologies and a growing number of agrochemicals. This indicates an increasing application of phase-out as a policy approach in new arenas.

In sum, the three periods of scholarly engagement with phase-out are not only characterized by different volumes and growth rates of scientific work, but can also be distinguished by different sets and varieties of phase-out targets. While early targets (ODS, nuclear) have all but disappeared from scholarly discourse, new ones (fossil fuel based arrangements, chemicals, etc.) have entered into academic debates, driving forward a productive line of research.

10.3.2 Engagement with phase-out in different academic fields

What communities have been at the core of academic debates surrounding phase-out? The subject area data drawn from Scopus shows that the academic discourse takes place in a variety of communities spanning social and natural sciences, engineering and medicine (see Figure 10.3a). More than half of the sampled output can be attributed to environmental sciences. Meanwhile, energy studies, social sciences and engineering also account for substantial shares of publications.

In terms of publishing venues, we find that scholarly debates are widely dispersed, testifying to the interdisciplinary nature of phase-out research. While our dataset covers publications from 406 different outlets, Figure 10.3b outlines a breakdown of the number of publications for those journals in which at least 1% of all surveyed works were published. We find that almost a third (30%) of scholarly work was published in 16 journals. Amongst these, *Energy Policy* figures most prominently with 50 publications, followed by two leading journals from the environmental sciences (*Environmental Science and Technology* and *Science of the Total Environment*) and *Climate Policy*.

Our periodization introduced above prompts the question whether there have been shifts with respect to the level of engagement by different communities over time. As seen in Figure 10.4, environmental sciences represent the dominant field throughout all periods. That said, energy studies, which accounted for a mere 15% of output during the emergence period, accounted for 43% during the acceleration period. We attribute this to increasing discussions of phase-out in the context of reducing carbon dioxide emissions from energy technologies and infrastructures. The share of scholarly work from the social sciences has increased also markedly, rising from 12% to 24%. This suggests that the social and socio-political dynamics of phase-out have become an important focus more recently. Interestingly, this tendency is not matched by the fields of economics and business studies. While economics has somewhat strengthened its level of engagement over time, the discipline accounted for only 37 publications (or 8%) during acceleration. Business, management and accounting decreased in prominence, contributing 20 papers (or 4%) over the same period. Another notable shift occurs with the relative loss of importance of engineering. While the field accounted for 22% of publications in the formative period, its share dropped to 14% in the acceleration period. This decrease provides another indicator of scholarly attention to phase-out expanding from technical to social dimensions. In summary, while environmental science continues to play a central role, there has been an increasing place for the social sciences and energy-related fields.

10.3.3 Phase-out in the context of various sustainability challenges

Which sustainability challenges have been discussed in scholarly work on phase-out, and how has their prominence shifted over time? Figure 10.5 illustrates the

(a)

Subject area	Number I share of papers
Environmental Sciences	495 I 57.0%
Energy	270 I 31.1%
Social Sciences	167 I 19.2%
Engineering	113 I 13.0%
Earth and Planetary Sciences	97 I 11.2%
Agricultural and Biological Sciences	85 I 9.8%
Chemistry	63 I 7.3%
Medicine	60 I 6.9%
Economics, Econometrics & Finance	53 I 6.1%
Business, Management & Accounting	46 I 5.3%

(b)

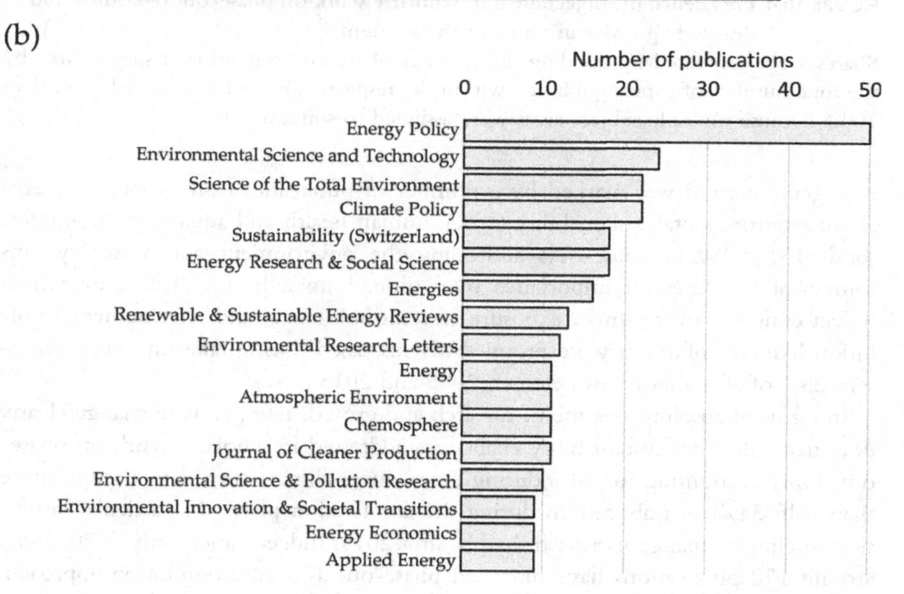

FIGURE 10.3 (a) Allocation of papers to the ten most frequent subject areas. Shares relate to the number of papers classified as relevant (n = 868). The sum of shares is higher than 100% because multiple subject areas were assigned to some papers. The dataset covers 26 subject areas in total. (b) Academic journals most prominently publishing work on phase-out. The figure includes all journals with at least 1% of the total relevant output across all surveyed works

prevalence of ten sustainability challenges discussed throughout the periods of emergence, take-off and acceleration.

During emergence, ozone depletion stands out as the most prominent driver of phase-outs, mentioned by 29% of all publications. However, its relative importance decreases substantially over time, with only a fraction of publications (6%) dedicated to the ozone problem during acceleration. Next to the ozone crisis, the

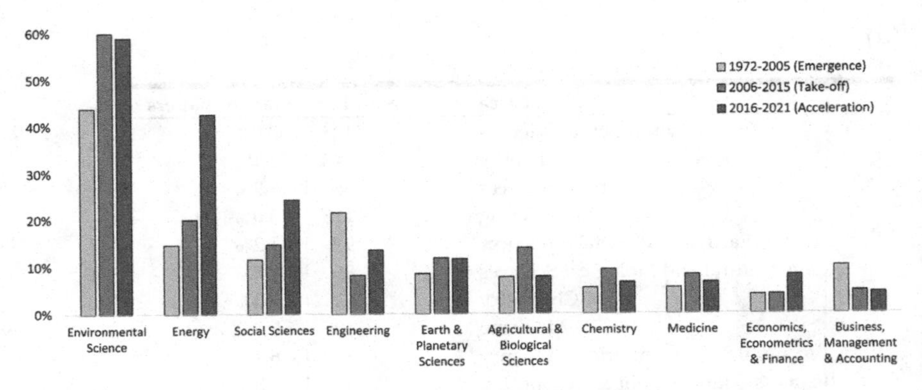

FIGURE 10.4 Prevalence of subject areas in scholarly work on phase-out over time and in descending order in terms of their volume

Shares were generated by dividing the number of papers assigned to a subject area by the total number of papers published within the respective period. Sums are higher than 100% because multiple subject areas were assigned to some papers.

emergence period was marked by a number of publications on general concerns about environmental sustainability (18%), human health and aquatic environments (both 15%). While phase-outs addressing the environment and water systems somewhat decreased in importance over time, human health challenges, which reflect concerns over human exposure to toxic substances and environmental pollution like air pollution, were prominently discussed during take-off. They represent 25% of all publications between 2006 and 2015.

In terms of absolute volume of research and growth rates, climate change clearly dominates all other sustainability challenges addressed in scholarly work on phase-out. Only accounting for 14% during emergence, climate surged to occupy more than half (54%) of publications during the acceleration phase. In particular, attention to climate change increased sharply after 2015. Indeed, since early 2020 alone, around 130 publications have discussed phase-out as a decarbonization approach. Our analysis hence documents that phase-out is increasingly being discussed as an approach to confront climate change.

10.3.4 Policy instruments

To systematically explore policy instruments, our analysis concentrates on three sustainability challenges: ozone, human health and climate. These three challenges are not only among the most widely discussed, but also maximize variance with respect to the temporal evolution of scholarly engagement with phase-out. As illustrated in Figure 10.5, ozone represents the dominant challenge during emergence, but its share has shrunk to 5% recently. The pattern is the reverse for climate change, which represents one among several challenges during emergence but dominates the contemporary phase-out research. Phase-outs attached to human

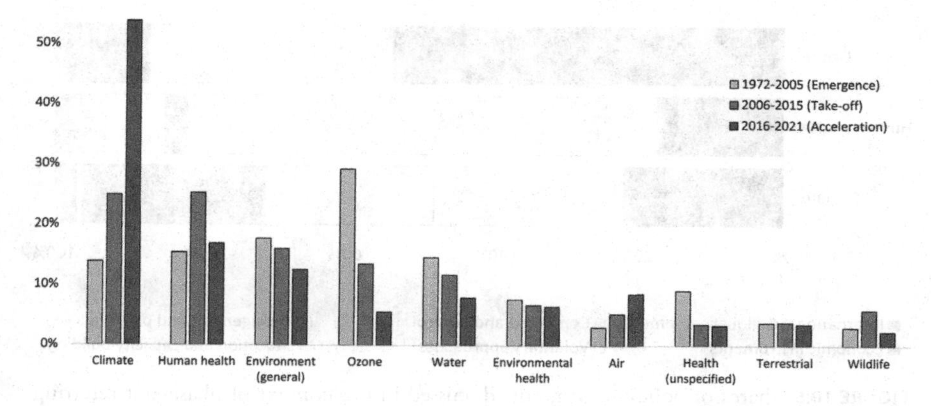

FIGURE 10.5 Sustainability challenges discussed in scholarly work on phase-out: percentage of publications highlighting challenges over time

Based on 868 articles. Shown in descending order in terms of their volume.

health challenges account for the highest share of publications during take-off and remain among the top sustainability challenges later on. The following analyses are based on 99 ozone-related, 163 health-related and 311 climate-related phase-out publications.

Based on a slightly adapted classification of environmental policy instruments initially put forward by the OECD (2001) and used in policy research (e.g., Jordan et al. 2003), Figure 10.6 illustrates the relative distribution of instruments for the three environmental sustainability challenges. According to our inductive coding approach, publications discussing several instruments received several codes, while publications mute on instruments were not coded. The bar chart indicates some variation in the importance of specific instruments for each challenge: while ozone is dominated by international and global agreements, phase-outs for health reasons are predominantly associated with command-and-control approaches. For climate change, no single policy instrument appears to dominate. In what follows, we unveil these patterns in more depth.

International and global agreements account for more than half (53%) of instrument mentions in the context of the ozone challenge. This reflects strong scholarly interest in the Montréal Protocol on Substances that Deplete the Ozone Layer, signed in 1987, and subsequent amendments. Setting the global framework to guide the phase-out of various ODS, this regime sets precise schedules for the stepwise reduction of their production and use. Based on the principle of common but differentiated responsibilities, developing countries are granted more leniency, receiving about 10–18 years more time than developed countries to complete their phase-out (Gonzalez et al. 2015). Early contributions emphasizing this global governance instrument were specifically concerned with the consequences of phasing out refrigerants (mainly CFCs) on industry (Dunn 1990) and the environmental impact of substitutes (McCulloch 1999). But there was little systematic scholarly engagement with the novelty or functioning of this policy approach, let alone an

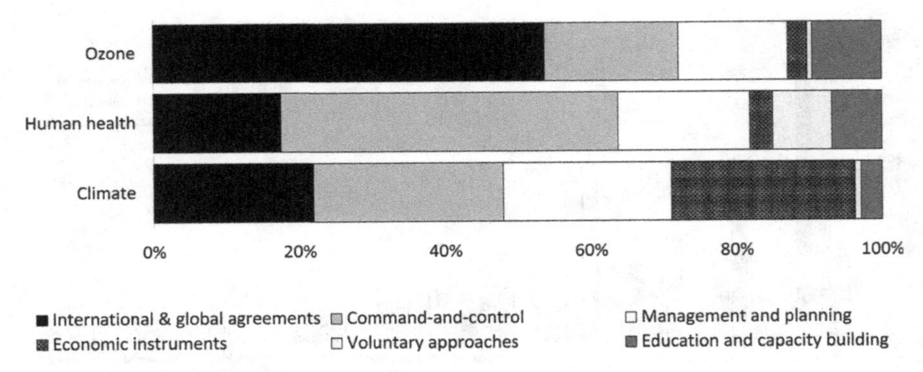

FIGURE 10.6 Shares of policy instruments discussed in the context of phase-out referring to three sustainability challenges

ambition to draw lessons for other environmental challenges. This has changed more recently. An effort to investigate the success factors of this phase-out framework, which is widely considered to be a landmark success of global environmental governance, was undertaken in a special issue of the *Journal of Environmental Studies and Sciences* (e.g., Andersen 2015; Gonzalez et al. 2015; Shende 2015).

In the context of human health challenges and climate change, international and global phase-out agreements are less commonly discussed (18% and 24% of instrument mentions, respectively). With respect to health challenges, the Stockholm Convention on Persistent Organic Pollutants (POPs; signed in 2001) and the Minamata Convention on Mercury (signed in 2013) gain some prominence. These international environmental agreements, which entail specific provisions for global substance phase-outs, set the agenda for a number of studies exploring economic implications. These include the cost of phasing out hazardous chemicals in developing countries (Zhu et al. 2016) and practical implications regarding the recycling and disposal of substances targeted by phase-out (Li et al. 2019). Given the more localized nature of health challenges, it is not surprising that international and global approaches play a less prominent role in this context.

When it comes to climate change, scholarly work occasionally refers to the United Nations Framework Convention on Climate Change (UNFCCC), the Kyoto Protocol and the Paris Agreement. Despite the transboundary nature of climate change and calls to phase out fossil fuels being raised as early as 1990 (Krause et al. 1990), the technology-neutral and policy-neutral paradigm driving climate policy within the UNFCCC had, for a long time, neglected phase-out as a policy approach. Although it does not explicitly state the need to phase out fossil fuels, the Paris Agreement and its goal to limit global temperature increase to well below 2°C are considered to have stimulated efforts to phase out fossil fuels and associated technologies at the domestic level (van Soest et al. 2017). The emergence of phase-out initiatives such as the launch of the Powering Past Coal Alliance in 2017 were driven by desires to fill the remaining gap (Blondeel et al.

2020). Finally, the Glasgow Climate Pact is the first UN-level climate policy framework that explicitly incorporates the notion of phase-out by stating the importance of "accelerating efforts towards the phasedown of unabated coal power and phase-out of inefficient fossil fuel subsidies" (UNFCCC 2021). Somewhat unexpectedly perhaps, the most prominently discussed global policy instrument in the climate context, however, is the Montréal regime. Although originally designed to halt ozone layer depletion, the Montréal Protocol has evolved into a tool of climate change mitigation as well. This occurred after the realization that several ODS and their substitutes contribute to the warming of the atmosphere, prompting their addition as phase-out targets in later amendments to the treaty (Graziosi et al. 2017).

Across the three sustainability challenges, command-and-control approaches are the second most frequently discussed class of instruments. While the majority of international and global agreements discussed in the phase-out literature refers to legally binding institutions, and thus could be classified as command-and-control instruments as well, we use this category to capture *domestic and subnational* policies, such as regulations and environmental standards, and those emanating from the EU. In the ozone context, early studies highlight domestic regulations towards a phase-out of certain ODS, for instance in the US (Stryker 1991), and considerations of countries like Australia to unilaterally phase out substances earlier than prescribed by the Montréal Protocol (Fraser & Bouma 1990). Later work highlights domestic legislation in various jurisdictions (e.g., Japan and the EU) to limit or phase out the use of substances not (yet) targeted by the Montréal regime. These regulatory approaches often serve as a pacesetter for later amendments of the Montréal Protocol (Velders et al. 2015).

Command-and-control approaches to phase-outs are emphasized frequently as a means to address human health challenges. Accounting for 46% of instrument mentions in this context, they comprise a broad range of country contexts and phase-out targets, with leaded gasoline, pesticides and other chemicals being among the most prominent targets. Scholarly work also examines instances of subnational phase-out regulation. For instance, methyl tertiary butyl ether (MTBE) was phased out as a fuel additive via state regulations in several American states (Thornburg 2016) before it was eventually phased out in the US altogether. The literature also hints at the dynamics that can affect policies at multiple jurisdictional levels. For instance, recent work on the phase-out of mercury pesticides in Australia (adopted in 2020) highlights how a country that rejects ratifying an international treaty (the Minamata Convention) can nevertheless be pressured to enact strong domestic phase-out legislation (Schneider 2021).

In the context of climate change, command-and-control-based approaches to phase-outs were scantly discussed in the literature until 2015. The recent scholarship, however, reverses this tendency. In particular, publications from the social sciences examine the making and functioning of coal phase-outs in jurisdictions that have enacted legislation for this purpose, such as Ontario (Rosenbloom 2018) and the UK (Brauers et al. 2020). Moreover, recent work emphasizes the role of regulatory instruments for phase-outs targeting fossil fuels in specific industries such as construction (Williamson & Finnegan 2021) or fossil fuel based industries and infrastructures in their entirety (Furnaro 2021). Much of this recent work is explicitly prescriptive in calling for command-and-control approaches to phase-out carbon-intensive arrangements domestically, be it coal-fired power generation (Kittel et al. 2020), cement manufacturing (Feng et al. 2018) or fossil fuels in general (Murombo 2021). This prescriptive turn is likely due to a combination of increasing perceptions of climate urgency among scholars (Wilson & Orlove 2021) and frustrations around the fact that other policy instruments like international agreements and economic instruments have largely failed to curtail the extraction and combustion of fossil fuels in line with internationally agreed temperature targets (Piggot et al. 2020).

Management and planning approaches (Persson 2006) are also used to drive phase-outs. These rely less on authority and more on the organizational resources of states (see Hood 1983). In the ozone context (11% of instrument mentions), they mostly comprise national-level timetables and targets for ODS phase-outs that deviate from the international ozone regime (e.g., Schmidt 1991). Similarly, the literature sheds light on environmental management tools in the context of health-related phase-outs, such as the so-called National Action Plans for reducing the human health and environmental risks of pesticide use, which are mandated by a directive for all EU member states (Schulte-Oehlmann et al. 2011). With respect to climate-related phase-outs, management and planning approaches (23% of instrument mentions) capture a variety of initiatives to phase out carbon-intensive arrangements. Some of these refer to existing policy frameworks such as Hong Kong's Climate Action Plan (Durmaz et al. 2020), under which fossil fuel based technologies are to be phased out. Others refer to scholarly proposals for phase-outs, such as the idea to progressively discontinue the use of internal combustion engine vehicles in Canada (Bahn et al. 2013). Moreover, the Paris Agreement has given rise to demands that nationally determined contributions (i.e., the non-binding mechanism allowing each country to set its own mitigation and adaptation objectives) should include precise roadmaps and long-term commitments for the phase-out of fossil fuels (Rauner et al. 2020).

The three remaining classes of phase-out instruments—economic instruments, voluntary approaches, and education and capacity building—are discussed much less prominently. However, there is interesting variation concerning their association with the three sustainability challenges under scrutiny (Figure 10.6). Economic instruments are almost absent in the phase-out literature about ozone and health, but account for 25% of mentioned policy instruments in the climate context. About half of these mentions relate to the removal or reform of subsidies for carbon-intensive arrangements, covering a variety of geographic and institutional contexts (Burniaux & Chateau 2014; Gençsü et al. 2020), with much of this work being prescriptive. Besides, emission pricing and emission trading systems are frequently theorized to contribute to climate-relevant phase-outs. For instance, Orthofer et al. (2019) show how an ambitious carbon pricing scheme would lead to a coal phase-out in South Africa's power sector. Meanwhile, using state-level data to examine the impact of the Regional Greenhouse Gas Initiative on coal and natural gas consumption in the electricity sector, Yan (2021) identifies this cap-and-trade program as a direct driver of coal and natural gas phase-outs within regulated US states. In general, the higher prevalence of economic ideas in the climate context appears to be in line with the relatively high level of engagement of scholarly communities from economics and business with climate change (see also Hulme et al. 2018).

Educational policy instruments and capacity-building approaches are sometimes used as complementary means to support the implementation of primary phase-out instruments. They play a non-negligible role in the context of ozone (10% of instrument mentions) which reflects scholarly interest in various flanking measures that support the global ODS phase-out. For example, scholarship discusses domestic schemes for technology cooperation and transfer, as well as the Regional Networks of Experts, an institution established by the United Nations Environment Programme (UNEP) to "provide a regular forum for the exchange of knowledge and experience, as well as development of skills among officers of National Ozone Units" (Shende 2015). Education and capacity building also appear occasionally in the context of health-related phase-out discussions, but less so in the climate context.

In contrast to the ozone and climate cases, discussions of phase-outs targeting human health challenges sometimes evoke voluntary approaches (8% of instrument mentions). This mainly reflects specific voluntary initiatives and corporate policies to phase out hazardous substances such as phthalate plasticizers (Nardelli et al. 2015) or polybrominated diphenyl ethers (Shaw et al. 2010). By voluntarily committing to a phase-out, a firm can try to position itself as a sustainability champion, based on the calculation to gain a competitive edge (e.g., Paska 2010). Alternatively, voluntary approaches may emerge in an effort to address voids left by missing or weak regulatory instruments (e.g., Tickner et al. 2005). Yet, the role of voluntary approaches has not been examined more systematically so far.

To conclude, our results demonstrate that multiple policy instruments are considered as drivers of phase-outs in practice and scientific discourse. Furthermore, patterns can be distilled by examining the specific phase-out instruments associated with different sustainability challenges. Accordingly, discussions of ODS phase-outs are associated mainly with international and global governance instruments, while phase-outs concerning human health issues are more frequently associated with command-and-control approaches. Finally, phase-outs in the context of climate change mitigation tend to be associated with a broader variety of instruments.

10.3.5 Phase-out functions

Most scholarly work discusses the phase-out of socio-technical elements primarily as a way to mitigate or ameliorate certain environmental sustainability challenges. Beyond that, a number of publications offer broader reflections on the ways in which phase-out may shape the evolution of socio-technical systems. Inspired by research on technological innovation systems (Kivimaa & Kern 2016), we refer to these as "functions"; i.e., processes caused by phase-outs that implicate a certain development within socio-technical systems. These are frequently discussed as a rationale for phase-outs. While we do not make the claim to comprehensively and accurately "measure" such functions in academic publications, our inductive coding led to the identification of three distinct sets of logics guiding phase-outs: substituting, signaling and transforming.

Of these, substituting is associated with the smallest degree of change in a socio-technical system. It captures the idea of using a phase-out intervention to remove an undesirable element and replace it with an alternative (Rinscheid et al. 2021). Signaling, by contrast, relates to the notion of using phase-out as a long-term direction-setting device for socio-technical change, simultaneously delegitimizing ongoing investment in the established system and reinforcing the general characteristics of the ultimate end point of a transition (Meckling & Nahm 2019). It emphasizes the role of phase-out as a trigger for innovation in terms of novel technologies, business practices and potentially societal arrangements more broadly. Finally, transforming captures the idea that phase-outs can target a broader range of technologies, infrastructures or even institutions such as entire fossil fuel based industries, often combined with additional policy interventions. In this case, phase-out is used as a lever to incite transformative change across the broader web of interlinked technologies, practices and rules that comprise socio-technical systems (Semieniuk et al. 2021).

In the surveyed literature, substituting is evoked regularly across the three challenges studied above (ozone, climate change and human health). It figures most prominently in the context of ozone, where several works put forward the phase-out of ODS as a means to trigger a relatively seamless switch from processes or technologies containing these substances to environmentally more benign replacements (e.g., Agrawal & Shrivastava 2010). The same logic is also evoked in several works relating to health and climate challenges, for instance in the context of replacing coal-fired power plants with natural gas.

Signaling and transforming, by contrast, are absent from the literature on ODS phase-out (see Figure 10.7). In some health- and climate-related publications, however, the function of signaling is evoked. Here, it is sometimes discussed as a way to provide sufficient preparation time to industry and stakeholders, aiming to reduce the cost and pain of transitioning from incumbent technologies, materials and business models to more sustainable alternatives (Howarth & Rosenow 2014; Meckling & Nahm 2019). A prototypical contribution to understanding the role of signaling in phase-outs is Rogge and Johnstone's (2017) paper. This work postulates—and demonstrates empirically—that phase-out policies in the energy sector cause manufacturers' innovation expenditures for new technologies to rise, outpacing other political interventions by far. Similarly, Bretschger and Zhang (2017) argue that a nuclear phase-out induces innovation and increases investment in renewable energies, an argument reiterated by Carrara (2020). While these

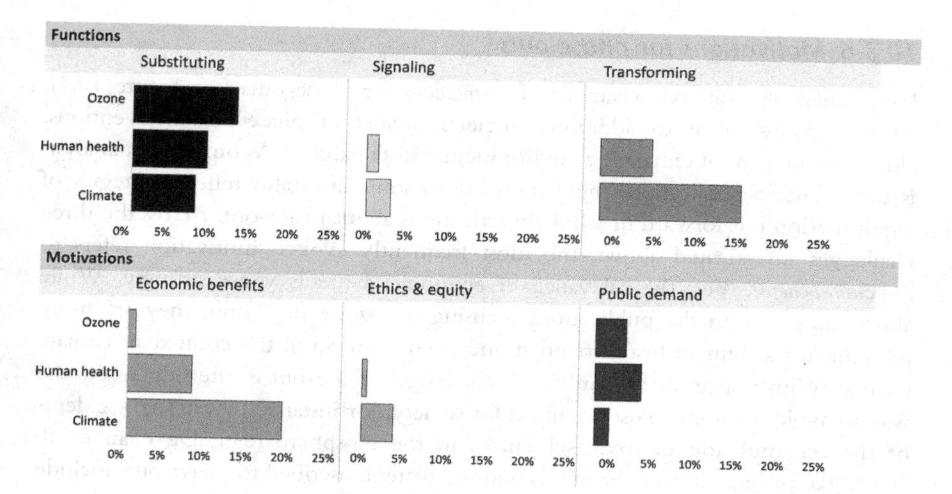

FIGURE 10.7 Functions and motivations ascribed to phase-outs in the context of three environmental sustainability challenges

The horizontal axes indicate the share of publications evoking a function or motivation within the three subsamples.

contributions emphasize the key role of policy signals in shaping market actors' expectations about visions for future arrangements that may add up to systemic change, signaling does not necessarily entail destabilization of incumbent regimes. Rather, through processes like creative accumulation (Bergek et al. 2013), recombination and diversification (Andersen & Gulbrandsen 2020), incumbents may be able to internalize a phase-out and remain among the shapers of socio-technical systems.

Transforming presumes far-reaching implications of phase-out in that such interventions can unsettle incumbent systems, trigger or accelerate broader change processes, and thereby fundamentally transform entire socio-technical systems (Rinscheid et al. 2021). An example associated with health challenges involves phase-out interventions targeting the use of toxic materials, thereby advancing the transformation towards a circular economy in the building sector (Wiprächtiger et al. 2020). Similarly, in the climate context, phase-out interventions targeting coal mining or power generation are frequently associated with fundamental transformations towards a sustainable society (e.g., Vögele et al. 2018). Focusing on the phase-out of lignite in Greece, Nikas et al. (2020) consider phase-out as a means to overcome multiple lock-ins and enact multidimensional changes in the Greek economy. Moreover, a number of works emphasize how the removal of fossil fuel subsidies may have far-reaching implications for energy supply systems and foster sustainable development by rupturing the tendency to follow historically established carbon-intensive development paths (Mutanga et al. 2018; Xu 2021).

10.3.6 Motivations for phase-outs

Researchers also observe a variety of *motivations* for phase-outs (see Figure 10.7). These broadly relate to additional societal drivers of phase-out interventions, alongside the goal of enhancing environmental sustainability. As our unit of analysis is the scientific publication, these motivations sometimes also reflect strategies of legitimization put forward in scholarly reflections about phase-out. Across the three challenges investigated above, the most frequently evoked motivation refers to *economic benefits*. Yet, the relevance of economic benefits varies strongly. While almost absent from the publications focusing on ozone depletion, they are more relevant in the human health context and even more so in the context of climate change. With respect to human health challenges, phase-out is often discussed as a way to avoid economic costs incurred for society, for instance by nuclear accidents or the accumulation of toxic substances in the biosphere (e.g., De Cian et al. 2014). Regarding climate change, economic benefits ascribed to phase-outs include multifaceted considerations. These range from economic savings (e.g., from phasing out fossil fuel subsidies) to employment creation (e.g., through a transition to decentralized renewable energy), to bolstering energy security, and finally to mitigating financial risks (e.g. stranded asset risks associated with coal-fired power plants). Motivations relating to *ethics and equity*, which comprise concerns for future generations or social equity more generally, are much less pronounced in the literature. Finally, *public demand* is discussed as a societal driver for phase-out interventions across all three challenges. While it plays a marginal role in the context of climate change related phase-out discussions, it figures more prominently in particular in relation to human health challenges (e.g., Viklund 2004).

10.4 Discussion and conclusion

The purpose of this chapter was to trace how scholarly work has mobilized phase-out, understood as a policy approach to terminate socio-technical elements with harmful or undesirable properties for environmental sustainability outcomes. Based on our systematic review, we derive five main insights.

First, after a slow accumulation of early works from the 1970s to the 1990s and a period of markedly higher academic interest between 2006 and 2015, scholarly work on phase-out has expanded rapidly since 2016.

Second, different academic communities from the social and natural sciences, engineering, medicine and others contribute to scholarly debates about phase-out. The shared interest in phase-out as a policy approach to address sustainability challenges and its prominence in current policy debates suggest that phase-out serves as a boundary object that mediates between diverse constituencies (Turnheim et al. 2020). It thus connects science with policymakers as much as it does differing strands of science.

Third, phase-out has been employed and conceived to tackle a varying set of sustainability challenges over time. Initially associated strongly with the challenge of atmospheric ozone depletion, phase-out has also been considered as a tool to tackle human health and other environmental challenges. More recently, phase-outs are increasingly discussed in the context of climate change mitigation.

Fourth, across the scholarly publications reviewed, a broad range of policy instruments are described to pursue phase-outs. While the ozone crisis is predominantly addressed with an internationally coordinated approach, human health challenges are often associated with domestic phase-outs using command-and-control instruments. Climate-related phase-outs rely on a range of instruments including command-and-control, international agreements, management and planning, and economic instruments.

Fifth, some of the more conceptually oriented literature associates phase-out with specific functions that entail implications for broader socio-technical systems. These range from phase-out as a means for substituting particular elements to phase-out as a signal for a potentially broader societal shift, to phase-out as a vehicle for enacting transformative change.

Finally, methodological choices and implications for future research deserve highlighting. A limitation of this investigation into phase-out was to rely on academic publications as units of analysis. Therefore, our findings may be as much a product of academic trends and analytical inclinations as they are a reflection of how phase-out is employed as a policy approach for confronting sustainability challenges. We thus caution against overinterpreting our scientometric analyses. While they reflect trends in the literature, they do not necessarily indicate the main targets and challenges addressed by phase-out policies at a given point in time. Moreover, our analysis does not distinguish between discussions of phase-out as a theoretical concept, an empirical case, a socio-technical process, a unit of a computational model or a normative goal. Another methodological limitation is that our systematic coding relied on the titles, abstracts and keywords of publications, but not the full texts.

Future research needs to deepen our knowledge on phase-out, both conceptually and empirically. Our review of 868 publications suggests that theory-building on phase-out has not progressed very far to date. While phase-out has become a major tool in confronting multiple sustainability crises, a widely known and accepted theoretical model of phase-out has not yet been developed. One step in this direction may be seen in our investigation of the functions ascribed to phase-out. Yet, our inductive analysis is not entirely unproblematic, since the functions identified in the literature partly overlap, may not capture the full array of logics guiding phase-outs, and are theoretically underspecified. Still, they may provide a useful starting point for more theoretically informed work on phase-out.

Empirically, we urge for systematic comparative analyses of the conditions that allow phase-out interventions to be considered and enacted as a tool for confronting sustainability challenges, the challenges that accompany its implementation in various contexts, and potential coping strategies. Important questions to be addressed include: Which policy instruments or mixes are most effective in phasing out unsustainable socio-technical elements? How can phase-out interventions be designed to ensure acceptance by affected stakeholders and regions without compromising their intended goals? How do phase-out interventions interact with other policy approaches steering innovation and decline in socio-technical systems? And to what extent can insights from "successful" phase-outs be transferred to other settings (i.e., other places, challenges, political contexts, etc.)? In tackling these questions, we hope that different academic communities keep contributing to this field of research, with the aim of generating knowledge that both advances the state of the art and informs decision-making for sustainable development.

Note

1 We thank Peter Stegmaier, Bruno Turnheim and Joshua Woodyatt for excellent comments on an earlier draft of this chapter. Adrian Rinscheid acknowledges funding from the Basic Research Fund (GFF) at the University of St. Gallen. Daniel Rosenbloom acknowledges the financial support of the Social Sciences and Humanities Research Council of Canada.

References

Agrawal, A.B. and Shrivastava, V. (2010) Retrofitting of vapour compression refrigeration trainer by an eco-friendly refrigerant. *Indian Journal of Science and Technology*, 3(4), 455–458.

Andersen, A.D. and Gulbrandsen, M. (2020) The innovation and industry dynamics of technology phase-out in sustainability transitions: Insights from diversifying petroleum technology suppliers in Norway. *Energy Research and Social Science*, 64, 101447.

Andersen, S.O. (2015) Lessons from the stratospheric ozone layer protection for climate. *Journal of Environmental Studies and Sciences*, 5, 143–162. https://doi.org/10.1007/s13412-014-0213-9

Andersen, S.O., Halberstadt, M.L. and Borgford-Parnell, N. (2013) Stratospheric ozone, global warming, and the principle of unintended consequences: An ongoing science and policy success story. *Journal of the Air and Waste Management Association*, 63(6), 607–647.

Ashford, N.A. and Hall, R.P. (2019) *Technology, Globalization, and Sustainable Development: Transforming the Industrial State*. Routledge.

Bahn, O., Marcy, M., Vaillancourt, K. and Waaub, J.P. (2013) Electrification of the Canadian road transportation sector: A 2050 outlook with TIMES-Canada. *Energy Policy*, 62, 593–606.

Bergek, A., Berggren, C., Magnusson, T. and Hobday, M. (2013) Technological discontinuities and the challenge for incumbent firms: Destruction, disruption or creative accumulation? *Research Policy*, 42(6–7), 1210–1224.

Blondeel, M., van de Graaf, T. and Haesebrouck, T. (2020) Moving beyond coal: Exploring and explaining the Powering Past Coal Alliance. *Energy Research and Social Science*, 59, 101304.

Bornmann, L., Haunschild, R. and Mutz, R. (2021) Growth rates of modern science: A latent piecewise growth curve approach to model publication numbers from established and new literature databases. *Humanities and Social Sciences Communications*, 8(1), 1–15.

Brauers, H., Oei, P.-Y. and Walk, P. (2020) Comparing coal phase-out pathways: The United Kingdom's and Germany's diverging transitions. *Environmental Innovation and Societal Transitions*, 37, 238–253.

Bretschger, L. and Zhang, L. (2017) Nuclear phase-out under stringent climate policies: A dynamic macroeconomic analysis. *Energy Journal*, 38(1), 167–194.

Burniaux, J.M. and Chateau, J. (2014) Greenhouse gases mitigation potential and economic efficiency of phasing-out fossil fuel subsidies. *International Economics*, 140, 71–88.
Carrara, S. (2020) Reactor ageing and phase-out policies: Global and regional prospects for nuclear power generation. *Energy Policy*, 147, 111834.
De Cian, E., Carrara, S. and Tavoni, M. (2014) Innovation benefits from nuclear phase-out: Can they compensate the costs? *Climatic change*, 123(3), 637–650.
Dewling, R.T. and Anderson, P.W. (1976) New York Bight I: Ocean dumping policies. *Oceanus*, 19(4), 2–10.
Dunn, J.R. (1990) Environmental concerns will widen auto use of specialty elastomers. *Elastomerics*, 122(2), 12–12.
Durmaz, T., Pommeret, A. and Tastan, H. (2020) Estimation of residential electricity demand in Hong Kong under electricity charge subsidies. *Energy Economics*, 88, 104742.
Feng, X.-Z., Lugovoy, O. and Qin, H. (2018) Co-controlling CO2 and NOx emission in China's cement industry: An optimal development pathway study. *Advances in Climate Change Research*, 9(1), 34–42.
Fraser, P.J. and Bouma, W.J. (1990) New ozone protocol and Australia. *Search*, 21(8), 261–264.
Furnaro, A. (2021) The role of moral devaluation in phasing out fossil fuels: Limits for a socioecological fix. *Antipode*, 53(5), 1442–1462.
Geels, F.W., Sovacool, B.K., Schwanen, T. and Sorrell, S. (2017) The socio-technical dynamics of low-carbon transitions. *Joule*, 1(3), 463–479.
Gençsü, I., Whitley, S., Trilling, M., *et al.* (2020) Phasing out public financial flows to fossil fuel production in Europe. *Climate Policy*, 20(8), 1010–1023.
Gonzalez, M., Taddonio, K.N. and Sherman, N.J. (2015) The Montreal Protocol: How today's successes offer a pathway to the future. *Journal of Environmental Studies and Sciences*, 5(2), 122–129.
Grant, M.J. and Booth, A. (2009) A typology of reviews: An analysis of 14 review types and associated methodologies. *Health Information and Libraries Journal*, 26(2), 91–108.
Graziosi, F., Arduini, J., Furlani, F., *et al.* (2017) European emissions of the powerful greenhouse gases hydrofluorocarbons inferred from atmospheric measurements and their comparison with annual national reports to UNFCCC. *Atmospheric Environment*, 158, 85–97.
Haddaway, N.R., Woodcock, P., Macura, B. and Collins, A. (2015) Making literature reviews more reliable through application of lessons from systematic reviews. *Conservation Biology*, 29(6), 1596–1605.
Hood, C.C. (1983) *The Tools of Government*. MacMillan.
Howarth, N.A.A. and Rosenow, J. (2014) Banning the bulb: Institutional evolution and the phased ban of incandescent lighting in Germany. *Energy Policy*, 67, 737–746.
Hulme, M., Obermeister, N., Randalls, S. and Borie, M. (2018) Framing the challenge of climate change in nature and science editorials. *Nature Climate Change*, 8, 515–522.
Johnstone, P. and Newell, P. (2018) Sustainability transitions and the state. *Environmental Innovation and Societal Transitions*, 27, 72–82.
Jordan, A., Wurzel, R.K.W. and Zito, A.R. (2003) "New" instruments of environmental governance: Patterns and pathways of change. *Environmental Politics*, 12(1), 1–24.
Kittel, M., Goeke, L., Kemfert, C., *et al.* (2020) Scenarios for coal-exit in Germany: A model-based analysis and implications in the European context. *Energies*, 13(8), 2041.
Kivimaa, P. and Kern, F. (2016) Creative destruction or mere niche support? Innovation policy mixes for sustainability transitions. *Research Policy*, 45(1), 205–217.
Koretsky, Z. (2023) Dynamics of technological decline as socio-material unravelling. In Koretsky, Z. *et al.* (eds) *Technologies in Decline: Socio-Technical Approaches to Discontinuation and Destabilisation*. Routledge.

Krause, F., Bach, W. and Koomey, J. (1990) *Energy Policy in the Greenhouse*. Earthscan.
Li, Y., Chang, Q., Duan, H., *et al.* (2019) Occurrence, levels and profiles of brominated flame retardants in daily-use consumer products on the Chinese market. *Environmental Science: Processes & Impacts*, 21(3), 446–455.
Lovei, M. (1998) *Phasing out Lead from Gasoline*. World Bank Technical Paper. https://elaw.org/es/system/files/worldbanklead.pdf
Markowitz, G. and Rosner, D. (2002) *Deceit and Denial: The Deadly Politics of Industrial Pollution*. University of California Press.
McCulloch, A. (1999) CFC and Halon replacements in the environment. *Journal of Fluorine Chemistry*, 100(1), 163–173.
Meckling, J. and Nahm, J. (2019) The politics of technology bans: Industrial policy competition and green goals for the auto industry. *Energy Policy*, 126, 470–479.
Murombo, T. (2021) Regulatory imperatives for renewable energy: South African perspectives. *Journal of African Law*, 66(1), 97–122.
Mutanga, S.S., Quitzow, R. and Steckel, J.C. (2018) Tackling energy, climate and development challenges in Africa. *Economics*, 12(1). doi:10.5018/economics-ejournal.ja.2018-61.
Nardelli, T.C., Erythropel, H.C. and Robaire, B. (2015) Toxicogenomic screening of replacements for Di(2-Ethylhexyl) Phthalate (DEHP) using the immortalized TM4 Sertoli cell line. *PLOS ONE*, 10(10), e0138421.
Nikas, A., Neofytou, H., Karamaneas, A., *et al.* (2020) Sustainable and socially just transition to a post-lignite era in Greece: A multi-level perspective. *Energy Sources, Part B: Economics, Planning, and Policy*, 15(10–12), 513–544.
Normann, H.E. (2019) Conditions for the deliberate destabilisation of established industries: Lessons from U.S. tobacco control policy and the closure of Dutch coal mines. *Environmental Innovation and Societal Transitions*, 33, 102–114.
OECD (2001) *Sustainable Development: Critical Issues*. OECD.
Orthofer, C.L., Huppmann, D. and Krey, V. (2019) South Africa after Paris: Fracking its way to the NDCs? *Frontiers in Energy Research*, 7, 1–15.
Paska, D. (2010) Facilitating substance phase-out through material information systems and improving environmental impacts in the recycling stage of a product. *Natural Resources Forum*, 34(3), 200–210.
Peñasco, C., Anadón, L.D. and Verdolini, E. (2021) Systematic review of the outcomes and trade-offs of ten types of decarbonization policy instruments. *Nature Climate Change*, 11, 257–265.
Persson, Å. (2006) Characterizing the policy instrument mixes for municipal waste in Sweden and England. *European Environment*, 16(4), 213–231.
Piggot, G., Verkuijl, C., van Asselt, H. and Lazarus, M. (2020) Curbing fossil fuel supply to achieve climate goals. *Climate Policy*, 20(8), 881–887.
Rauner, S., Bauer, N., Dirnaichner, A., *et al.* (2020) Coal-exit health and environmental damage reductions outweigh economic impacts. *Nature Climate Change*, 10(4), 308–312.
Rinscheid, A., Rosenbloom, D., Markard, J. and Turnheim, B. (2021) From terminating to transforming: The role of phase-out in sustainability transitions. *Environmental Innovation and Societal Transitions*, 41, 27–31.
Rogge, K.S. and Johnstone, P. (2017) Exploring the role of phase-out policies for low-carbon energy transitions: The case of the German Energiewende. *Energy Research and Social Science*, 33, 128–137.
Rosenbloom, D. (2018) Framing low-carbon pathways: A discursive analysis of contending storylines surrounding the phase-out of coal-fired power in Ontario. *Environmental Innovation and Societal Transitions*, 27, 129–145.

Rosenbloom, D. and Rinscheid, A. (2020) Deliberate decline: An emerging frontier for the study and practice of decarbonization. *Wiley Interdisciplinary Reviews: Climate Change*, 11(6), e669.

Rotmans, J., Kemp, R. and van Asselt, M. (2001) More evolution than revolution: Transition management in public policy. *Foresight*, 3(1), 15–31.

Schmidt, K. (1991) How industrial countries are responding to global climate change. *International Environmental Affairs*, 3(4), 292–315.

Schneider, L. (2021) When toxic chemicals refuse to die: An examination of the prolonged mercury pesticide use in Australia. *Elementa – Science of the Anthropocene*, 9(1), 1–18.

Schulte-Oehlmann, U., Oehlmann, J. and Keil, F. (2011) Before the curtain falls: Endocrine-active pesticides – A German contamination legacy. *Reviews of Environmental Contamination and Toxicology*, 213, 137–159.

Semieniuk, G., Campiglio, E., Mercure, J.-F., Volz, U. and Edwards, N.R. (2021) Low-carbon transition risks for finance. *Wiley Interdisciplinary Reviews: Climate Change*, 12(1), e678. https://doi.org/10.1002/wcc.678

Shaw, S.D., Blum, A., Weber, R., *et al.* (2010) Halogenated flame retardants: Do the fire safety benefits justify the risks? *Reviews on Environmental Health*, 25(4), 261–305.

Shende, R. (2015) Networking to save the world: UNEP's regional networks – conflict resolution in action. *Journal of Environmental Studies and Sciences*, 5(2), 138–142.

Stegmaier, P. (2023) Conceptual aspects of discontinuation governance: An exploration. In Koretsky, Z. *et al.* (eds) *Technologies in Decline: Socio-Technical Approaches to Discontinuation and Destabilisation*. Routledge.

Stryker, R.G. (1991) The regulatory phase-out of CFCs and halons. *Energy Engineering*, 88 (3), 22–27.

Thornburg, E. (2016) Public as private and private as public: MTBE litigation in the United States. In Hensler, D.R., Hodges, C. and Tzankova, I. (eds) *Class Actions in Context*. Edward Elgar.

Tickner, J., Geiser, K. and Coffin, M. (2005) The U.S. experience in promoting sustainable chemistry. *Environmental Science and Pollution Research*, 12(2), 115–123.

Trencher, G., Rinscheid, A., Rosenbloom, D. and Truong, N. (in review) The rise of phase-out as a critical decarbonisation approach: A systematic review. *Environmental Research Letters*.

Turnheim, B. (2023) Destabilisation, decline and phase-out in transitions research. In Koretsky, Z. *et al.* (eds) *Technologies in Decline: Socio-Technical Approaches to Discontinuation and Destabilisation*. Routledge.

Turnheim, B., Asquith, M. and Geels, F.W. (2020) Making sustainability transitions research policy-relevant: Challenges at the science–policy interface. *Environmental Innovation and Societal Transitions*, 34, 116–120.

UNFCCC (2021) *Glasgow Climate Pact*. https://unfccc.int/sites/default/files/resource/cop26_auv_2f_cover_decision.pdf.

van Soest, H.L., de Boer, H.S., Roelfsema, M., *et al.* (2017) Early action on Paris Agreement allows for more time to change energy systems. *Climatic Change*, 144(2), 165–179.

Velders, G.J.M., Fahey, D.W., Daniel, J.S., *et al.* (2015) Future atmospheric abundances and climate forcings from scenarios of global and regional hydrofluorocarbon (HFC) emissions. *Atmospheric Environment*, 123, 200–209.

Viklund, M. (2004) Energy policy options – from the perspective of public attitudes and risk perceptions. *Energy Policy*, 32(10), 1159–1171.

Vögele, S., Kunz, P., Rübbelke, D. and Stahlke, T. (2018) Transformation pathways of phasing out coal-fired power plants in Germany. *Energy, Sustainability and Society*, 8(1), 25.

Williamson, A. and Finnegan, S. (2021) Sustainability in heritage buildings: Can we improve the sustainable development of existing buildings under Approved Document L? *Sustainability*, 13(7), 3620.

Wilson, A.J. and Orlove, B. (2021) Climate urgency: Evidence of its effects on decision making in the laboratory and the field. *Current Opinion in Environmental Sustainability*, 51, 65–76.

Wiprächtiger, M., Haupt, M., Heeren, N., *et al.* (2020) A framework for sustainable and circular system design: Development and application on thermal insulation materials. *Resources, Conservation and Recycling*, 154, 104631.

Xu, S. (2021) The paradox of the energy revolution in China: A socio-technical transition perspective. *Renewable and Sustainable Energy Reviews*, 137, 110469.

Yan, J. (2021) The impact of climate policy on fossil fuel consumption: Evidence from the Regional Greenhouse Gas Initiative (RGGI). *Energy Economics*, 100, 105333.

Zhu, J., Liu, J.-G., Hu, J.-X. and Yi, S. (2016) Socio-economic analysis of the risk management of hexabromocyclododecane (HBCD) in China in the context of the Stockholm Convention. *Chemosphere*, 150, 520–527.

11

THE END OF THE WORLD'S LEADED PETROL ERA

Reflections on the final four decades of a century-long campaign

Peter Newman

11.1 Introduction

Leaded petrol began to be used in the United States of America in 1921 to enable Cadillac cars to run without engine knocking. The use of lead for this purpose spread rapidly across the world, setting up a battle between commercial interests and public health as the data grew rapidly showing that workers and then the public were receiving brain damage. The phase-out began in the US in the 1970s but Japan was the first to completely phase out leaded petrol in 1986, and in July 2021 Algeria was the last place it was sold for road use. This chapter will examine the last 40 years of this long battle to end the leaded petrol era and in particular how companies kept polluting our world for many decades after incontrovertible evidence of its health dangers. The chapter ends with lessons for governing the end of similar harmful and dangerous technologies.

The leaded petrol story relates to the themes of this book in four ways:

First, *Decline, Destabilisation, and Discontinuation* are all evident with regime resistance and the weight of incumbency preventing a quick process.

Second, *Structural Inertia and Lock-In* are evident due to automobile dependence and the continued insistence that changing vehicles was too expensive, so leaded petrol must be maintained, reinforced by the companies who were making money from this.

Third, *Active Construction of Ignorance* and indeed the weaponisation of disinformation was a dominant factor in slowing down the phase-out of leaded petrol, fogging of the truth through funded pro-lead research, and even publicly acknowledged bribery in some low income nations. The role of *Alternatives* will be outlined as part of this public fight.

Fourth, *Governance Failure* is shown through the lack of an adequately fast *Global Phase-Out* strategy and the necessity of grass-roots campaigning.

DOI: 10.4324/9781003213642-11

The chapter is written from the perspective of an activist as well as an academic involved in this issue and stresses the importance of civil society groups in public battles like this, against heavily embedded incumbency. The lessons learned for change processes associated with any technology or product that threatens humanity are suggested to be universally applicable. The industry lobby groups that ensured there was a need for lead in petrol will not be seen in any historical perspective to have been a constructive contributor to change. Companies today have ESG strategies to ensure they manage their environmental, social and governance responsibilities. From an ESG perspective today most companies would look back and cringe as we do over asbestos use and how we will do over climate change. But this period of industrial civilization continued using leaded petrol far beyond any sensible health and environmental policy would have allowed.

11.2 Why experts warned leaded petrol needed to be rapidly phased out

Lead is a toxic product that kills ordinary people decades after exposure. This has been known for centuries but was first recognised as being associated with lead in petrol in the 1920s. The importance of preventing lead poisoning and specifically of eliminating leaded petrol was well known, therefore, for over a hundred years. In 1990, the WHO's World Health Statistics Quarterly declared the addition of lead to petrol to be "*The* Mistake of the 20th Century" (Shy 1990).

According to environmental scientist Philippe Grandjean, the European Food Safety Authority declared there is no known safe exposure to lead for a developing foetus and "work should continue to reduce exposure to lead, from both dietary and non-dietary sources" (Grandjean 2010). He suggests that lead is the environmental pollutant with the largest toxicological database for good reason. Lead is seen by scientists now as the most common contaminant of the industrial age, and of the Anthropocene, so everyone is exposed to some degree and lead negatively affects every organ of every organism on earth (Fiałkiewicz-Kozieł et al., 2018).

The issue of lead in petrol is not one of widespread ignorance about lead's toxicity but is much more a story of widespread government failure to properly regulate, and of companies' failure to take on alternatives. It is a failure to manage technology transition.

As the historian Bill Kovarik puts it,

> The leaded gasoline story provides a practical example of how industry's profit-driven decisions—when unsuccessfully challenged and regulated—can cause serious and long-term harm. It takes individual public health leaders and strong media coverage of health and environmental issues to counter these risks.
>
> *(Kovarik 2021)*

The leaded petrol story is one of only "Eight Stories of Corporations Defending the Indefensible, from the Slave Trade to Climate Change" in the book *Industrial-*

Strength Denial (Freese 2020). Fred Pearce calls the inventor of leaded petrol, Thomas Midgley, "a one-man environmental disaster" (Pearce 2017), and environmental historian John McNeill describes Midgley as having "had more impact on the atmosphere than any other single organism in earth history" (Bess 2002).

The World Bank called the lead additive (tetraethyl lead) in petrol the "greatest environmental threat to health in many Third World cities". In June 1996 the UN recommended that governments eliminate as soon as possible the use of lead in gasoline and the World Bank supported them by saying this task is cheap and technically achievable for all cars.

> The world's biggest aid lender to low-income countries now puts banishing lead from petrol as its number one priority for Third World transport investment ... There is, says the World Bank, "no excuse for continuing to allow leaded fuels in any city".
>
> *(Pearce 1996)*

11.3 The growing science of health and environmental impacts from lead in petrol since the 1970s

Lead is a known neurotoxin with a long history of causing brain damage, most notably having a likely role in the fall of the Roman Empire (Nriagu 1983; Winder 1984; Gilfillan 1990). The idea that Western civilization had created a similar possible decline and fall scenario began when scientists focused on lead in petrol in the 1970s.

The need to remove lead in petrol was highlighted in the 1970s by research on neurobehavioural impacts of leaded petrol in children. Lead inhibits certain enzymes or catalysts that are important in brain function. The growing brains of children are the most susceptible to this toxic substance. Studies such as Needleman et al. (1979), Weiss et al. (1979), Garnys et al. (1979), Wibberley et al. (1977) and Bryce-Smith and Stevens (1981) found that high lead levels cause increased risk for spontaneous abortions, and high lead levels in children's milk teeth were associated with IQ reductions, lower average performance in attention spans, lower reaction times, behavioural disorders and generally lower levels of overall functioning in schools. Similar studies confirmed this in the UK, Australia and across the US. Leaded petrol was also linked to behaviour issues in adults, cancer, decreased resistance to and interference with drugs (see references in Newman and Kenworthy 1985).

The US Environmental Protection Agency estimated in 1978 that between 90% and 98% of lead emissions in cities was from car exhausts and from old sump oil and engines. Lead measured in Greenland ice shows that the pre-industrial levels were around 0.005 micrograms per kilogram but by 1940 this had risen more than ten times to 0.07 micrograms per kilogram. By 1980 this was 200 times the level of pre-industrial lead from atmospheric fallout. Lead had even increased in open-ocean concentrations by ten times since pre-industrial times by the late 1970s

(Train 1979) and this was the only heavy metal at that time to be measurably increased in the ocean. Lead from natural sources was measured at around 2,000 tonnes per year but 400,000 tonnes came from human sources with 280,000 tonnes (the biggest by far) from leaded petrol (Settle and Patterson 1980).

Awareness grew from this time that lead from motor vehicle exhausts included volatile compounds such as lead halides (caused by lead scavengers added to the petrol to reduce damaging lead deposition in the engine and exhaust system), lead oxides and unburnt tetraethyl lead. These particles are less than 0.1 micron in size and awareness grew that they would pass easily into lower parts of the lungs as well as pass directly through the skin when in direct contact with leaded petrol.

While the original concerns were about ambient levels of airborne lead, a paper in *Nature* in 1988 showed that consumption of leaded petrol was a major source of global pollution of not only the air, but also water and soils (Nriagu and Pacyna 1988). However, the biggest source of public health concern was how the dust on the side of roads built up lead to dangerous levels in house dust and play soil, and was ingested by children through normal activity.

The emerging concern among scientists led many to begin to make political statements: "Unlike many neuropsychiatric diseases, little remains to be discovered about lead poisoning. The etiology, some of the biochemical toxicology and the necessary steps to eliminate the problem are clearly spelled out. The larger challenges with lead lie in finding the will and the means to remove it from the human environment" (Needleman 1981: 7).

The claim of Octel/Innospec (Associated Octel 1995a) that lead pollution from cars was "not significant in terms of effects on human health" should have been seen as outrageous based on the scientific data available in the previous 20 years. However, the phase-out did not happen for decades.

11.4 A brief history of the thwarting of the phase-out of lead in petrol

The history of lead in petrol from the 1920s shows that its use became completely embedded in the emergence of the automobile, especially in the post-war period when automobile dependence became a structural phenomenon in cities across the world (Newman and Kenworthy 1989). Not only were vehicles built to use leaded petrol in their engines (valve seats were designed to use lead particles to seal them) but the oil companies refined petrol to fit this (lower octane ratings were needed). Technological lock-in was thus well established and the politics of shaping the automobile's future was not open to changing such a winning formula.

The public health process to remove lead from petrol began in the 1980s but still took 40 years for it to be fully adopted despite options being available. The first opportunity came when the global growth in awareness of car-related smog led to the invention of the catalytic converter which was very successful at removing hydrocarbons from vehicle emissions but could not work if lead was in the petrol as it poisoned the catalyst. Thus, countries began to legislate for unleaded petrol for new cars in the 1980s as these were what the Japanese manufacturers were providing and places like California were regulating to support the transition.

However, tetraethyl lead manufacturer Octel/Innospec was telling governments in the 1990s that unleaded fuel is only suitable for cars with catalytic converters, a claim described as "nonsense" by experts (Pearce 1996). The issue of leaded petrol was therefore clouded for most high-income governments presumably because of the powerful cultural and economic commitment to the continued expansion of automobility, as well as the vested interests and active lobbying to deflect challenges by companies.

The vehicle, oil and lead industries continued to assert that old vehicles needed lead or there would be financial outcomes that would not be politically feasible. The lobby that maintained this highly misleading approach was able to convince motor vehicle associations and governments that they should not worry about lead in petrol and that it was far too expensive to change to higher level octane ratings in the refinery process or better valves and pistons in old motor vehicles.

Throughout the 1980s and early 1990s this approach was perpetrated across the global motoring public. The public's infatuation with motor vehicles and their dependence on them for daily life meant that they were easy fodder for the deceptive companies with a vested interest in keeping lead in petrol (Newman and Kenworthy 1985).

The public and governments were unsure due to such lobbying. The academic and public health world were also confused by the lead and oil companies who helped to muddy the debate by funding lead research to make the issues seem to be less clear cut and thus just a matter of opinion. The Lead Fund, created by Octel/Innospec, continued throughout this period, creating a fog of scientific papers.

The beginning of the end of lead in petrol had arrived with catalytic converters but it still took decades to remove. The public health evidence became more and more persuasive (as outlined above) but the actions were limited by the public fog created by the industry lobbying activity. Politicians and decision-makers just stood back and watched without any kind of phase-out strategy, other than hoping that new vehicles would take over the market and slowly remove leaded petrol.

Old vehicles kept using lead in petrol for many decades as Octel/Innotec continued telling governments that "unleaded petrol should only ever be used in cars fitted with a catalytic converter", and "not all cars designed to run on leaded 4-star [petrol] can switch to super unleaded without running the risk of engine damage". They were printing this in a brochure (Associated Octel 1995a) and distributing it, along with a cancer fear-mongering brochure suggesting that higher refined petrol would contain more carcinogens (Associated Octel 1995b). These pamphlets were widely distributed to car clubs and the media and then further distributed around the world.

The phase-out began with new vehicles but continued with old vehicles unless governments intervened to remove all lead from petrol. This took a decade or so in the high-income world but took much longer elsewhere.

The worst part of this story is that low-income nations continued to use leaded petrol, and in their crowded cities the impacts were terrible. It took another 20 years after the high-income nations' phase-out for leaded petrol to be phased out globally, with Algeria in 2021 becoming the last country. Lead particles remain along main roads, especially in the slow phase-out cities. As Ritchie said in February 2022:

> Around one-in-three children globally suffer from lead poisoning… Lead poisoning is estimated to account for about 1% of the global disease burden… in some countries, the costs of lead exposure are equivalent to as much as 6% of GDP.
>
> *(Ritchie 2022)*

11.5 The campaign for phasing out leaded petrol in high- and low-income nations

The 1980s were characterised by a growing campaign by newspapers such as *The Guardian* and *The Observer* through civil society groups such as The Conservation Society. Much research by medical academics was gathering in the US. The growing weight of the scientific evidence and the horror of leaded petrol's impact on unsuspecting children began to emerge. The articles and writings from academics outlined the long history of lead from the 1920s, when Thomas Midgley showed Cadillac how adding lead to petrol was a cheap and easy way to prevent engine knocking and "had no side effects" (Needleman 1999). Although leaded petrol use had become widespread, its global acceptance was now coming into conflict with serious health issues.

Scientists and civil society began forming associations to fight leaded petrol in the 1980s (Newman and Kenworthy 1985). Elizabeth O'Brien in Sydney established The LEAD (Lead Education and Abatement Design) Group, along with three other families who were dealing with lead poisoning among their relatives. In the LEAD Group Newsletter, the role of academics in providing perspectives now was given an activist outlet, for example, *Pollution and the Poor* on leaded petrol and social justice (Newman 1993) and *Green Accounting and Car-Free Planning* (Newman 1998) were distributed widely in the LEAD Group Newsletter. The LEAD Group's leaded petrol phase-out campaign was part of a wider car use reduction campaign that is now focused on reducing greenhouse gas emissions but from the 1980s had a broad approach to multiple impacts from the over-use of automobiles (Newman and Kenworthy 1992).

Although the first steps to transition away from leaded petrol started in the US in 1972 with the governments' requirement that unleaded petrol be made available, the first country to complete the phase-out of leaded petrol was Japan in 1986 (Lovei 1998). As Japan had also required catalytic converters on cars manufactured for in-country sale, it was not unexpected that they would also put them in their exported vehicles thus ensuring that other countries had to follow by regulating away leaded petrol, at least in new cars. However, it was still very slow in some high-income countries. In Australia, the availability of unleaded petrol was required as of July 1985, and additionally all new cars sold from 1986 had to have a catalytic converter fitted. Australia finally completed the leaded petrol phase-out for all vehicles on 1 January 2002.

The LEAD Group continued with the campaign to remove lead in every low-income nation. This happened with help from global development agencies. In many low-income countries leaded petrol bowsers were never labelled with the octane number or even as "leaded"—so consumers had no choice but to buy leaded—let alone choose an octane rating, and in the poorest countries, leaded petrol was sold by the litre from unmarked portable cans from a stall by the side of the road.

The phase-out was slowed globally for as long as possible by the same lobby groups as had slowed it in the high-income world. As a result, it took nearly five decades for the transition to unleaded petrol to be completed. Readers can see a map[1] with a sliding timeline (at the bottom) from 1986 until 2021 showing how the transition occurred country by country, year by year. This process had various global agencies like the UN, World Bank and WHO playing various roles but the LEAD Group in Australia provided most of the detailed assistance that enabled many of the low-income nations to work against the incumbent companies who insisted leaded petrol was essential.

A detailed *Chronology—Making Leaded Petrol History* (1921 to 2011)—including steps needed to complete the global phase-out—is provided by the LEAD Group (O'Brien et al. 2011). And the UN's leaded petrol timeline[2] and numerous other UN articles are provided by the LEAD Group in the *LEAD Action News* August 2021 issue (O'Brien 2021).

Leaded petrol has now stopped being sold globally, nearly 50 years after the phase-out began. Lead continued to grow in urban dust throughout this period, causing lead poisoning to growing brains, so the impact will continue for generations.

The phase-out of leaded petrol from the 1980s taught us how deeply political and anti-science were such environmental and health issues in transport. They helped show us that the socio-technical approaches to change had to be more central in winning the politics at every level.

11.6 The lessons learned from the leaded petrol phase-out

The leaded petrol story relates to the themes of this book in four structural ways:

1. Decline, Destabilisation and Discontinuation are all evident with regime resistance and the weight of incumbency preventing a quick transition process. The evidence for this is apparent in the way that the lead and oil company regime created a (dis-informational) strategy to string out the removal of lead in petrol. They based their strategy around the political power of motor vehicle associations, who supported the lack of intervention as it enabled no

extra costs to be incurred, and weak governments that did not see that the health impacts warranted a mandated change to unleaded fuel.

2. Structural Inertia and lock-in is evident due to automobile dependence reinforced by the companies who were making money from this. Automobile dependence was created as a term in 1989 by Newman and Kenworthy to explain the large differences in fuel use in US and Australian cities compared to European and Asian cities. This led to significant changes in the priority given to funding just highways instead of transit and active transport systems, but these American and Australian cities continue to have substantial auto-dependence though 'peak car' has begun (Newman and Kenworthy 1999, 2015). In Europe this notion was called lock-in and inertia and in recent decades this has grown through peri-urban processes (Newman et al. 2019). Lead was an issue in the early debates but mostly fuel vulnerability dominated the debate as to why car use should be constrained. Now it is climate-related. These structural issues are real and deep and in the early days of the lead campaign it was the lead and oil company lobbies that used this dependence to incite fear of change and higher costs in car owners.
3. Active Construction of Ignorance and the weaponisation of disinformation was a dominant factor in slowing down the removal of leaded petrol through corporate marketing strategies, fogging of the truth through funded pro-lead research, and even bribery in some low-income nations. Octel/Innospec successfully convinced millions of drivers that their cars needed lead, and their health and the environment would be better off if they used leaded petrol. The role of *Alternatives* in overcoming this ignorance was part of the campaign and these were constantly demonstrated to be simple and cheap to implement. However, the oil and vehicle lobbies only chose to phase out lead in new vehicles when the catalytic converter was invented and found to be poisoned by lead. Older vehicles continued to use lead until vehicle owners discovered they could fix the problem with better valves and pistons and hence did not need lead. Governments did not step in to hasten this change and indeed leaded petrol lasted for decades more in the low-income world.
4. Governance Failure. Governance failure is shown through the lack of an adequately fast Global Phase-Out strategy by a single lead agency from beginning to end and the necessity of grass-roots campaigning to fill the gaps. The WHO barely had a role in leaded petrol phase-out, the OECD were active from 1991 to 2012, the World Bank were the most focussed from the early 1990s to 2002, the UN finished the job (painfully slowly) from 2002 to 2021. This changing scene of Alliances was chronicled throughout by the LEAD Group in Australia, which acted as a clearinghouse for information and a conduit between individuals and organisations who didn't know where to turn to speed the phase-out in their own country or region. Small bits of funding from the Australian Government occasionally assisted the LEAD Group to conduct this global role but they had no power apart from their ethical influence based on their accumulated knowledge.

The next stage of lead activism is phasing out leaded aviation fuel "AvGas". Decision-makers across the globe need to realise that the same story as outlined above applies to lead in AvGas. There is a long story of regime resistance outlined above due to automobile dependence in our cities that has influenced the slow removal of leaded petrol. Aviation dependence in our regions has some structural hold over remote areas but such structural change must not be allowed to override health issues again. It will prevail if we continue to believe that lead is needed to keep aircraft flying. Total removal of lead from fuel should be a part of climate strategies and other global commitments like the Sustainable Development Goals.

11.7 Looking ahead

Leaded petrol has now been phased out throughout the world. This has been a very long and painful process with millions of children impacted in the functioning of their brains and millions of adults destined for early death. The reality is that most concerned scientists and civil society groups with an interest in the issue were sure it would not take as long as it did. The weight of incumbency, the structural inertia surrounding the automobile and the power of corporate ignorance-construction allowed the profit-makers to be the major winners over and over.

There are lessons to be learned from this battle to phase out a toxic product that impacts health and environments across the globe. We are beginning to see more action now on climate change, not because the science is any clearer, not because governments have gotten braver, but because the technology of solar, batteries and electric vehicles are now superior and the tech giants are now seeing that they can invest profitably whilst averting climate disaster (Gates 2020).

The 50-year global process to eliminate leaded petrol is not a story where government leadership could be seen. Civil society such as the LEAD Group persevered in the campaign for 30 years. The role of civil society groups in such battles against heavily embedded incumbency cannot be minimised. They are now turning their attention to leaded AvGas as it remains in many parts of the world. The hope for a complete eradication of this modernist scourge of leaded fuel remains.

The lessons from phasing out lead in petrol can be learned for the change processes associated with any technology or product that threatens humanity. Perhaps plastic will be the next focus as it has growing impacts from its breakdown products, some similar to lead (Zaman and Newman 2021). The same oil companies are responsible for these products.

As shown by Franta (2018), Shell and Exxon already knew in the 1980s that global warming was caused by their products but concluded that government and consumers should take the leadership role on this and indeed completely prevented anyone knowing about their internal reports which were finally leaked in 2015. The world cannot afford any further delay on phasing out fossil fuels and all companies as well as governments and cities must have phase-out strategies that fit the IPCC advice. These should now include the phase-out of oil and gas-derived plastics. The battle continues.

Notes

1 https://ourworldindata.org/grapher/lead-petrol-ban?time=latest.
2 https://leadsafeworld.com/wp-content/uploads/2021/09/08-Leaded-Petrol-Timeline.pdf.

References

Associated Octel (1995a) *Associated Octel Brochure: Petrol and the Air You Breathe.*

Associated Octel (1995b) *Associated Octel Brochure: Petrol – Facing Reality.*

Bess, M. (2002) *Something New Under the Sun: An Environmental History of the Twentieth-Century World*, by J.R. McNeill (2001), New York: Norton, xxvi, 421 pp. Reviewed by Michael Bess, History Department, Vanderbilt University, Nashville, TN. *Journal of Political Ecology: Case Studies in History and Society*, 9. https://web.archive.org/web/20040328013401/http://dizzy.library.arizona.edu/ej/jpe/volume_9/1101bess.html.

Bryce-Smith, D. and Stevens, R. (1981) *Lead or Health*, 2nd edn. Conservation Society.

Fiałkiewicz-Kozieł, B., De Vleeschouwer, F., Mattielli, N., Fagel, N., Palowski, B., Pazdur, A. and Smieja-Król, B. (2018) Record of Anthropocene pollution sources of lead in disturbed peatlands from Southern Poland. *Atmospheric Environment*, 179, 61–68.

Franta, B. (2018) Shell and Exxon's secret 1980s climate change warnings: Newly found documents from the 1980s show that fossil fuel companies privately predicted the global damage that would be caused by their products, *The Guardian*, 19 September. www.theguardian.com/environment/climate-consensus-97-per-cent/2018/sep/19/shell-and-exxons-secret-1980s-climate-change-warnings.

Freese, B. (2020) "How wrong one can be": Bias, tribalism, and leaded gasoline. In *Industrial-Strength Denial: Eight Stories of Corporations Defending the Indefensible, from the Slave Trade to Climate Change.* University of California Press.

Garnys, V.P., Freeman, R. and Smythe, L.E. (1979) *Lead Burden of Sydney Schoolchildren.* Department of Analytical Chemistry, University of NSW.

Gates, B. (2020) *How to Avoid a Climate Disaster: The Solutions We Have and the Breakthroughs We Need.* Penguin. Audible Books © Bill Gates, Penguin Audio. Read by Wil Wheaton and Bill Gates.

Gilfillan, S.C. (1990) *Rome's Ruin by Lead Poison.* Wenzel Press.

Grandjean, P. (2010) Comment: Even low-dose lead exposure is hazardous. *The Lancet*, 376. www.thelancet.com/journals/lancet/article/PIIS0140-6736(10)60745-3/fulltext.

Kovarik, B. (2021) A century of tragedy: How the car and gas industry knew about the health risks of leaded fuel but sold it for 100 years anyway, *The Conversation*, 9 December. https://theconversation.com/a-century-of-tragedy-how-the-car-and-gas-industry-knew-about-the-health-risks-of-leaded-fuel-but-sold-it-for-100-years-anyway-173395.

Lovei, M. (1998) *Phasing out Lead from Gasoline: Worldwide Experience and Policy Implications* (Vol. 397). World Bank Publications.

Needleman, H.L. (ed.) (1981) Quoted in CALIP Newsletter No 13 March–May, from The Conservation Society, London.

Needleman, H.L. (1999) *History of lead poisoning in the world.* Conference on Lead Poisoning Prevention & Treatment, Bangalore, India, February 8–10. www.lead.org.au/history_of_lead_poisoning_in_the_world.htm.

Needleman, H.L., Gunnoe, C., Leviton, A., Reed, R., Peresie, H., Maher, C. and Barrett, P. (1979) Deficits in psychologic and classroom performance of children with elevated dentine lead levels. *The New England Journal of Medicine*, 300(13), 689–695. https://afitch.sites.luc.edu/Articles/Needleman%20NEJM%201979.pdf.

Newman, P. (1998) Green accounting and car-free planning – extracts from a speech called the implications of the environmental agenda for the future development of Australian human settlements. *LEAD Action News*, 6(3), 28 October. https://lead.org.au/lanv6n3/lan6n3-11.html.

Newman, P. (1993) Pollution and the poor. *Murdoch News*, 1 December. Reprinted by The LEAD Group in *LEAD Action News*, 7(1). www.lead.org.au/lanv7n1/L71-16.html.

Newman, P. and Kenworthy, J. (1985) Lead in petrol: An environmental case study. *Environmental Education and Information*, 4 (1), 1–56. www.researchgate.net/publication/295967540_Lead_in_petrol_an_environmental_case_study.

Newman, P. and Kenworthy, J. (1989) *Cities and Automobile Dependence: An International Sourcebook*. Gower.

Newman, P. and Kenworthy, J., with Robinson, L. (1992) *Winning Back the Cities*. ACA (Australian Consumers' Association) and Pluto.

Newman, P. and Kenworthy, J. (1999) *Sustainability and Cities: Overcoming Automobile Dependence*. Island Press.

Newman, P. and Kenworthy, J. (2015) *The End of Automobile Dependence: How Cities Are Moving Beyond Car-Based Planning*. Island Press.

Newman, P., Thomson, G., Helminen, V., Kosonen, L. and Terama, E. (2019) Sustainable cities: How urban fabrics theory can help sustainable development. *Reports of the Finnish Environment Institute*, 39, 1–33. https://helda.helsinki.fi/handle/10138/305336

Nriagu, J.O. (1983) Saturnine gout among Roman aristocrats: Did lead poisoning contribute to the fall of the Empire? *New England Journal of Medicine*, 308(11), 660–663.

Nriagu, J.O. and Pacyna, J. (1988) Quantitative assessment of worldwide contamination of air, water and soils by trace metals, *Nature*, 333, 134–139. www.nature.com/articles/333134a0.

O'Brien, E. (2021) Celebrating the end of the leaded petrol era! *LEAD Action News*, 21(3), 31 August. https://leadsafeworld.com/wp-content/uploads/2021/08/LANv21n3-Celebrating-the-End-of-the-Leaded-Petrol-Era.pdf.

O'Brien, E., Roberts, A. and Cooper, D. (2011) Who will end the leaded petrol death trade? *LEAD Action News*, 11(4), June. www.lead.org.au/lanv11n4/LEAD_Action_News_Vol_11_No_4.pdf.

Pearce, F. (1996) A heavy responsibility – Removing lead from petrol is cheap and easy, so why are poor countries resisting attempts to phase out a substance that threatens the health of millions? *New Scientist*, 27 July. www.newscientist.com/article/mg15120402-400-a-heavy-responsibility-removing-lead-from-petrol-is-cheap-and-easy-so-why-are-poor-countries-resisting-attempts-to-phase-out-a-substance-that-threatens-the-health-of-millions/

Pearce, F. (2017) Inventor hero was a one-man environmental disaster: From poisonous cars to the destruction of the ozone layer, Thomas Midgley almost single-handedly invented a global environmental crisis, finds Fred Pearce. *New Scientist*, 7 June. www.newscientist.com/article/mg23431290-800-inventor-hero-was-a-oneman-environmental-disaster/#ixzz7LZE19Bxg.

Ritchie, H. (2022) Around one-in-three children globally suffer from lead poisoning: What can we do to reduce this? Our World in Data. https://ourworldindata.org/reducing-lead-poisoning.

Settle, D.M. and Patterson, C.C. (1980) Lead in Albacore: Guide to lead pollution in Americans. *Science*, 207, 1167–1176.

Shy, C.M. (1990) Lead in petrol: The mistake of the XXth century. *World Health Statistics Quarterly*, 43(3), 168–176.

Train, R. (1979) *Quality Criteria for Water*. Castle House.

Weiss, G. *et al.* (1979) Hyperactives as young adults: A controlled prospective ten-year fol low-up of 75 children. *Archive General Psychiatry*, 36, 657.

Wibberley, D.G., Khera, A.K., Edwards, J.H. and Rushton, D.I. (1977) Lead levels in human placentae from normal and malformed births. *Journal of Medical Genetics*, 14(5), 339–345.

Winder, C. (1984) *Developmental Neurotoxicity of Lead*. MTP Press.

Zaman, A. and Newman, P. (2021) Plastics: Are they part of the zero-waste agenda or the toxic-waste agenda? *Sustainable Earth*, 4, 4. https://doi.org/10.1186/s42055-021-00043-8.

12

CONCLUSIONS AND CONTINUATIONS

Horizons for studying technologies in decline

Zahar Koretsky, Peter Stegmaier, Bruno Turnheim and Harro van Lente

At the start of this book we asked how technologies decline and how this entails re-negotiating relationships with technology in a social context. The need to seriously discuss and practise endings of established socio-technical systems comes from the policy and research worlds alike. Bolder claims are being made to prohibit certain ways of doing things that are being framed as outdated, unsustainable or undesirable. Cities like Amsterdam, Barcelona, Brussels, Helsinki, Paris, and Birmingham have been taking it upon themselves to ban cars from city centres. In 2019, an alliance of private, public and non-governmental organisations emerged with the aim of Powering Past Coal.[1] In 2020, the UK government requested public advice about phasing out petrol, diesel and hybrid cars.[2] While these initiatives are only slowly bearing fruit—and some may be primarily symbolic—in the academic literature, the topic of active abandonment of technological trajectories is attracting growing interest.

The climate crisis is among the main existential threats to humanity in this century. To address it, innovations such as clean technologies have been gaining significant momentum in recent years. But, at the same time, emissions continue to grow. The large, overarching trajectory towards more environmental degradation, inequality and conflict that stems from this seems difficult to understate and undermine. Currently, sustainability options are largely discussed through the lenses of cost–benefit, risk, and anticipation in attempts to shed light on how certain paths may be purposely avoided or reoriented. In this book, we instead discussed and illustrated the possibilities for abandoning, discontinuing, destabilising established socio-technical trajectories as they become undesirable or misaligned with environmental and social justice objectives. The contributors to this book sought to uncover the characteristics of active governance (how and under which conditions are technologies and systems abandoned?) as well as the means for active governance (how can discontinuation be pursued, under which constraints?).

In the theoretical Part, we introduced and expanded on the notions of decline, destabilisation, and discontinuation—not as *the* theory of decline, but as stepping

DOI: 10.4324/9781003213642-12

stones for further conceptual and empirical investigations. In the empirical explorations Part, we opened up the floor and brought in insights from cases across the world. The third Part of the book explored important governance aspects of declining technologies, showing both current efforts and achieved results. Taken together, we hope the Parts of this book can be read as a timely guide to some of the current research, useful for policymakers, researchers, and students, and offering an agenda for future research endeavours.

Much of mainstream research, such as economics and policy studies, still tends to neglect decline and related phenomena. The contributions of the present book demonstrate that destabilisation, discontinuation, exnovation, phase-out, withdrawal, and unravelling offer relevant entry points for research and governance strategies for addressing challenges related to sustainability and problematic technology–society relationships. They also help show how the processes of decline are more complex and more intricate than often assumed. Seen in this light, it is obvious to consider decline as an object of inquiry in its own right.

In this conclusion, we will not offer a unified theory or analytical scheme of 'technologies in decline', but rather take stock of how convening a variety of emerging perspectives may enable the collective opening up of what appears as a particularly important, timely yet challenging research problem. The phenomenon we are dealing with here cannot simply be put into the familiar categories. It is not even clear whether it is a phenomenon or a whole class of them, how it relates to other phenomena and which categories fit best. First, appropriate categories and notions have to be developed, tested, and sharpened in empirical cases. We therefore, firstly, propose a whole range of perspectives and, secondly, see their task as serving as heuristics for discovering what constitutes the phenomena of decline, destabilisation and discontinuation in the first place.

The thrust of the book is that the question of technologies in decline requires a plurality of approaches. The contributions bear witness to the significant variety of approaches that may be employed to address such problems—and we are aware that we have barely scratched the surface. In the following, we identify a series of themes and questions stemming from the contributions to the volume, before concluding with a central conundrum around which many studies of decline are articulated, both here and in the wider literature—the relationship between decline and innovation.

12.1 Studying decline: Five themes

In the present volume we see five overlapping and interconnected themes emerging, relating to geography, governance, temporality, stabilisation work, as well as terms and concepts of technologies in decline.

12.1.1 Geography

From the contributors we learn that a given socio-technical trajectory is not globally homogeneous but differs from region to region and country to country. This was already an important theme in the work of Thomas P. Hughes, notably around the notions of regional variety and technological style, but may take a new flavour in the context of decline. Technologies tend to have key regions or countries from where the large-scale trajectory can be influenced (Weiss & Scherer, 2023). These influences can occur via supply chains, for example (ibid.), and be achieved with the help of powerful large-scale actors (e.g., national governments, industry associations), but also with the help of small-scale actors, such as cities, which can join forces and, in this way, increase their collective power (Callorda Fossati et al., 2023). The most effective interventions seem to come from regions that are central to the technology (e.g., based on patent activities) and are possible with the help of coalitions of smaller actors. Further, the regional embedding of socio-technical systems has significant influence over the specific inertial ties developed, the opportunities for the unmaking of such ties, and the resulting experiences of loss that discontinuation governance is having to deal with (Turnheim, 2023).

More generally, geography matters, and the regional scale is likely to become foregrounded as societies engage more fervently with discontinuation governance.

12.1.2 Governance

Deliberate decline interventions range from soft to hard, e.g., economic incentives, rules and sanctions, management and planning, international agreements, but also command-and-control policies (Rinscheid et al., 2023; Turnheim, 2023; Stegmaier, 2023a). The spectrum of governance pathways entails weakening and ending, which means that not all efforts will always lead to (or seek) complete abandonment. It also means that a ladder of discontinuation steps can be observed that goes from control and observation via reduction of use to restriction of production (Stegmaier, 2023a). If the path of weakening is followed and legacy elements remain, discontinuation governance can even include caring for such remnants for a longer period of time. Discontinuation governance is not only about getting rid of something, but also about caring for the remnants that are difficult to eradicate (Stegmaier, 2023b). Care may also involve the things that are to be preserved intact like musealised technology (van de Leemput & van Lente, 2023), special purpose technology (e.g., incandescent light bulbs in ovens), or waste (e.g., nuclear). Here, in addition to discontinuation governance, we may productively think of the spaces and practices of aftercare, maintenance, or protected exceptions.

There tends to be a large variety of opponents and proponents of a technology (Turnheim, 2023; Markard et al., 2023; van de Leemput & van Lente, 2023; Stegmaier, 2023a), ranging, typically, from NGOs, activists, scientists and political parties on the one side, to incumbents such as utilities and industry associations on the other (Markard et al., 2023). Users should not be forgotten, either, who choose to withdraw from products, or who cling to what industry or politics would prefer to see removed.

The rhetorical framing by critics and supporters of a given technology demonstrates a pattern (ibid.). The framing by the critics tends to focus around the climate crisis and different options (substitutes) to address it, such as diffusion of renewables. The supporters of the technology tend to flag cost-efficiency and reliance of the technology, as well as negative consequences of switching from it. Discontinuation efforts entail a nuanced mix of measures that use existing governance momentum (where previously consolidated assemblages, arrangements, regimes are already beginning to destabilise) and create such momentum (in order to push and pull where things are still too stuck), while anticipating that and from where resistance will come and where there could be setbacks (Stegmaier, 2023a). Momentum can come not only from governance itself, but from technology development, social innovations, new business models, new scientific ways of looking at and proving things, shifts in expert discourse, public issue careers, and much more. The burden of proof tends to lie on the proponents of phase-out policy intervention (Callorda Fossati et al., 2023). These efforts are often met with strong opposition by incumbent actors with significant power, resources and vested interests in maintaining things as they are (Turnheim, 2023) who can engage in campaigns of obfuscation and ignorance construction, as happened in the case of leaded petrol (Newman, 2023), or contribute to capturing and co-opting political agendas of transformation. In the rarest of cases, there is likely to be a straight path, with governance executed as planned (Kuhlmann et al., 2019). Processes drag on and are subject to the influences of changing circumstances.

12.1.3 Temporality

The intensity of pressures for intervention is likely to change over time (Turnheim, 2023). While it may be hypothesised that destabilisation is a cumulative and self-reinforcing process—with a typical progression from relatively balanced debates during early societal discussions on phase-out, to a strengthening of pro-decline discourses as phase-out becomes more of a reality (Markard et al., 2023)—it is also important to note the scope for reversals, backlashes and reactive sequences (Turnheim, 2023). This is a matter for empirical investigation, and no doubt future research will further analyse a variety of possible patterns.

More generally, *technologies in decline* is an inherently processual research focus that conjures up questions of temporality, such as the confrontation of temporalities of socio-technical inertia and change with those of social and political mobilisation.

12.1.4 Stabilisation work

Work directed at preventing decline, whether for preservation or resistance motives, tends to involve much stabilisation, maintenance, care and preservation, often involving constructing (new) networks, learning new competences, and a large variety of actors (van de Leemput & van Lente, 2023; Koretsky, 2023). Substitutes and alternative technologies can also play an important role in avoiding combating decline as they can serve the purposes of either decline or continuity (Goulet, 2023). In the case of leaded petrol, for example, substitutes played a key role in ending the controversy and largely closing down debates (Newman, 2023). Stabilisation is a sort of antidote (and hence obstacle) to destabilisation. They are opposite attractors, both require the combination of active work and the activation of structural dynamics. In turn, additional work on the stabilisation–destabilisation dichotomy may prove fruitful (see also Turnheim, 2023).

From some theoretical perspectives decline is seen as the end phase of a life cycle; from others, a breakdown of innovation (Latour, 1992, 1996); from others yet, a latent tendency for chaos precariously resisted by ordering work. Decline often does not mean complete abandonment or eradication, but rather subsequent existence of the socio-technical system, its maintenance, repair and continuity over a longer period of time. Instead, we observe contractions or retreat into protected spaces of technology use or special applications. This implies that in a way a decline process becomes stabilised at some point. For the question about where the point of no return is reached or the last stabilising constellation ceases to keep a dying technology alive, it is important to note here that not all remnants hold the potential for stabilisation. It depends on the degree of destabilisation or the relationship between stabilisation and destabilisation. Decline that has gone too far is hardly reversible (or only at high cost (Koretsky, 2023)). Conversely, there may be an interest in keeping largely banned technology or dangerous remnants alive and stable for special uses in residual forms (e.g., DDT pesticide or nuclear waste).

More generally, decline and emergence, destabilisation and stabilisation, discontinuation and continuation are inherently related as radically opposed horizons of development for the socio-technical system, yet it is important to resist thinking of them in terms of reverse or symmetrical processes. Better understanding how things die out no doubt promises important lessons for the understanding of how things continue living on—and vice versa.

12.1.5 Terms and concepts

The authors of the chapters have also contributed to qualifying certain adjacent concepts to decline, destabilisation and discontinuation. Namely, phase-out is elaborated as a boundary object that is often used to link different strands of literature across disciplines, and to link the research and policymaking worlds (Rinscheid et al., 2023). In this volume, we proposed 'decline' as an umbrella term that could be used to refer to both coordinated and non-coordinated processes, but perhaps 'phase-out' is indeed a better term. That said, 'phase-out' implies a phased, gradual process, and thus attempts to conceal—or at least project control over—abrupt forms of decline, sudden turnarounds, hard exits and abandonments.

Contributors to this volume are also in active conversation with additional notions, such as transition, transformation, substitution, ban, termination, removal, dismantling, obsolescence, maintenance, divestment, exit, extinction, and more. Other notions are not central parts of the contributors' framings, but are nonetheless relevant to the topic. A non-exhaustive glimpse of this broader constellation includes disruption, crisis, collapse, senescence, sunset industries, and decommissioning.

More generally, it is hardly surprising that such an interesting and relevant problem is being addressed in a variety of ways, convening different intellectual traditions, language, and metaphors. This plurality is welcome and stimulating, but requires additional work of positioning and clarification.

12.2 Continuing to study decline: Unsolved puzzles

We also find a number of unsolved puzzles which we think could be points of departure for future research. Studies of decline are picking up speed. What is so exciting about the current moment is that when the first questions are (almost) clarified, other not necessarily simpler questions always arise. So, Rinscheid et al. (2023) call for a theoretical model of phase-out, with which effective phase-out policies and their design could be studied and with which heuristics could capture key pathways and practices of discontinuation governance (Stegmaier, 2023a). Other contributors suggest that in the future, studies of decline will need to focus more on the emergence of substitutes as a way to understand decline (Goulet, 2023), notably to understand their co-constitution. At the same time, multiple contributors suggested that decline should not be equated with substitution because it encompasses broader questions about how to live (Callorda Fossati et al., 2023; Rinscheid et al., 2023), or because substitution is but one possible outcome of destabilisation processes (Turnheim, 2023). Other case studies would register additional aspects and needs, it can be assumed. We cannot be exhaustive here, but explore in different directions to give further impulses. From the above themes, we propose several key questions for future studies of the topic—what constitutes decline and how it occurs, how it can be dealt with, and how it can be better conceptualised.

12.2.1 What constitutes decline and how does it occur?

Which patterns of decline are recurring across individual cases? A whole number of further comparative case studies will be needed before one can indicate forms of decline that can be generalised to some extent. The same applies to the active pursuit of decline through public politics, non-, lesser-, or no-longer-use, or business management. We hope to contribute to the initiation of further studies that go far beyond the topics of this book and begin to get to grips with the wealth of decline phenomena. A look at the contributions suggests that the circumstances of decline are indispensable in any comparison and generalisation. Likewise, engaging with various conceptual lenses, carried out in a non-eclectic but factual manner, is likely to be illuminating, since a new complex of phenomena is unlikely to be grasped in old terms alone. In this context, it is also important to further study the role of impulses and triggers.

What is the relation between decline and persistence? Sometimes things persist because other things decline. One could easily assume substitution relationships, but some technology exits and some replacements also change the structure itself, so that rather a reconfiguration takes place. For instance, LEDs allowed for design and use contexts, new types of light sources and lighting designs not possible with incandescent bulbs (Stegmaier et al., 2021; Koretsky, 2021). Another example is the ban of DDT, which led to a stabilisation of the pesticide regime (Pellissier, 2021) via sacrifice of a part (DDT) to enable the continuity of the whole (a pesticides regime) (Joly et al., 2022). It can be observed that something declines against the background of something stable, such as the ban on internal combustion engines which doesn't necessarily put into question automobility in general but one form of motorising the car (Newman, 2023; Hoffmann et al., 2017). This raises the questions of the multiple scales and temporalities of decline.

12.2.2 How to deal with decline?

How to influence decline? Decline can occur without active intervention, but could be triggered, pushed further, or channelled in certain directions through more active engagement. How and who does this, under what conditions and with what prospects of achieving the desired effects, requires many more studies that also descend to the mundane side of governance practice and get involved instead of just outlining the big picture and the abstract connections. The latter is also important in order not to get lost. In more general terms, the question of influence could also help to clarify how far both declines as such and influencing them are intentional or unintentional. It seems decline policies—such as initiating decline, dealing with decline and aftercare—could show a 'career pattern' when a full trajectory is traced. Or these policies could at least be the alternative main starting points of active decline policies.

How to achieve technology decline in times of globalisation, which implies fast and global communication, global value chains, and, in some cases, harmonised legislation? The contributions in the book are unfortunately still geographically focused on the global North, but international linkages are sometimes addressed, albeit insufficiently. Research on decline urgently needs studies that explore developments 1) in the global South, 2) in global linkages, and 3) in global impacts. We expect the triggers of decline to present themselves differently, to play out differently, in other places in the world and under conditions other than those of the rich North. We expect alternative decline cases, different choices, different implications for societies, political entities, economies and ecologies. There are likely to be spillover effects when technologies are abandoned in lead economies, on the one hand, or in vulnerable emerging economies, on the other.

How to keep a technology in decline? Not only can existing socio-technical systems meet with resistance and see their existence called into question, but decline and active phase-out efforts also meet with opposition, countermeasures, and in some cases significant backlashes (Pel 2021). We still know too little and have not sufficiently systematically followed up on how exits are attacked, how the legitimacy and desirability of the declines themselves are delegitimised, and when, where or how such struggles play out. Massive interests are associated with many declining technologies; it would be astonishing if there were not organised and routine as well as reflexive-spontaneous resistances and reactions. At the same time, decline is not just about resistance, but about complicated processes and measures that take time and energy. We would not be surprised if not only incumbencies and novelties were interrupted (see section 12.3 below), but also destabilisation processes and discontinuation efforts.

What are the dynamics of care and aftercare for technology in decline? When it is clear that parts of a technology cannot be made to disappear, the question arises of how to deal with the remnants and what dynamics unfold around these remnants. These need to be seen in relation to habitualised practices of preservation, maintenance and care. Sustainability, to take this example, can result both from ending technology that destroys the basis of life and from the preservation as finished use of existing technologies and systems, insofar as this saves resources that would have to be spent on replacement. Let us think, for example, of the internal combustion engine car, which can be replaced by a more energy-saving and less emission-emitting electric car, or which can be kept because of reluctance to simply dispose of the resources used. It is important to look at how such conflicts of means and goals are negotiated and decided.

12.2.3 How to further elaborate conceptions of decline?

Is decline of a system or regime qualitatively different from decline of a product or substance? When it comes to what is given over to decline, the essentialist question of the difference between large systems and small devices is raised again and again. In this way, one can ask—or can try to find out—in which relations a questionable object of decline observation is embedded and what the relationship to disappearing and remaining elements looks like, how decline changes the wider fabric and what possible structural differences look like when it comes to very massive, widely ramified technologies or very confinable, small-scale objects that look like they are easy to handle. It remains to be seen whether such or other categorisations will work.

What conceptual fruits can a meta-theoretical comparison of the established transitions approaches on emergence and innovation (such as TIS and MLP) offer? If sufficient high-quality studies are available and relevant theoretical cornerstones become visible, an analysis of the adequacy of prevailing theoretical offers should also be undertaken at the theoretical level. This should say more than the theory comparisons and critiques already available, namely as offers of abstraction for the particular subject matter itself.

What can be learned about technologies in decline from other disciplines? The authors who contributed to this book come from different disciplines, but are in turn united by the focus on technology and innovation and the theories and methods common in the field of science, technology, innovation policy studies. Other subjects and researchers not represented in this broader context can make important contributions. Questions from psychology, geography, philosophy, ethics, political and administrative science, economics, history, cultural anthropology, historical studies, ecology and more could—and, it is our conviction, should—be considered.

12.3 Decline and the limits of innovation

What this volume is about is captured in many different terms. We have learned that we need to be careful with the terms that are used. It is important to look carefully at what is being referred to, what it is about, what horizon a term has in its context of use. We noticed this on two levels: It can be seen in the subtleties and inconveniences with which the whole battery of terms around decline, such as destabilisation and discontinuation, phase-out and exit, termination and exnovation, are handled. It has not yet been decided which terminology or theoretical tradition will prevail, which research direction and its language will become dominant and sediment in various research traditions. It can also be seen in how they relate to the notion of 'innovation'. As it seems decline, destabilisation, and discontinuation aren't simply the reverse of innovation.

One understanding of 'innovation' involves scientific, technological, organisational, financial, and commercial activities around implementing technologically new products and processes, as well as significant technological improvements for them (OECD, 2018), rooted among others in Joseph Schumpeter's (1934) idea that firms seek advantage and opportunity against competitors and thereby create something later framed as 'innovations'. In this account, there is almost no 'other side' mentioned, that something new would replace something old, or destroy it. What it hints at is 'interrupted innovation', an activity that can be 'aborted, discontinued, or put on hold, for instance when activities to develop an innovation are stopped before implementation' (OECD 2018: 80). We have discussed such cases in this volume (cf. Turnheim, 2023; Stegmaier, 2023a; Koretsky, 2023).

Schumpeter (2003/1943) originally presented capitalism as an evolutionary process, as 'a form or method of economic change' (p. 82) that is fuelled by novelty: 'The fundamental impulse that sets and keeps the capitalist engine in motion comes from the new consumers' goods, the new methods of production or transportation, the new markets, the new forms of industrial organization that capitalist enterprise creates' (Schumpeter, 2003/1943: 82–83). At the same time, Schumpeter, however, also emphasised the destructive character of capitalism by referring to the counterpart which he calls the 'process of Creative Destruction': a 'process of industrial mutation... that incessantly revolutionizes the economic structure from within, incessantly destroying the old one, incessantly creating a new one' (ibid.: 83). Novelty creation and incumbency destruction are here seen as two sides of the same coin.

We would like to suggest a third view. Especially in the face of pressing ecological and economic crises, but also in the face of incoming revolutionary new technologies, the destabilisation and discontinuation of technologies can be seen as a *double innovation*. The new is not only the replacing technology, but also the letting go of the old. When an existing technology is deeply embedded in everyday routines, it is a new turn, an act of renewal and rethinking, when existing technology is actively terminated or passively left to decline. This impression is reinforced by the fact that active discontinuation actually cannot do without new or adapted old instruments that make the turning away operational.

Another lesson the studies teach us is not to think that destabilisations, discontinuations, however much they may be innovations, are easy or quick problem solvers. It is about nothing more or less than negotiating the relationship with familiar (old) and unfamiliar (new) technologies and social uses.

12.4 The urgency of studying decline

Getting away from technologies and thus ways of life that create them or build upon them and that no longer seem acceptable is a growing theme of the times. We have noted this not only in the introduction, but also in many of the contributions. We are following this development with curiosity and interest. Here are some reasons why we are so fascinated by it, and we hope that this volume will help to carry this fascination further.

Since the modern project has been discussed over the past 300 years (more recently rethought by Berger et al. (1973), Giddens (1991), Lash and Friedman (1992), Latour (1993), Beck (1996)), not only progress, improvements, intensifications, expansions, and technisations of human impact on the world we live in have been diagnosed, but also shedding, purging, overcoming, breaking with traditions, replacing old with new, both destruction and creation, and manifold hybridities and hybridisation. Often enough, this happened in counter-arguments of what modernisation is and what one should value or reject about it—if at all ascribing to such a reading of contemporary history. We are just learning that both belong together. They are not either/or, but there is *both/and*. If we are really serious about understanding change, we see no way out of considering the ruptures and break-offs. Change is not only constructive in its effect, but also destructive, not only desirable, but also challenging to the point of repulsion. Those who promise change without ending are deceptive. Those who explore change without ending are blind in one eye. We have opened the narrowed eye and are beginning to glimpse what the change concepts and narratives lacked. In this sense, a focus on decline, destabilisation, discontinuation and related conditions of loss can also be seen as an emancipatory programme because destabilisation developments and discontinuation efforts by definition attempt to challenge the established order and undermine incumbent control. We have no doubt that this also applies to theories and research on socio-technical relations.

Notes

1 www.poweringpastcoal.org
2 www.gov.uk/government/consultations/consulting-on-ending-the-sale-of-new-petrol-diesel-and-hybrid-cars-and-vans/

References

Beck, U. (1996) *The Reinvention of Politics. Rethinking Modernity in the Global Social Order*. Cambridge University Press.

Berger, P.L., Berger, B. and Kellner, H. (1973) *The Homeless Mind: Modernization and Consciousness*. Random House.

Callorda Fossati, E., Pel, B., Sureau, S., Bauler, T. and Achten, W. (2023) Implementing exnovation? Ambitions and governance complexity in the case of the Brussels Low Emission Zone. In Koretsky, Z. *et al.* (eds) *Technologies in Decline: Socio-Technical Approaches to Discontinuation and Destabilisation*. Routledge.

Giddens, A. (1991) *The Consequences of Modernity*. Polity.

Goulet, F. (2023) The role of alternative technologies in the enactment of (dis)continuities. In Koretsky, Z. *et al.* (eds) *Technologies in Decline: Socio-Technical Approaches to Discontinuation and Destabilisation*. Routledge.

Hoffmann, S., Weyer, J. and Longen, J. (2017) Discontinuation of the automobility regime? An integrated approach to multi-level governance. *Transportation Research, Part A*, 103, 391–408.

Joly, P.-B., Barbier, M. and Turnheim, B. (2022) Gouverner l'arrêt des grands systèmes sociotechniques. In Goulet, F. and Vinck, D. (eds), *Faire sans, Faire avec Moins: Les Nouveaux Horizons de L'innovation*. Presses des Mines.

Koretsky, Z. (2021) Phasing out an embedded technology: Insights from banning the incandescent light bulb in Europe. *Energy Research and Social Science*, 82, 102310. https://doi.org/doi.org/10.1016/j.erss.2021.102310.

Koretsky, Z. (2023) Dynamics of technological decline as socio-material unravelling. In Koretsky, Z. *et al.* (eds) *Technologies in Decline: Socio-Technical Approaches to Discontinuation and Destabilisation*. Routledge.

Kuhlmann, S., Stegmaier, P. and Konrad, K. (2019) The tentative governance of emerging science and technology: A conceptual introduction. *Research Policy*, 5, 1091–1097. https://doi.org/https://doi.org/10.1016/j.respol.2019.01.006.

Lash, S. and Friedman, J. (eds) (1992) *Modernity and Identity*. Blackwell.

Latour, B. (1992) Where are the missing masses? The sociology of a few mundane artifacts. In Bijker, W.E. and Law, J. (eds) *Shaping Technology / Building Society*. MIT Press. doi:10.2307/2074370.

Latour, B. (1993) *We Have Never Been Modern*. Harvard University Press.

Latour, B. (1996) *Aramis and the Love of Technology*. Translated by C. Porter. Harvard University Press.

Markard, J., Isoaho, K. and Widdel, L. (2023) Discourses around decline: Comparing the debates on coal phase-out in the UK, Germany and Finland. In Koretsky, Z. *et al.* (eds) *Technologies in Decline: Socio-Technical Approaches to Discontinuation and Destabilisation*. Routledge.

Newman, P. (2023) The end of the world's leaded petrol era: Reflections on the final four decades of a century-long campaign. In Koretsky, Z. *et al.* (eds) *Technologies in Decline: Socio-Technical Approaches to Discontinuation and Destabilisation*. Routledge.

OECD (2018) *Osolo Manual: Guidelines for Collecting, Reporting and Using Data on Innovation*. https://doi.org/10.1787/9789264304604-en.

Pel, B. (2021) Transition 'backlash': Towards explanation, governance and critical understanding. *Environmental Innovation and Societal Transitions*, 41, 32–34.

Pellissier, F. (2021) *Tuer les pestes pour protéger les cultures: Sociohistoire de l'administration des pesticides en France* (Doctoral dissertation, Université Gustave Eiffel).

Rinscheid, A., Trencher, G. and Rosenbloom, D. (2023) Phase-out as a policy approach to address sustainability challenges: A systematic review. In Koretsky, Z. *et al.* (eds) *Technologies in Decline: Socio-Technical Approaches to Discontinuation and Destabilisation*. Routledge.

Schumpeter, J.A. (1934) *The Theory of Economic Development: An Inquiry into Profits, Capital, Credit, Interest, and the Business Cycle*. Harvard University Press.

Schumpeter, J.A. (2003/1943) *Capitalism, Socialism, and Democracy*. Routledge. https://doi.org/10.4324/9780203202050.

Stegmaier, P. (2023a) Conceptual aspects of discontinuation governance: An exploration. In Koretsky, Z. *et al.* (eds) *Technologies in Decline: Socio-Technical Approaches to Discontinuation and Destabilisation*. Routledge.

Stegmaier, P. (2023b) Aftercare, or Doing Less with Discontinuation Niche Governance. In Goulet, F. and Vinck, D. (eds.) *Doing with less, doing with out. New horizons of innovation*. Edward Elgar.

Stegmaier, P., Visser, V.R. and Kuhlmann, S. (2021) The incandescent light bulb phase-out: Exploring patterns of framing the governance of discontinuing a socio-technical regime. *Energy, Sustainability and Society*, 11(1), 1–22. doi:10.1186/s13705-021-00287-4.

Turnheim, B. (2023) Destabilisation, decline and phase-out in transitions research. In Koretsky, Z. *et al.* (eds) *Technologies in Decline: Socio-Technical Approaches to Discontinuation and Destabilisation*. Routledge.

UK Department of Transport (2020, April 9) Consulting on ending the sale of new petrol, diesel and hybrid cars and vans. Gov.uk. www.gov.uk/government/consultations/consulting-on-ending-the-sale-of-new-petrol-diesel-and-hybrid-cars-and-vans/consulting-on-ending-the-sale-of-new-petrol-diesel-and-hybrid-cars-and-vans.

van de Leemput, D. and van Lente, H. (2023) Caring for decline: The case of 16mm film artworks of Tacita Dean. In Koretsky, Z. *et al.* (eds) *Technologies in Decline: Socio-Technical Approaches to Discontinuation and Destabilisation*. Routledge.

Weiss, D. and Scherer, P. (2023) Mapping the territorial adaptation of technological trajectories—The phase-out of the internal combustion engine. In Koretsky, Z. *et al.* (eds) *Technologies in Decline: Socio-Technical Approaches to Discontinuation and Destabilisation*. Routledge.

INDEX

Page numbers in **bold** refer to tables and those in *italics* refer to figures.

Printed in the United States
by Baker & Taylor Publisher Services